# INTEGRATED DIGITAL NETWORKS

L. S. LAWTON

SIGMA PRESS
Wilmslow, England

**First published in 1993**
Sigma Press, 1 South Oak Lane, Wilmslow, Cheshire SK9 6AR, UK

**First printed 1993**

**ISBN: 1-85058-181-9**

**British Library Cataloguing in Publication Data**
A CIP catalogue record for this book is available from the British Library

**Typeset and Designed** by  Sigma Press, Wilmslow, UK

**Cover Design** by Design House, Marple Bridge.

**Printed by:** Interprint Ltd, Malta

**Acknowledgements**
In this book, many of the designations used by the manufacturers and sellers to distinguish their products may be claimed as trademarks. Due acknowledgement is hereby made of all legal protection.

**General disclaimer**
Any computer programs or instructions described in this book have been included for their instructional value. They are not guaranteed for any particular purpose and the publisher does not offer any warranties or representations, nor does it accept any liabilities with respect to the user of the programs.

# *Foreword*

The operation, maintenance and management of telecommunications networks is no longer the sole remit of the national PTTs. Following liberalisation, the way has been opened for others to provide communications services to meet a variety of user needs.

Many organisations, previously used to owning or leasing a PABX for intra-office communications, now own and operate their own private voice and data networks to provide inter-site communications. This trend is set to continue in the future. However this does not mark any reduction in the importance of public telecommunications networks. Private networks will continue to depend upon leased PTT circuits and be subscribers to an increasing variety of PTT provided services, in order to deliver essential communications to their customers and suppliers in the organisation.

That the world of digital telecommunications is changing rapidly is not in doubt. 1990 saw the introduction of British Telecom's Integrated Services Digital Network offerings, ISDN2 and ISDN30 and a continuation of massive investment in digital switching and transmission equipment by the world's major PTTs. World wide, the PABX manufacturers are offering ISDN type services on their digital switches.

As a result, within industry and commerce in general, there is a growing requirement for personnel, other than those employed by the traditional network operators, to be knowledgeable in aspects of digital communications. Within the military, for example, we are increasingly dependent upon a wide variety of information systems to support command and control, logistics, personnel and a number of other administrative functions in peace and war, as well as to provide essential data for weapons systems.

These require a comprehensive array of digital communications networks, including satellite communications, UHF and SHF Radio Relay, HF and VHF Radio and leased PTT circuits to support them. As a result the British Army in the UK now operates one of the largest private leased digital networks in the country. This is linked to other national and NATO military systems on the continent and elsewhere to provide comprehensive data communications at all levels of command.

By taking the ISDN route both military and civilian users will be able to take advantage of the increased speed and flexibility, and easier interworking with other networks so that it will be possible to more easily provide voice and data services, such as desktop conferencing, with organisations and outside to other parties and customers.

This book clearly describes the technology of modern digital communications networks to those without the benefit of a formal communications engineering background and goes a long way to meet the requirement for greater knowledge in this area. The reader is taken stage by stage through the concepts of digital transmission, digital control, digital switching, common channel signalling, System X, ISDN and optical fibre systems.

By reading this book, you will understand the basic building blocks of PCM multiplexers and digital switches. You will also learn how communications networks are built from these basic blocks using high capacity optical fibre digital transmission systems, and complex digital signalling systems into full public or private ISDNs.

During his long military service Captain Lewis Lawton has worked on a wide range of communications systems from analogue FDM networks to mobile satellite communications systems.

During his time as a lecturer he helped to develop much of the course material covering this subject area, and he is widely known and highly regarded as an instructor in this field by the Officers and Senior NCOs of the Royal Corps of Signals whom he has taught. I am delighted to commend this book which I hope will attract a wide readership among the increasing numbers now becoming involved in modern digital telecommunications.

*Major General R F L Cook MSc MPhil CEng FIEE,*

*Director General Command, Control, Communications and Information Systems (Army) and Signal Officer-in-Chief (Army),*

*Ministry of Defence, London.*

# CONTENTS

# 1

# *Introduction to Telephone Communications Systems*

The first telephone networks were analogue. Since the late 1960s, digital equipment has been introduced into the telephone network, initially digital transmission equipment, then digital computer controlled exchanges, followed in the late 70s by digital switching and signalling systems. The aim of this chapter is to provide an introduction to the subject of telephony by briefly describing the analogue telephone network. Readers who are familiar with such things as Strowger exchanges and FDM multiplex may wish to skip this chapter. However if you are new to the subject or have limited knowledge, it should provide you with an understanding of the basic concepts of telephony. Most of the content is still valid as a large proportion of all telephone networks throughout the world are still analogue.

## 1.1 The Functions of the Telephone Network

Ever since there were more than two people owning telephones there has been a requirement for a telephone network. Today there are something like 20 million telephones in the United Kingdom, and 600 million world-wide. Although it is unlikely that any one telephone user would want to call every other user, it is true to say that with a few exceptions, one could call any telephone anywhere in the world simply by pressing several buttons on the keypad. We have come to expect a lot of this huge network, the telephone system.

The three basic functions of the telephone system are listed in Table 1.1.

*Table 1.1 Functions of a telephone system*
Connect two telephones together for a two-way conversation
Provide a *private* connection
Provide an instant connection

That is to say, a user should be able to call any other user connected to the same telephone system, or to another telephone system connected to the caller's system, and be sure of a reasonably noise free connection to the called user. The caller can also expect that no other user is connected either by malicious action (phone tapping or other unauthorised eavesdropping) or by faulty equipment routing the call to the wrong number, or worse still, to two numbers simultaneously such that the caller is unaware of the fact that a third (unwanted) person is listening to the call.

The required connection should be provided automatically by the system within a few seconds of the caller completing dialling the required number. Once established the connection should support two-way simultaneous communications.

To carry out these basic functions, the telephone system requires the four main items of equipment (or plant), listed in Table 1.2.

---

*Table 1.2   Components of telephone networks*

Telephones
Switching Systems (or Exchanges)
Transmission Systems (or Lines)
Electrical Power

---

The telephones are the terminal equipments of the telephone network, and are connected to the telephone exchanges of the network by cables which normally consist of just two copper wires. The exchanges are in fact no more than highly complex switches which are able to connect one telephone line to another.

The exchanges themselves are interconnected by a variety of different types of transmission system. In the simplest case, the transmission system may be a cable of many pairs of copper wires, with one telephone call being connected over each pair. In some cases the cable may be a coaxial or optical fibre system, both of which are capable of handling many thousands of calls simultaneously. The inter-connection between exchanges may also include Ultra High Frequency (UHF) radio links. For very long distance links where it is impractical to erect radio towers, for example across the oceans, these radio links may be relayed via a communication satellite.

Within the telephone network, the human voice is carried from telephone to telephone in the form of electrical signals. The equipment within the telephone exchanges which route calls from one telephone to another are electrically operated. Thus the system requires a source of electrical power.

Traditionally this power has been provided by large banks of batteries located in the telephone exchange building. The batteries are constantly on charge from mains rectifiers, so that should the mains supply to the exchange fail, there will be enough reserve power in the batteries to carry on operating the system for several hours until the mains supply is restored.

It is not uncommon nowadays to find small private exchanges which are only mains operated, especially in offices where the provision of a battery supply may be difficult, and where it is possible to tolerate the inconvenience of being without a telephone system should the mains supply fail. There will be other equipment within the telephone system that requires electrical power, for example radio transmitters and receivers.

The subject of electrical power will not be covered in detail in this book.

## 1.2 The Telephone

The telephone is the device which transmits voice sounds, as electrical signals, into the network and receives electrical signals from the network. The received signals are used to reproduce the original voice sounds as faithfully as the system will permit. The telephone must therefore be capable of converting sound energy into electrical energy and vice versa.

The two components within the telephone which carry out this energy conversion are the microphone and the receiver (or earpiece).

The principles of these two components have changed little over the last century, although successive designs have increased their efficiency. This chapter describes the basic principles of operation of the carbon granule microphone, and the rocking armature receiver.

### 1.2.1 The Carbon Granule Microphone

Figure 1.1 shows the construction, circuit symbol and principles of this microphone.

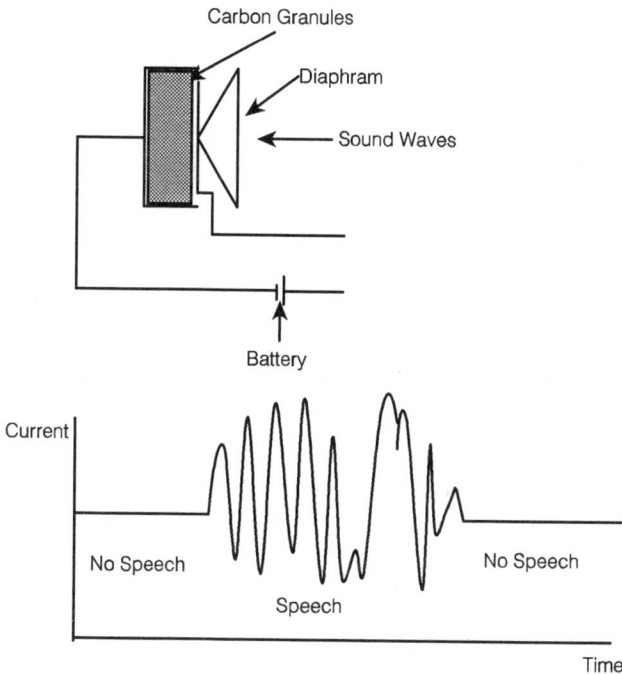

*Figure 1.1 The carbon granule microphone*

The microphone consists of a diaphragm which vibrates when struck by sound waves. The diaphragm is connected to a chamber containing carbon granules. The vibrations of the diaphragm cause the pressure on the granules to vary, and this pressure variation causes a variation in electrical resistance. If a source of direct current (DC) electricity is connected to the carbon granule microphone, the DC current flowing varies in sympathy with the sound waves. The effect is that an alternating current (AC) directly proportional to the volume of the original sound is superimposed upon the DC current. It is this AC current which is the electrical signal transmitted through the network to be converted back to sound in the receiver of a distant telephone. There are still many millions of this type of microphone in use today, although generally they are being replaced by dynamic microphones in new telephone instruments.

## 1.2.3 The Rocking Armature Receiver

At the distant telephone, the incoming electrical signal must be converted back into an audible sound. The small AC current generated by the microphone may have travelled hundreds of miles through the network, and will have suffered from some attenuation (reduction in amplitude). The receiving device has therefore to be very efficient if most use is to be made of the available energy. All receivers are electromagnetic devices constructed in a similar way to the moving coil loudspeaker. The principles of the rocking armature receiver are shown in Figure 1.2.

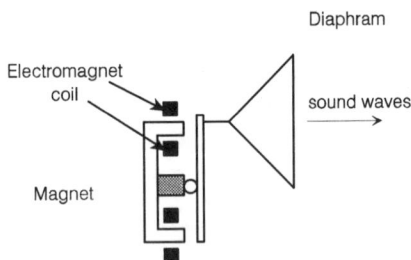

*Figure 1.2 The principle of the rocking armature receiver*

The receiver consists of a permanent magnet fixed to a ferrous core. Around this are wound two coils connected in series. An armature is pivoted above the magnet and is attracted equally by both poles of the core. Connected to one end of the armature is a driving pin, which is itself connected to a diaphragm. The coils are wound so that when they are energised, they produce opposing flux. In one of the pole pieces the flux due to the current will add to the flux due to the magnet, while in the other pole piece, the fluxes will tend to cancel.

The overall effect of this arrangement is that when an AC current is applied to the receiver the armature is attracted to one pole piece during the positive half cycles, and to the other during the negative half cycles. As the armature rocks on its pivot the diaphragm vibrates at a rate equal to the frequency of the applied current, and with an amplitude proportional to the magnitude of the current.

## 1.2.4 A Basic Telephone System

A functional, although not practical, telephone could be made by directly connecting a microphone in series with a DC battery and two terminals. A receiver could then be connected to two other terminals. The telephone would then have four terminals. A simple one-to-one link could be made by connecting the microphone and battery of one telephone to the receiver of the other as shown in Figure 1.3 on the next page.

Although this arrangement would work it has several problems which make it impractical in use. The main problems are:

❑ When the phones are idle, there is no method of alerting the user at the distant end of an incoming call.

❑ Four wires are required for the connection. This is expensive especially on long links.

❑ Two batteries are required. If these are dry cells they will need replacing from time to time

❑ The batteries are in circuit constantly even when the phones are not in use, thus battery life will be short

❑ The constant DC current through the receiver may cause it to become polarised, which will decrease the efficiency of the device

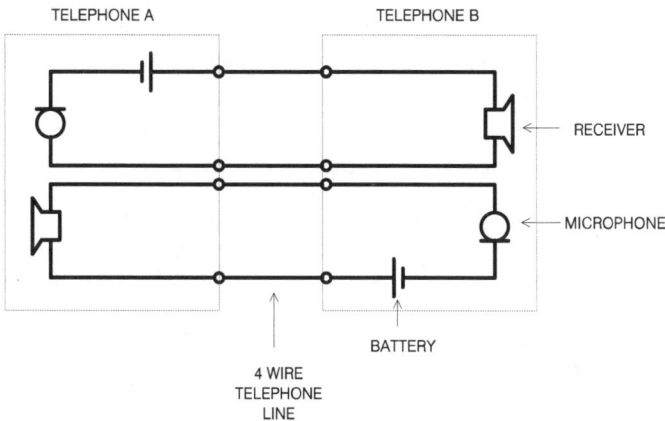

*Figure 1.3 A basic, but impractical telephone system*

## 1.2.5 Telephone Signalling

The first problem is solved by introducing a signalling system into the telephones. Both phones must be capable of sending a signal (signalling out) to the other, and receiving signalling (signalling in). In the early telephones signalling out was achieved by incorporating a hand operated current generator. To make a call the user had to turn the generator several times, to produce a low frequency, high voltage AC current which would activate an electric bell in the telephone at the distant end. This method of signalling is known as magneto signalling; it is still in use today for some applications, for example, in the military where there is often a requirement for one-to-one telephone links.

Except in the case mentioned above, telephones today are rarely connected directly to anything other than an automatic telephone exchange. The signalling system in use must reflect the requirements of an automatic rather than a manual system. Signalling from the exchange to the telephone is still the magneto system already described. To alert a user of an incoming call, the exchange connects a low frequency (17Hz to 25Hz), high voltage (50Volt to 80Volt) current to the line, which rings the bell in the phone.

The method of signalling from the telephone to the exchange is more complex, as there are two types of signal to be sent. Firstly, it is necessary to alert the exchange that the user wishes to make the call, and secondly it is necessary to tell the exchange which number is required. The initial *call request* signal is sent when the user picks up the handset. This causes a switch in the telephone to operate and connect a resistive load to the line. In telephony terms this is known as a *loop*, due to the fact that the telephone line is looped back to the exchange. This loop is detected by the exchange and interpreted as a request for a new call. The exchange then sends a dial tone to the caller, who then sends an *address* signal by dialling the required number.

## 1.2.6 Loop Disconnect Signalling

There are two basic methods of signalling between telephones and exchanges. For many decades a system known as Loop Disconnect (LD) signalling has been employed.

Operation of the rotary dial causes the resistive load to be successively disconnected from the line, and then reconnected several times according to the number dialled. For example dialling 2 causes two disconnect pulses, dialling 3 causes three disconnect pulses, and so on up to 0, which causes 10 disconnect pulses. As these pulses are used to operate electro-mechanical switches within the exchange, the speed at which they can be sent is determined by the mechanical design of the switches.

The usual rate at which these disconnect pulses are transmitted from the telephone dial is 10 pulses per second (pps), and hence this signalling system is often called 10pps, instead of Loop Disconnect.

To allow the exchange to discriminate between the LD pulses of one digit and the next, a short period of time during which no pulses can be sent is inserted. The construction of the rotary dial is such that the pulses are transmitted only during the return period after the user has dialled the digit. Figure 1.4 shows the fundamental principles of LD signalling. It should be pointed out that the telephone networks of different countries use slightly different pulse ratios, so telephones designed for one system may not operate satisfactorily on others.

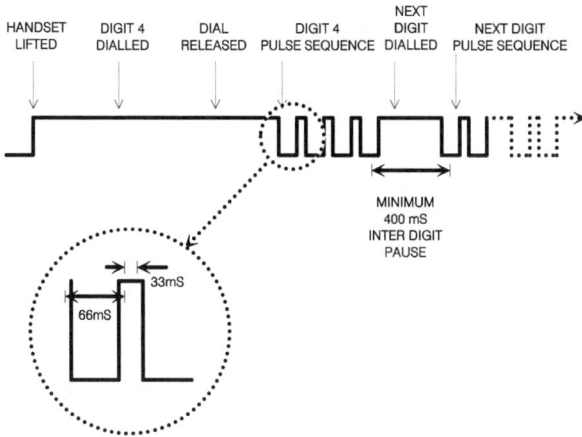

*Figure 1.4 Loop Disconnect (10 pps) signalling*

## 1.2.7 Multi-Frequency Signalling

Most modern telephone exchanges are designed to accept LD signalling so that they are compatible with older telephones. But as the LD system is slow, modern exchanges, for example British Telecom's System X exchanges, and most Private Automatic Branch Exchanges (PABX) use a faster signalling system known as Dual Tone Multi-Frequency (DTMF). In this system, which is particularly suited to push button telephones, a unique pair of voice frequency tones is sent to the exchange for each digit dialled, including the symbols # (Hash or Square) and * (Star).

A particular advantage of DTMF signalling, apart from being significantly faster than Loop Disconnect signalling, it that it is fairly standardised. This implies that telephones can be designed to operate on any manufacturer's exchange in virtually any country of the world.

Chapter 2 includes a description of the DTMF system.

## 1.2.8 Summary of Telephone Functions

The design of the telephone must overcome all the problems mentioned as economically as possible, both in terms of manufacture and maintenance. Today's telephone manufacturers are producing instruments with many features to make use of the telephone simpler and more efficient. The basic functions of all telephones are summarised in Table 1.3.

*Table 1.3 The Basic functional components of a telephone*

| Function | Components |
|---|---|
| Transduction | Microphone and receiver |
| Signalling out | Cradle switch, dial, MF keypad |
| Signalling in | Bell, electronic sounder |

# 1.3 The Telephone Exchange

## 1.3.1 The Functions of a Telephone Exchange

The exchange is rather like a very complex switch. In fact, the first automatic exchanges contained many thousands of electrically operated switches operated by the exchange control system as it responded to the loop disconnect pulses of each digit dialled. Exchange technology has changed dramatically since Britain's first automatic exchange opened in Epsom over 75 years ago, but the main functions of an automatic exchange remain the same. As an illustration of these functions we describe the sequence of events required to connect a local call. The flow chart in Figure 1.5 shows, starting from a line idle state, the main events required to connect the call.

**A.** In the line idle state, the exchange is waiting to receive a CALL REQUEST signal from the telephone.

**B.** On receipt of call request, the exchange must arrange to connect suitable equipment to the line to receive the dialled digits. This will take a short but variable amount of time, so the exchange sends a READY signal to the user in the form of dial tone, when the equipment is connected. Normally if dialling takes place before dial tone is heard the first digits will not be correctly recognised by the exchange and result in the call being incorrectly connected.

**C.** After the first digit has been received, there is no further need to send a dial tone, so this is removed from the line. The first digit is stored and checked to ascertain if it is valid. For instance if no number, or inter-exchange connection starts with a 3, a first digit 3 will cause the exchange to send a number invalid or NUMBER UNOBTAINABLE (NU) tone to the caller.

**D.** The dialling sequence continues until either:

❏ A complete valid number is dialled, or

❏ An invalid number is dialled

**E.** If an invalid number is dialled, NU tone is returned to the caller.

**F.** If a complete valid number is dialled, the exchange then commences the actions necessary to complete the call.

**G.** The exchange interrogates the called line to ascertain whether it is free, i.e. not busy on another call. If the line is busy, an engaged tone is returned to the caller. Should the line become free while the engaged tone is being sent, there is normally no way for the exchange to connect the call, and the caller must re-dial.

**H.** If the line is free, the exchange sends two signals. The first is the Alert Called Party signal, which of course is the ringing current to operate the telephone bell. The second is an information signal to the caller, telling him that the called line is free and that the called party has been alerted, i.e. the exchange sends a ring tone to the caller. Since ring tone and ring current are very different electrical signals, these two signals are not sent from the same source within the exchange. It should be pointed out that very few exchanges have the ability to determine whether or not a phone is actually connected to the line, and so it is possible to hear a ring tone even if the telephone at the distant end has been temporarily disconnected.

## LEGEND

| | | | |
|---|---|---|---|
| ⬭ | START | ⬠ (signal shape) | SIGNAL FROM SUBSCRIBER TO EXCHANGE |
| ▭ | ACTIVITY | ⬠ (signal shape) | SIGNAL FROM EXCHANGE TO SUBSCRIBER |
| ▭ | IDLE OR WAITING | ◁ | TEST DIALELD DIGITS |
| Ⓒ | CALLER CLEARS | ◇ | DECISION |

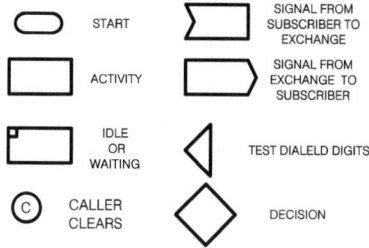

*Figure 1.5a Exchange flow chart legend*

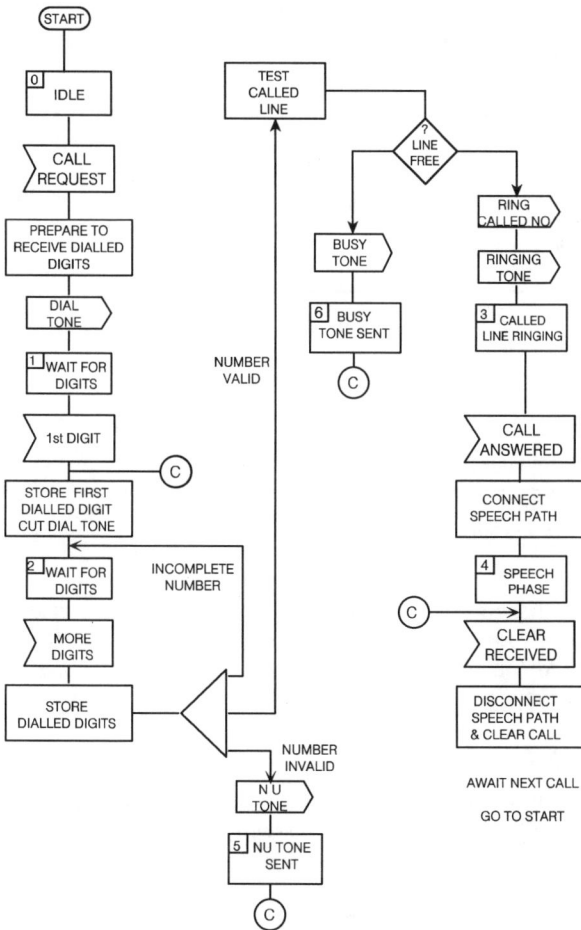

**START**

0 IDLE

CALL REQUEST

PREPARE TO RECEIVE DIALLED DIGITS

DIAL TONE

1 WAIT FOR DIGITS

1st DIGIT

Ⓒ

STORE FIRST DIALLED DIGIT CUT DIAL TONE

2 WAIT FOR DIGITS

MORE DIGITS

STORE DIALLED DIGITS

INCOMPLETE NUMBER

NUMBER VALID

NUMBER INVALID

NU TONE

5 NU TONE SENT

Ⓒ

TEST CALLED LINE

? LINE FREE

BUSY TONE

6 BUSY TONE SENT

Ⓒ

RING CALLED NO

RINGING TONE

3 CALLED LINE RINGING

CALL ANSWERED

CONNECT SPEECH PATH

4 SPEECH PHASE

Ⓒ

CLEAR RECEIVED

DISCONNECT SPEECH PATH & CLEAR CALL

AWAIT NEXT CALL

GO TO START

*Figure 1.5b Exchange operation flow chart*

**I.** Two events are now possible: either the called party answers, or the caller gives up waiting and hangs up. In the latter case, the exchange simply disconnects ringing current and ring tone and restores both lines to their idle state.

**J.** When the called party answers, ringing current is disconnected immediately to prevent damage to the telephone microphone, and stop high level unpleasant sounds in the receiver, which may also damage the device. The speech path between the two telephones is only made at this stage. In a public network, a charge for the call will be made, so a START CHARGING signal is sent to the relevant equipment within the exchange.

**K.** During the conversation phase, no further action takes place. The exchange however is monitoring both lines awaiting a CALL TERMINATED signal which will occur when either party replaces his handset. This monitoring is known as SUPERVISION in telephony terms. On receipt of a call terminated message from one of the two parties, the lines are restored to their idle state, and in a public exchange a STOP CHARGING signal is sent to allow the charging equipment to establish the charge for the call.

**L.** At any time, the caller could decide that he no longer wishes to proceed with the call, and clears down. The exchange will react to this and restore the caller's line, and all equipment used, to the idle state. This is represented by the symbol C in the diagram. The main functions of the exchange, as outlined in the flow chart are summarised in Table 1.4.

---

*Table 1.4   The main functions of an exchange*

Signalling in
Signalling out
Switching
Control (or intelligence)

---

## 1.3.2 Signalling in

The signalling in function of the exchange is to receive signalling out from the telephone. The exchange must be able to recognise the call request signal from the telephone. It must also be able to interpret the dialled digits whether they are in Loop Disconnect or DTMF form. Thus the telephones and the exchange must use the same signalling system to be compatible.

## 1.3.3 Signalling out

Similarly, the signalling out function is to transmit signalling in to the telephone. We have already seen that this consists of sending ringing current to the bell in the telephone, so magneto signalling is used for the signalling out function.

## 1.3.4 Inter-Exchange signalling

A very large percentage of telephone calls are connected via more than one exchange.

To enable each exchange to know how to connect the call there must be a signalling system between exchanges. There are many different types of inter-exchange signalling system in use today, although as we move into the digital era the number of systems is likely to be reduced to a few standard types. The basic functions of inter-exchange signalling are summarised in Table 1.5.

---

*Table 1.5  The functions of an inter-exchange signalling system*

Selection of inter-exchange circuit

Transmission of call request signal to distant exchange

Transmission of address information (i.e. dialled digits)

Transmission of call answered signal

Reception of signals

Detect end of call

---

To explain these functions, consider a call routed through two local exchanges. The number of circuits provided between the two exchanges will depend upon the amount of traffic between them. Of these circuits the originating exchange has to select one of those not already in use, and send a call request signal to the destination exchange. This message informs the destination exchange that a new call is to be processed on this circuit and that it can not itself select this circuit for another call.

In some systems a call acknowledge signal is passed back, however this is far from always the case. A short time after sending the call request, the originating exchange assumes that the destination exchange has received it, and proceeds to send the required address digits. The destination exchange uses these digits to route the call within the exchange.

When the called number answers, the destination exchange sends a CALLED SUBSCRIBER ANSWER signal to the originating exchange so that if necessary, charging for the call can be commenced. At this stage, the speech path between the two telephones is completed.

The exchanges are transparent to the speech signals passing in both directions. The subscribers' loops are used to hold all the required connections in the exchanges in position. When either user replaces his handset, his loop is disconnected. This is detected by his exchange which clears the call by returning all the switches used to their idle state, and informs the other exchange that the call is terminated. The call is then cleared at the distant exchange, and from the inter-exchange circuit which is then free to be used for another call.

## 1.3.5 Exchange Switching

Until recently the switching function of an exchange was implemented using electromechanical switches. Ever since the invention of the transistor by The Bell Telephone Company, telephony engineers have sought a switching system based on semiconductor technology. This technology is now available and used in most new telephone exchanges, but as an introduction to switching it is useful to take a look at the old electromechanical switching systems.

## 1.3.6 Manual Exchanges

The first telephone exchanges were manually operated, and did not employ switches in the modern sense of the word. Each telephone line was connected to a line jack socket and an indicator. When the user wished to place a call, he rang the exchange using the telephone's magneto generator. At the exchange the ringing current operated a latching indicator, which remained operated until the operator answered the call by inserting a jack plug into the socket. After ascertaining which number or line the caller required, the operator would insert another plug into the appropriate line jack and ring down to the called user's telephone. The plug used to answer the call, was connected to the plug used to place the call by a cable (or cord). The cord was in fact the switching system and provided a temporary connection between the two telephone lines. The call was cleared by the operator removing the plugs after having checked that the call was no longer in progress.

The manual system was slow, labour intensive and by its very nature devoid of privacy. One of its major demerits was that the operator could connect your call to some one other than the person required. For example a user could request a call to a particular trader, only to find that the call had been connected to another. The inventor of the first practical automatic exchange realised this when he believed that people calling his undertaker's business were having their calls routed to one of his competitors.

There was one facet of the manual system which even today has not been provided in even the most sophisticated telephone system. The user was not required to know the number of the person he wished to call, in most cases just providing his name and business was sufficient for the operator to perform a mental translation into the required number. In an automatic system much of the responsibility for correctly connecting the call is placed on the user. This implies that any automatic system must be as easy to use as possible. We will see in later chapters how various manufacturers have tried to achieve this aim, especially on the more modern exchanges which offer many features to users.

## 1.3.7 Automatic Switching

The undertaker Almon B Strowger patented the first practical automatic switch. His rotary switch had a single inlet connected to a wiper which could be switched to any one of 10 outlets by the action of a solenoid on a ratchet mechanism. A single such switch connected to every telephone line would thus allow a connection between the line and 10 others. In Figure 1.6 we show that the number of possible connections can be increased by cascading switches such that each of the outlets of the first switch is connected to the inlet of a second switch, giving 100 possible connections. This scheme, which can be expanded almost indefinitely using further switching stages, has become known as the Strowger system, after its inventor.

SINGLE MOTION SELECTOR

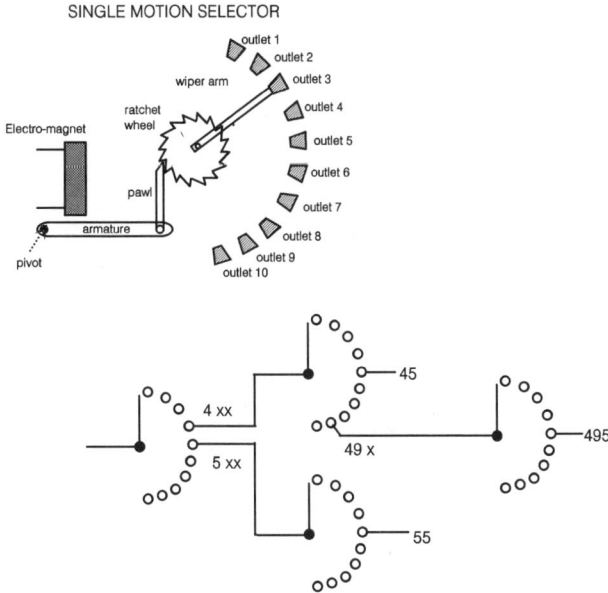

*Figure 1.6 The Strowger switching principle*

## 1.3.8 Strowger Switching

Thousands of exchanges have been built on the Strowger concept and many are still in service today. The design of the switching mechanisms has improved since the early rotary switch which moved in one plane only. To increase the capacity of a single switch, a design in which the wipers moved in two dimensions was introduced. The simplified diagram of this switch in Figure 1.7 shows how the wipers of this switch first move vertically to any one of 10 possible banks. Each bank is fitted with 10 outlets. Having moved vertically to the correct bank, the wipers then move horizontally to the correct outlet on that bank. The capacity of this two-motion switch was thus 100 outlets.

A Strowger switch consists essentially of two units: a switch unit that makes the required physical connection, and a control unit, that contains several relays.

The control unit responds to the loop disconnect dial pulses sent by the telephone and operates the switch unit to cause the wipers to move to the required outlet. The block diagram symbol for a two-motion selector is also shown in Figure 1.7. The 10 lines beneath the control unit represent the 10 banks of contacts onto the which the wipers can step vertically. Each bank contains 10 outlets, but these are not shown individually.

## 1.3.9 Strowger Exchanges

To illustrate the principles of Strowger switching, Figure 1.8 shows the block diagram of a 10,000-line exchange. The exchange uses single motion switches, known as uni-selectors, and two motion selectors (or switches) to provide a switching system in which any of the 1,000 lines can be connected to any other.

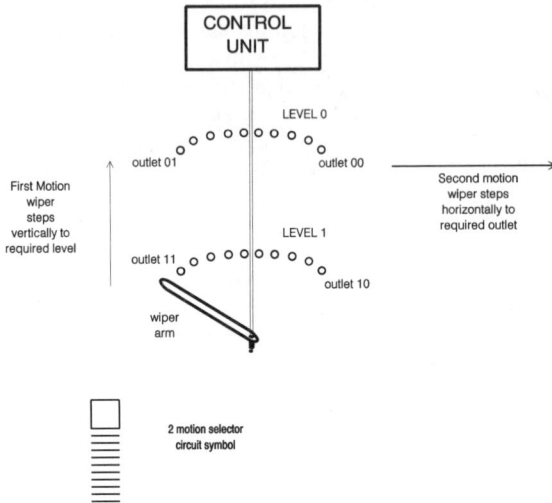

*Figure 1.7 Strowger two-motion selector*

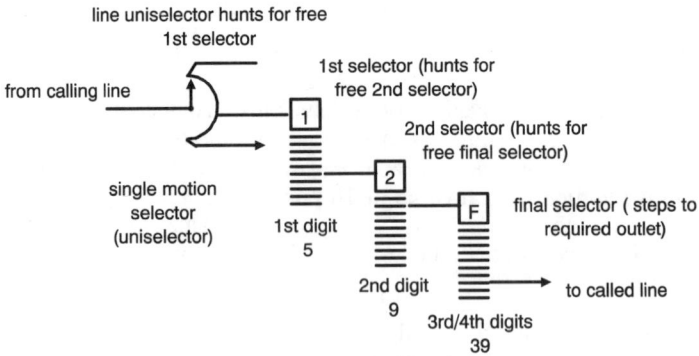

*Figure 1.8 Strowger exchange block diagram*

Each telephone line is connected to the inlet of a uni-selector, which in practice has up to 25 outlets. Each outlet is connected to the inlet of a two-motion selector. Since not all telephones will be in use at the same time it is possible for a large number of uni-selectors to be connected to the same 25 two-motion selectors simply by wiring all the corresponding outlets in parallel and to the inlet of the two-motion selector.

When the uni-selector circuit detects a calling loop, the wipers are moved over the outlets testing each one in turn hunting for a two-motion selector that is not in use. When a free selector is found, the line is connected via the outlets of the uni-selector to the inlet of the free two-motion selector. Of course it may be that no selector is free, in which case the uni-selector circuit connects equipment busy tone to the caller's line.

The two-motion selector is the first switching stage within the exchange, and will respond only to the first digit dialled by the subscriber. For this reason it is called a First Group selector. It is from this stage that dial tone is sent back to the caller, as now the exchange is ready to receive the first digit. When the first digit is dialled, the control unit of the selector will operate the wipers vertically, such that they step up one bank of contacts for each loop disconnect pulse. In other words if the user dials 5, the wipers step up to the fifth bank of contacts.

Each of the 10 outlets on this bank is connected to a similar two motion selector, which, as it will respond to the second dialled digit, is called the Second Group selector. The first group selector having stepped the wipers up to the required bank of contacts then changes from vertical stepping under control of the dial, to horizontal self-drive, stepping round each outlet in turn, searching for a free second group selector from the 10 that are connected.

When a free second group selector is found, the dial is then connected to the control unit of this selector, so that the second digit may be dialled. All this must happen during the time between one digit being dialled and the next. At this stage dial tone is removed from the caller's line.

The second group selector operates identically to the first group selector and will connect the dial to a third selector. This responds to the last two digits to be dialled, and is known as a Final Selector.

When the user dials the third digit, the final selector control unit steps the wipers up the appropriate number of banks, and then, instead of self driving round, as in the case of the first and second group selectors, waits for the user to dial the fourth digit. The control unit now steps the wipers round such that they end up on the outlet connected to the required subscriber's line.

The final selector control unit then tests the required line to see whether it is free or not, and completes the required actions to connect the call. Once the call is connected, relays in the control unit of the final selector supervise the call, waiting for either user to *hang up* at the end of the call. When the final selector control detects this call terminated signal, it returns the wipers of the switch to their idle (or Home) position, and via a control circuit linking all the selectors used for the call, instructs the subscriber's uni-selector, first and second group selectors to do the same. These selectors are now free to be used to set up another call.

## 1.3.10 Other Electromechanical Exchanges

Two other switching technologies should be mentioned, not only for completeness but also because they were part of the evolution from the Strowger exchange to the modern digital exchange.

## 1.3.11 Crossbar Exchanges

The Swedish telephone company, Erricson, designed a very reliable switching system, called crossbar, that had fewer moving parts than the Strowger switch. The essence of this switch was a matrix of contacts operated by electromechanical relays. Unlike the Strowger system, in which each selector had its own control unit, a single control unit would operate a large number of switches in a crossbar unit.

The main advantages of the crossbar switching system were its speed of operation and a saving in capital costs. The savings accrued from having a control unit that was common to a large number of switches. We will look at this concept in more detail towards the end of this chapter.

## 1.3.12 Electronic Exchanges

Strowger and Crossbar exchanges are electromechanical switching systems, which employ relay logic for the control functions. The first step towards a truly electronic exchange was taken with the introduction of transistor logic control units. While the transistor has applications as a high-speed switch, it has proved difficult to build exchanges using the transistor as replacement for the Strowger switch. One of the main reasons for this is that, although the impedance of the transistor in its *off* state is very high, in an exchange application it is necessary to have many switches in parallel. The effect of the parallel connections is to substantially reduce the total *off* impedance, and cross-talk or overhearing between circuits will occur.

In the electronic exchanges introduced by the British Post Office, the switching unit was a reed relay matrix. The reed relays being operated by a transistorised control unit. Just as in the crossbar exchange, the control unit of the electronic exchange is able to control many reed relay switches.

## 1.3.13 Switching Networks

This section describes how the various types of switches may be connected within the exchange. If we consider an exchange to be a simple matrix of switches, for example reed relay contacts, even for a small exchange, say 1,000 lines, a very large number of contacts is required. Figure 1.9 shows a 1,000-line exchange, It has 1,000 inlets and 1,000 outlets, with a switch contact at each intersection, so that any incoming line can be connected to any out going line. Neglecting the fact that line 1 inlet would not need to be connected to line 1 outlet, since they would be connected to the same telephone line anyway, this simple switch requires a million (1,000 x 1,000) reed relay contacts. Since each line comprises two wires, and at least one control wire, the figure for the number of reed relay contacts actually approaches 3 million. As well as making this a costly system, the complexity of such a system is awesome.

## 1.3.14 Concentration and Expansion

The simple matrix just described has the ability to connect every inlet to the required outlet, irrespective of the number of calls already in progress on the exchange. The control system would ensure calls were not connected to lines already busy. But when we consider that the maximum number of calls in progress at any one time is 500, (since a call involves two lines), and thus the number of contacts in use is just 1,500. From a total of 3 million, this appears a very inefficient system. If the maximum number of contacts required is only 1,500 a switching system with fewer contacts would be more efficient, less complex and thus less costly. The first step to achieving this aim is to use two smaller switching matrices rather than a single large unit. In Figure 1.10 we show the same 1,000-line exchange as two smaller matrices, each of 1,000 x 100 contacts.

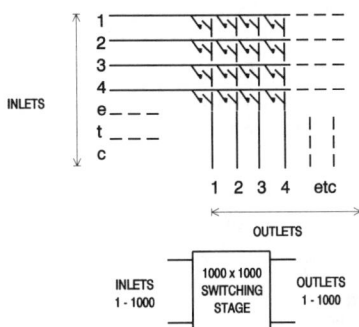

*Figure 1.9 Single stage switching network*

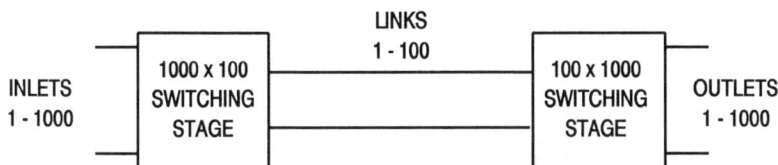

*Figure 1.10 Two stage switching network*

The matrices are arranged so that any one of the thousand inlets of matrix A can be connected to one of its hundred outlets. Each outlet of matrix A is connected to an inlet of matrix B. Each inlet to matrix B can be connected to any one of the thousand outlets of matrix B. Using such a switching network, we provide the ability to connect a call between any two of the thousand lines, at reduced cost in terms of the number of switches employed (300,000) but at the expense of reducing the maximum number of calls that can be in progress at any instant from 500 to 100.

As there are only 100 links between the two switching stages, this sets the limit on the number of calls in progress. If 100 calls were in progress and another call was attempted it could not be connected, and is said to be blocked as there is insufficient equipment within the exchange to connect the call. Since it is highly unlikely that more than 100 users would wish to place calls at the same time, this is a feasible solution. The cost of switching equipment has been cut by 80%. This has been achieved for a just small increase in the complexity of the control unit, and introducing the probability of a call being blocked during busy periods.

The term, CONCENTRATOR is used to describe the function of the first switching stage, Matrix A in the diagram. The switch concentrates calls from a large number of lines to a rather smaller number of links between the first and second switching stages. Matrix B performs the inverse function and is known as an EXPANDER stage. The block symbols used at the bottom of the diagram are attempts to show the functions of concentration and expansion graphically.

## 1.3.15 Grouping

The number of switch contacts in the concentrator and expander stages can be further reduced by a technique known as grouping. In this example, shown in Figure 1.11, the 1,000 inlet and outlet lines are divided into groups of 100. Each 100-line group terminates on a small switch matrix having 100 inlets and just 10 outlets. The total number of contacts has now been reduced to 60,000, some 2% of the original 3 million.

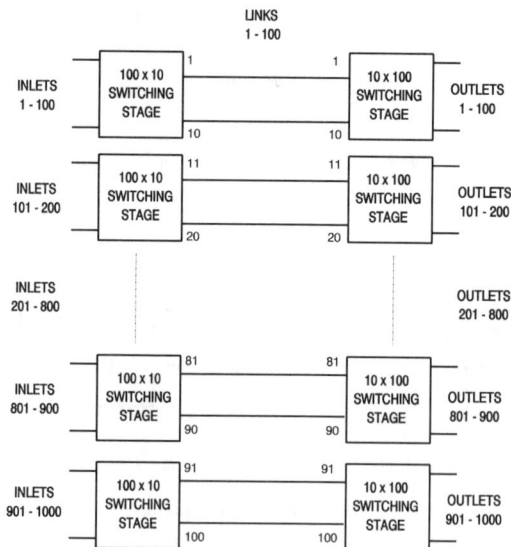

*Figure 1.11 Grouping*

The technique of grouping as shown in this example does introduce a few problems which need to be overcome. From the diagram it can be seen that lines in the group 1 to 100 (group 1) can only be connected to other lines in the same group, because in our diagram there are no links between group 1 and any other groups. This is obviously unsatisfactory and can be overcome without adding further switch contacts by connecting the first outlet of the group 1 concentrator matrix to an inlet of the group 1 expander matrix as shown in Figure 1.12. The second outlet of the group 1 concentrator matrix can then be connected to an inlet of the group 2 expander matrix, and so on. This provides the ability to connect any of the 1,000 lines to any other, but only one call can be connected between any pair of groups, since there is only one link connecting all the matrices.

LINKS
1 - 100

INLETS
1 - 100

100 x 10
SWITCHING
STAGE

1

1

10 x 100
SWITCHING
STAGE

OUTLETS
1 - 100

10

10

INLETS
101 - 900

OUTLETS
101 - 900

INLETS
901 - 1000

100 x 10
SWITCHING
STAGE

91

91

10 x 100
SWITCHING
STAGE

OUTLETS
901 - 1000

100

100

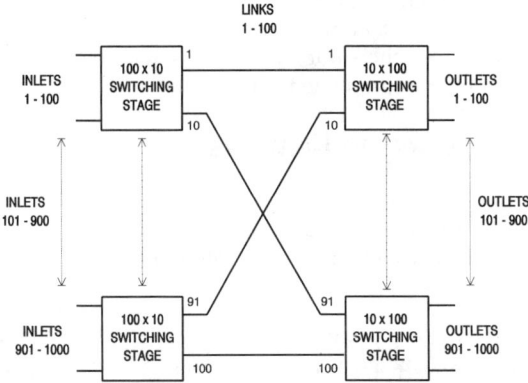

*Figure 1.12 Grouping with interconnections between groups*

This is still not satisfactory, as a situation could arise in which only one call was in progress through the exchange when a second call, involving the same groups as the first, is attempted. The system shown could not connect the second call, which is thus blocked despite the fact that many switch contacts in the exchange are idle. To overcome this problem, a technique known as distribution is employed.

## 1.3.16 Distribution

So far two switching stages, the concentrator and expander, have been discussed. Distribution adds a third stage between the first two. The cost in terms of numbers of switch contacts is increased slightly. There is also a corresponding increase in the complexity of the control unit to service three rather than two switch stages. Figure 1.13 shows how the distributor stage allows calls to be connected between one group and another, up to the maximum number of outlets from a group matrix (in this case 10).

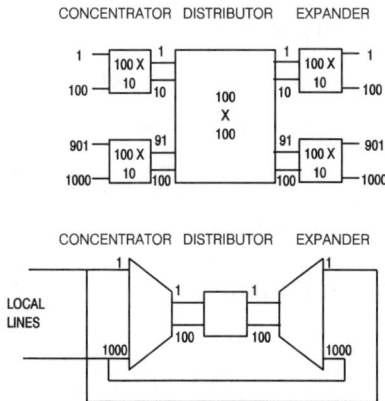

CONCENTRATOR   DISTRIBUTOR   EXPANDER

1

100 X
10

1

10

1

100
X
100

10

1

100 X
10

1

100

901

100 X
10

91

100

91

100

91

100 X
10

901

1000

1000

CONCENTRATOR   DISTRIBUTOR   EXPANDER

LOCAL
LINES

1

1

1

1

1000

100

100

1000

*Figure 1.13 Distribution*

The distributor in this diagram is a 100 inlet by 100 outlet matrix with full availability, i.e. any inlet can be connected to any outlet. The block diagram shows the three switching stages and their interconnections with circuits to other exchanges being terminated on the distributor switch stage.

### 1.3.17 Summary of Switching Networks

The techniques of concentration, expansion and distribution can be applied to any switching technology to reduce the total number of switch contacts required while still maintaining the ability to make a connection between any two lines on the exchange.

The main disadvantage of using these techniques is that new calls are blocked when all the links through the exchange are already in use. The design of the exchange switching network is thus a compromise between cost on the one hand and service level on the other. Although we have used a small 1,000-line exchange as an example, the techniques become more important in larger exchanges where there is a potential for extremely large numbers of switch contacts. In a later chapter we will see how these techniques are applied to digital exchanges.

# 1.4 Exchange Control

## 1.4.1 Control Functions

In the introduction to exchanges, it was stated that the four main functions of an exchange were signalling in, signalling out, switching and control. In this section we look at the function of control within the exchange.

In the early manual exchanges the control function was vested in the intelligence of the operator. As we have already said the manual system is very slow, but it was extremely flexible. Some of the features of such a system are its ability to:

❑ Retry a number found to be engaged

❑ Interrupt a call with one of a higher priority

❑ Try several inter-exchange routes when normal routes are busy, or faulty

❑ Interpret vague instructions from the user.

Some of these facilities are only now becoming available on automatic systems employing computerised control. We will be considering such control systems in a later chapter, but for now, the main principles of exchange control are introduced.

## 1.4.2 The Basic Control Functions

We can define the basic control functions of the control system as those activities necessary to recognise new call request signals, interpret address signals, send required tones, test called lines, connect and supervise calls, and of course clear calls when complete. We have already mentioned these functions under the headings of signalling in, signalling out and switching. The control system ensures all the required activities take place in the correct sequence.

## 1.4.3 Step-by-step Control

The control system of the Strowger exchange is made up of the control units of the selectors. As a call is processed from the initial call request through to the cleardown, control is passed from the subscriber's line circuit to the first group selector, then to the second group selector, and ultimately to the final selector. Because control passes from one control unit to another in this way, the Strowger system was said to use step-by-step control.

Another way of considering this system is to view the control elements as being distributed throughout the exchange. There are two particular advantages to this approach. Firstly, as the control system is built up of many similar small units, these units tend to be simple. Secondly, as each control unit handles one switch, and one switch only, should a fault develop in either the control unit or the switch itself, the faulty selector may be taken out of service. This leaves all the other selectors in service, and except under conditions of very heavy loading, the fault will not be noticed by the users.

The main disadvantages of step-by-step control are listed in Table 1.6.

---

*Table 1.6 Disadvantages of step-by-step control*

Lack of speed

Inflexibility

Lack of intelligence

Costs

---

The lack of speed is due to the fact that the system is implemented in electro-mechanical relays and switches. The loop disconnect dialling system was developed around this constraint. Some DTMF dialling systems are available on step-by-step exchanges. They are only a little faster than the loop disconnect systems as a translation process is involved, and the translator must itself operate the relays.

The lack of flexibility is inherent in the design of the control units. They are designed to use a particular type of signalling system, and a particular type of switch system. It would be prohibitively expensive to redesign the control unit to use another signalling system or switch, to take advantage of newer technologies.

The lack of intelligence manifests itself in several ways. The control system can not for example take account of the fact that the call can't be completed when the called number is engaged. The attempt to set up the call ties up the first, second and final selectors until the caller replaces his handset. This represents a waste of exchange equipment which could be used for other calls.

Generally it can be said that the lack of intelligence prevents the control system from being capable of anything other than basic call set up and release operations.

The cost of the control system is high. Although each control unit is itself simple and relatively cheap, there is one control unit per selector. Since there are many selectors, the overall capital cost is high. The operating costs for power and maintenance are also high due to the nature of the switching technology in use.

Consider the amount of time each control unit is actually in use. It is operational only for a short period during call set up and release, being completely idle during the conversation phase. This active time is only a small proportion of the total lifetime of the exchange, and this has the effect of making the costs appear even higher.

## 1.4.4 Common Control

Common control was an attempt to overcome some of the problems of the step-by-step system. The concept is to employ a control unit which is not tied to a single switch unit or selector. The diagram in Figure 1.14 is based upon Figure 1.13. The concentrator, distributor and expander may employ any type of switching system. The control unit may be relay or electronic logic, and is connected to various units in the exchange by dedicated control and signalling paths. The basic concept is that the control system carries out its functions by giving instructions or commands to each part of the exchange as it sets up a call. Once it has operated the required switches and so on for one call, the same control unit is then free to set up another call. A block diagram of a common control exchange is given in Figure 1.14, with a description of the system in Section 1.4.5. This description is based on the flow chart of Figure 1.5b, to which you may wish to refer.

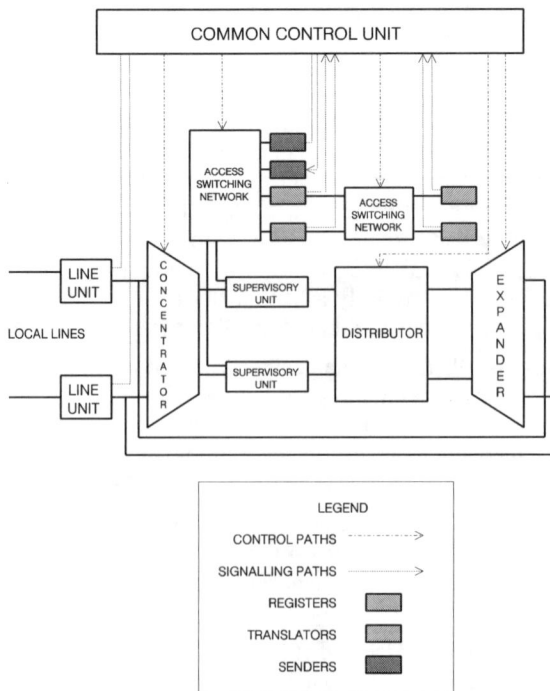

*Figure 1.14 Concepts of common control*

## 1.4.5 Operation of Common Control Systems

Line units are provided on a one per line basis. The line unit detects the presence of a calling loop when the subscriber lifts the handset. Via an internal signalling path, the line unit informs the control unit that a new call has been originated.

The control unit is aware of all supervisory units that are currently in use, and selects a free supervisory unit for this call. The control unit then operates the necessary switches within the concentrator to connect the call to the selected supervisory unit. Sufficient supervisory units will be provided to handle the expected traffic load during the busiest period of the day.

The control unit must also select a free register in which to receive the dialled digits. Since the register will only be in use during the dialling phase, there will be substantially fewer registers than supervisory units.

The control unit must arrange to switch the selected supervisory unit to the selected register by controlling appropriate switches in the access switching network. At this stage the caller will receive dial tone from the register, and can proceed with dialling. The dialled digits are stored in the register, which informs the control unit as each digit is dialled. If the first digits indicate that the call will be to another exchange, it will be necessary to connect a translator to the register in order to translate the dialling code digits into the necessary digits to route the call through the rest of the network. This allows each local exchange to be allocated a standard dial code, which may be used from anywhere in the country.

The translator carries out the conversion between the code dialled and actual digits needed to route the call. The routing digits are then passed to the senders. The control unit will select a suitable circuit to another exchange in the network and connect the sender to the selected circuit, and issue a suitable instruction, causing the sender to dial out the translated digits.

If the call is an own exchange call, the translator will not be required. When the complete number has been dialled, the control unit releases the register which is now free to be used on another call. The control unit now interrogates the called subscriber's line unit to ascertain whether or not it is engaged. If the line is busy, busy tone is switched into the caller's line via a relay in the supervisory unit.

If the called line is free, the control unit operates a relay in the relevant line unit to send ringing current to the called subscriber. Simultaneously the control unit operates a relay in the supervisory to send ringing tone to the caller. The control unit also selects a free link from the distributor to the expander and prepares to operate the necessary switches in these stages to connect the call to the called line.

The called subscriber's line unit detects the DC loop when the call is answered and informs the control unit, which then causes ringing current and ringing tone to be disconnected and then operates the selected switches in the distributor and expander. While the call is in progress the supervisory unit monitors both calling and called lines, waiting for the disconnection of the DC loop when either subscriber clears down. On completion of the call the supervisory unit informs the control unit, which then releases the switches operated in the three switching stages, and frees the supervisory unit for a new call.

## 1.4.6 Advantages of Common Control

The common control system overcomes the main disadvantages of step by step control in the following ways.

❑ The system can be substantially faster than step-by-step especially if an electronic control unit and tone dialling are used.

❑ In concept the system is more flexible because the control unit is designed independently of the switching technology. In practice common control techniques have been applied to crossbar and more effectively to reed relay exchanges. By rights, so long as the control and signalling interfaces are standardised, the same control unit can be used with any type of switching system. Equally it should be possible to redesign the control unit to incorporate new technology without having to modify the switching system. In practise this ideal is not achieved, however the problems of redesigning various parts of the exchange are substantially reduced when compared to step-by-step systems.

❑ Because there is only one control unit, time and effort can be afforded to ensure that this unit has a high degree of intelligence. This intelligence can more effectively deal with problems such as alternative routing round busy or failed parts of the network.

❑ The capital cost of an exchange using common control is smaller than an equivalent step-by-step exchange. There are several reasons for this. Firstly the overall cost of the control system is reduced. Although the control unit is far more complex, there is only one unit. Also the control system is used more efficiently than the step-by-step system. The common control unit is in use constantly, whereas the control units of Strowger selectors are idle for much of the time.

The second reason for the reduced costs is that as equipment such as supervisories, registers and translators are only in use when required. The total number of these required for any given exchange is fewer than in a step-by-step system.

The third main reason for reduced costs is due to the intelligence within the control unit being able to ascertain whether a call can be completed or not without the need to tie up switches. The exchange can thus be designed with smaller switching units, and fewer link circuits, e.g. a smaller distributor stage, for a given service level than would otherwise be the case.

## 1.4.7 Common Control Reliability Considerations

The most important aspect to be considered when applying common control techniques is that of reliability or security of operation. As the exchange includes several supervisories, registers and translators, a fault in one of these will not cause a catastrophic loss of service, neither will failure of a few switch contacts. The intelligence in the control unit can be designed to ignore faulty equipment, and not try to use it to process calls.

However the complexity and component count of the control unit make it vulnerable to failure. Since the operation of the whole exchange depends on the control unit, steps must be taken to improve its overall reliability. In the simplest case this will entail duplication of the control unit itself, with some form of automatic switch over between one control unit and the standby when a fault in the main unit is detected.

The control unit is constantly working to set up and release calls – it may often be processing several calls simultaneously. For example while many calls will be in the conversation phase concurrently, there will be several at different stages of the dialling phase. The process of setting up a call can not therefore be considered as one single operation, but as a number of sub-processes. The control unit must be designed to be able to deal with all calls offered to the exchange during the busiest period of the day, without users having to wait for dial tone when they pick up their handset.

Obviously electronic logic is faster than electrical relays, but the use of computers as the basis of a common control unit permits even faster, more flexible and more efficient control units to be designed. The application of digital computers using purpose built programs stored in memory to control exchange operation is known by the term *stored program control.*

## 1.4.8 Stored Program Control (SPC)

Almost all exchanges manufactured today are controlled by digital computers. The control systems within the exchange have become very complex, and are able to offer many facilities to the user. Facilities such as call transfer and call back when free would not be possible with the limited intelligence of relay or electronic logic control units. The diagram in Figure 1.14 is reworked slightly to show how a computer may be used for the control function. A description of the operation of the stored program control exchange in Figure 1.15 is not necessary at this stage, since it would be very similar to the previous description. You may like to try to describe the operation for yourself. The advantages of SPC are increased speed and flexibility, while all the points raised about the reliability of common control systems are also applicable.

We will be examining computer control of telephone exchanges in more detail in Chapter 7.

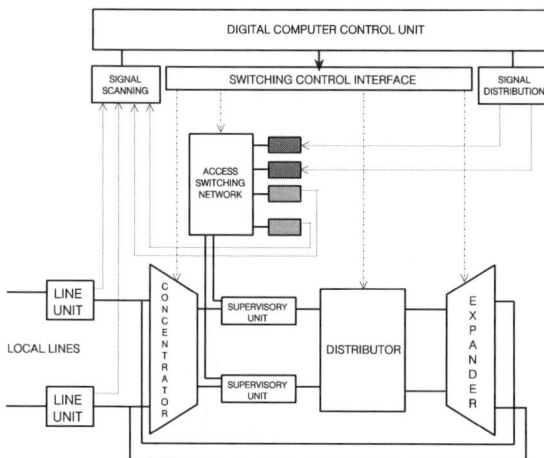

*Figure 1.15 Stored program control exchange*

## 1.4.9 Summary of Exchange Control

The control functions are those activities associated with call processing. These include recognising the initial call request signal, interpreting the dialled number and controlling the operation of those switches and tone supplies necessary to set up the call. Strowger exchanges use step-by-step control, in which the very limited intelligence is distributed to the many control units associated with each selector.

Common control, in which a single central control unit co-ordinates all the exchange processes, was introduced to overcome the disadvantages of Strowger exchanges. In the main common control was applied to electronic reed relay exchanges, such as the British Post Office's TXE4, many of which are found in the British public switched telephone network.

Stored Program Control was a natural successor to common control. SPC exchanges use a digital computer in place of an electronic logic control unit. Today the digital computer controls digital switching technology rather than analogue switches such as reed relays.

# 1.5 Transmission

## 1.5.1 What is Transmission?

In the first three sections of this chapter we have looked at the two components of the network of which most people are aware. In this section we look at the third component, the cables joining telephones to exchanges, and the cables which join one exchange to another. In telephony terms this subject is called transmission, and tends to be divided into two discrete areas.

One area of concern is the local cable network. That is the network of cables in underground ducts, on overhead poles and so on, linking every telephone to a local telephone exchange. The other area is the trunk network, that is the network of coaxial cables, microwave radio links, optical fibre cables and the like, which join all the telephone exchanges together, and to exchanges in other countries. In this introductory chapter we will briefly look at some of the early development in both the local and the trunk networks.

## 1.5.2 The Local Cable Network

There are approximately 22 million telephones in the UK (comparative figures for other countries are given in Table 1.7). Each of these telephones is connected to one of 6,000 local exchanges by a pair of copper wires. The average cable distance between telephone and exchange is around two to three kms. Some are far shorter than this especially in city areas such as London. On the other hand some cable lengths are greater than 10kms, especially in rural areas.

To a first approximation there are 50 million kilometres of telephone cable in the local cable network. In fact these cables account for a larger proportion of British Telecom's capital investment than the telephone exchanges.

*Table 1.7 Sizes of various major telephone networks*

| Country | Millions of lines | Country | Millions of lines |
|---------|-------------------|---------|-------------------|
| USA | 118 | Italy | 17 |
| Japan | 46 | Canada | 13 |
| USSR | 28 | Spain | 9 |
| W Germany | 26 | Brazil | 7 |
| France | 23 | Australia | 7 |
| UK | 22 | Netherlands | 6 |

When you consider just how much cable is involved, it becomes obvious that the cable costs must be kept to a minimum. This is the reason for selecting a two-wire cable of relatively light gauge and accepting the associated problems of signal attenuation and interfacing telephones and exchanges at each end.

Over this two-wire cable, it is necessary to transmit a DC current to power the carbon granule microphone, receive from the telephone a series of DC loop disconnect pulses representing the dialled number, transmit high-voltage high-current ringing, and finally telephony speech as low voltage electrical signals. Although this aspect of telephony is often overlooked, the requirement to transmit these vastly different signals over the telephone line means that the interfaces at the telephone and the exchange are quite complex.

## 1.5.3 The Subscriber's Loop

The term Subscriber's Loop (or local loop) refers to the telephone line and the telephone connected at the distant end. There is currently a lot of work being done to improve the transmission qualities of the subscriber's loop by introducing digital transmission and signalling into this area of the telephone network.

Also on the cards is the prospect of a single optical fibre bringing to the house, not only the telephone but TV and other services.

## 1.5.4 Transmission in the Trunk Network

Circuits which link one exchange to another are known as trunk circuits.

The number of trunk circuits between any two particular exchanges will depend upon how much traffic is expected to pass between the two exchanges in question. For example between London and Birmingham we can expect a very large number of trunk circuits since at any one time there will be many people in London wishing to talk to people in Birmingham, and vice versa.

In the discussion on common control techniques in Section 3 of this chapter, it was stressed that equipment such as supervisories and registers were selected and used for calls, then immediately released, becoming available for use on new calls. This is exactly the case with trunk circuits. When a trunk call is originated at a local exchange, the control system selects a free trunk circuit from those available and processes the call over that circuit. When the call is complete, the trunk is released ready for another call. Thus if one was to dial from London to Birmingham several times in succession, the probability of the call being connected over the same actual trunk circuit is very remote.

## 1.5.5 Multiplexing

In only a few cases is each individual trunk circuit actually routed over its own two-wire cable. Unlike the local cable network, where each pair terminates in a different location, all the trunk circuits between one exchange and another start and finish at the same place. This allows some economies to be made by allowing a number of trunk circuits to share the same cable.

It is obviously not possible to simply connect several circuits to the same cable at one exchange and expect the distant exchange to sort them out. How would the exchange know which set of signals belonged to which call, when they all occupy the same frequency spectrum?

Multiplexing is a process which allows several calls to be transmitted along the same pair of wires, coaxial cable, or optical fibre cable without interfering with one another. The process of sorting out the calls at the distant end is known as demultiplexing. Multiplexers are used to reduce the physical number of cables between one exchange and another, and thus reduce the overall cost of providing these trunk circuits.

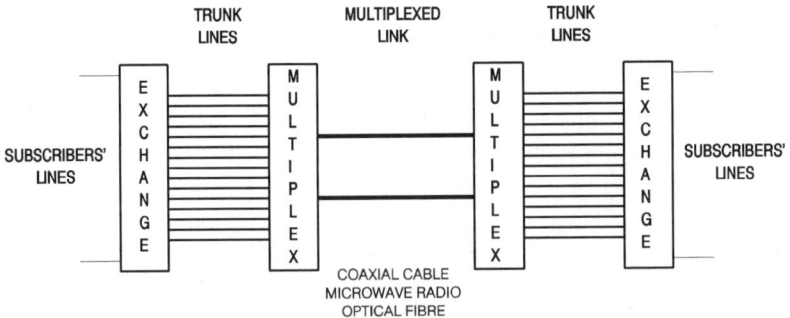

*Figure 1.16 The use of Multiplex equipment on trunk routes*

Multiplexing then, is a technique for combining a number of calls in order that they may all be transmitted on the same single cable. In Figure 1.16 we show the concept of multiplexing in which all the trunk circuits from one exchange on a particular route are connected to a multiplex equipment. The single output from the multiplex is connected by coaxial cable, optical fibre or radio relay link to the input of a similar equipment (co-located with the distant exchange) which is able to separate the individual circuits. These are then connected to the distant exchange. Each multiplex equipment carries out the multiplexing and demultiplexing process.

There are two main multiplexing techniques currently in use for voice communications systems: frequency division multiplexing (FDM), and time division multiplexing (TDM).

## 1.5.6 Frequency Division Multiplexing

FDM was the first technique to be used extensively in the telephone network. It is an analogue system based on the concept of dividing the available bandwidth of any transmission medium into slots. Each individual trunk circuit is allocated one of these slots, or channels. Many FDM systems are still in existence today, although they are being slowly replaced by the digital systems we will be looking at later.

## 1.5.7 The Principles of FDM

In order to explain the basic principles of FDM, we will look at one of the standard FDM systems as recommended by the CCITT. In this system each telephone channel is deemed to require a bandwidth ranging from 300 Hz to 3.4KHz. For reasons which will become obvious later, the actual bandwidth allocated to each channel is 4KHz. Twelve telephone channels are combined giving a total bandwidth of 48Khz. In the CCITT system this 48KHz band lies between 60KHz and 108KHz.

The process of taking a telephone channel from the audio frequency band up to the correct 4KHz slot in the multiplexed system is known as frequency translation. This process is normally carried out using single side band amplitude modulation techniques, in which the lower side band is selected.

For example, consider that a 1KHz tone is applied to channel 1 of an FDM equipment. The tone is used initially to modulate a carrier of 108KHz using double sideband suppressed carrier modulation. This produces two sidebands at 108+1Khz, and 108-1KHz. Since only one sideband is required the upper sideband (at 109KHz) is removed by filtering leaving only the lower sideband. The same process occurs for channel 2, except the carrier frequency is 104KHz, and a signal at 103KHz is produced.

All channels are similarly processed using carriers ranging from 64KHz for channel 12, to 108KHz for channel 1. The spacing between carriers giving rise to the 4KHz slot allocated to each channel. The outputs from all channel modulators are simply combined, and since they all occupy different frequency slots, they do not interfere with each other. In Figure 1.17 we represent the audio bandwidth of a telephone channel by a triangle, the apex of the triangle representing the high frequencies in the audio band. Since the lower sideband is selected, a frequency inversion takes place. i.e. the high frequencies of the original signal become the low frequencies of the translated signal, and this is represented by showing the triangles reversed in the output spectrum diagram.

At the de-multiplex equipment, each 4KHz channel is separated out by filtering and demodulated to reproduce the original signal applied to the channel.

## 1.5.8 Signalling in FDM Systems

We have described how a number of telephone channels are multiplexed using FDM. The description is valid for voice signals and any other voice frequency signal, such as the analogue output of a data communications modem. But since exchange to exchange signalling is often in a DC form, not dissimilar to the LD system used in the subscriber's loop, some extra processing is required before the signalling can be connected to the FDM multiplexer.

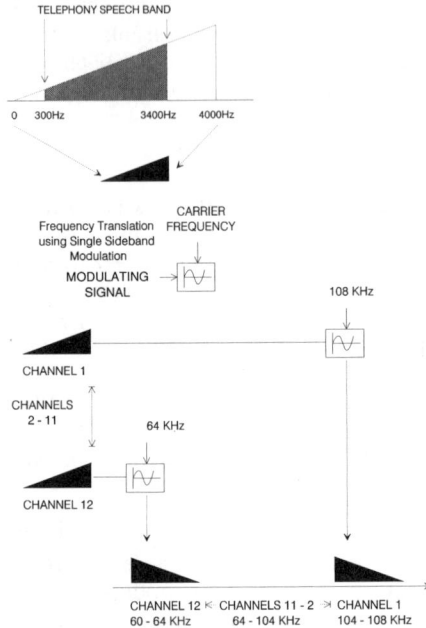

*Figure 1.17 Frequency division multiplexing*

There is not room in this book to describe fully the several techniques which are available. The basic concept is to convert the DC signalling conditions to voice frequency tones which can then be multiplexed as for voice signals. At the distant de-multiplexer, special circuits detect the signalling tones and reconvert them to the required DC signalling conditions.

The frequencies selected for signalling may be within the speech band of 300 Hz to 3.4KHz. For example 2280 Hz is used in many FDM systems. Such systems are called in-band systems and require a speech immunity circuit in the demultiplex to prevent actual speech signals at the signalling frequency operating the signalling circuits. When the selected signalling frequency lies outside the speech band, for example 3825 Hz, the systems are known as out-of-band, and do not require speech immunity circuits.

## 1.5.9 Higher Capacity FDM Systems

When the number of channels provided by the basic (or primary) 12-channel FDM multiplex is considered insufficient, capacity can be increased using secondary multiplexing without having to lay further cables. The 12-channel system is known as a basic group, and is the basic building block in a hierarchy of multiplexers which can provide up to 2,700 channels.

To provide a system capable of handling 60 channels, five separate 12 channel basic groups are each frequency translated by using each group to modulate one of the high frequency carriers as indicated in Figure 1.18. The five frequency translated groups can then be combined on to a single cable without interference to form a 60-channel super-group. To provide even higher capacity systems, it is possible, for example, to frequency translate five super-groups to produce a 300-channel system.

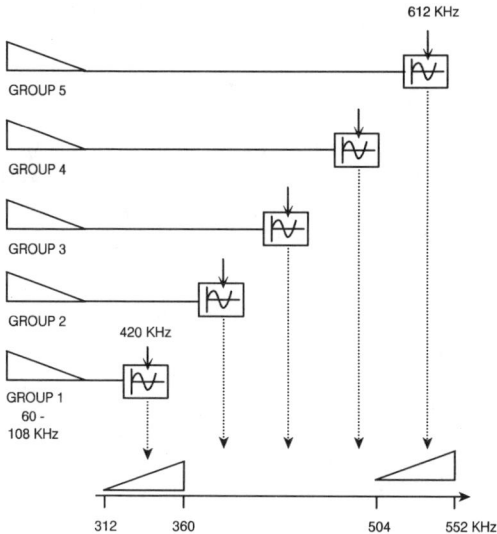

*Figure 1.18 The structure of a 60-channel FDM super-group*

## 1.5.10 Time Division Multiplexing

FDM systems divide the available bandwidth into a number of frequency slots. TDM systems on the other hand divide the time the bandwidth is available into a number of discrete time slots. Each audio channel is allocated a time slot, rather than a frequency slot, and during each channel's allocated time slot sufficient information is sent to allow the receiving de-multiplex equipment to reproduce the original audio signals. Since TDM systems are digital in nature the whole of Chapter 5 is dedicated to them.

# 1.6 Network Terminology

In this section, we explain a few of the terms commonly used when discussing telephony networks.

### Office of Telecommunications (OFTEL)

This non-ministerial Government department has several responsibilities. The main areas of OFTEL's concern are:

❏ Monitoring the Public Telecommunications Operators' (PTO) licences

❏ Monitoring Non-PTO licences

❏ Overseeing the approval of apparatus to be connected to the network

❏ Consumer affairs

While OFTEL controls type approval for such things as telephones, it is the Department of Trade and Industry (DTI) that issues operating licences to the telecommunications companies.

## Public Switched Telephone Network (PSTN)

The term PSTN refers to the normal public telephone network operated in most countries of the world. In many countries the PSTN is operated as a state owned monopoly. However in the UK, recent legislation has removed the monopolies and a state run organisation (the DTI) grants operating licences to the telephone companies. In the UK there are three PSTNs:

❏ British Telecom (British Telecommunications plc)

❏ Kingston Communications (Hull) plc

❏ Mercury Communications Limited

All three are interconnected, thus enabling subscribers on one network to call subscribers on either of the others.

## Subscriber

The term subscriber refers to the customers of the PSTN operators. The subscriber's apparatus refers to any terminal equipment connected to the PSTN.

## Trunk

A circuit between two Trunk exchanges in a PSTN.

## Junction

A circuit between two Local exchanges in a PSTN.

## Private Automatic Branch Exchange (PABX)

Most large office complexes are equipped with an internal telephone system which is normally based on a PABX. The PABX provides communications between all offices in the building, and may be owned and maintained by the organisation itself, or leased from a PABX supplier approved by OFTEL. In either case the maintenance must be carried out by staff approved by OFTEL.

In most cases the PABX will also be connected to the PSTN via a local telephone exchange, so that the PABX users are able to make calls to, and receive calls from the PSTN. In some cases the PABX will be connected to more than one PSTN, with software within the exchange determining call routing on a least cost basis.

## Exchange lines

The circuits connecting a PABX to a PSTN local exchange. Until recently, these lines have been very similar to normal analogue telephone lines, but there is now also the option of using digital lines.

## Tie line

This term refers to circuits linking PABXs. Since these circuits are normally leased from the PSTN operator, and used solely by the organisation concerned, the terms Private Wire (PW) or Leased Line are often used to mean the same thing. Until fairly recently these were predominantly analogue, but now digital links are mainly used.

## PABX extension

Any telephone or other equipment connected to a PABX telephone line is defined as an extension.

## Attendant's console

Although there is no longer a need for an operator to switch calls, most PABX installations include an attendant's console which is operated by a telephonist or other member of staff. Generally all incoming calls from the PSTN are routed by the PABX to the attendant's console. The telephonist can then direct the call to the requested extension, or assist the caller in some other way.

The attendant, can also make calls for those extensions of the PABX which are not provided with direct access out to the PSTN.

## Direct dialling in (DDI)

The DDI facility allows incoming calls to PABX from the PSTN to be routed directly to the required extension. The caller dials the PSTN number of the PABX, followed by the extension number. This facility is not available on all local exchanges at present. However most PABX are designed to operate DDI, so that it is available as soon as the local exchange is upgraded to provide the service. One of the major problems that has to be addressed when installing a DDI system is the requirement to ensure that the numbering scheme is compatible with the existing public network, without the need to redesign the numbering scheme of the PABX This is usually achieved by translating the digits sent by the PSTN to the PABX into a valid PABX extension number.

## Private networks

Many organisations operate a number of PABX in different sites around the country, and overseas. Since the PABX are connected to the PSTN, calls between offices can be made via the PSTN. However in circumstances where many calls are made between offices of the same organisation, the costs of these calls is high.

A private network consists of a number of PABX connected together over tie lines or private dedicated circuits. Inter-PABX calls are not routed over the PSTN but over these dedicated circuits, thus reducing costs.

## Circuit switching

The telephone network is an example of a circuit switched voice network. Telex is a circuit switched telegraph network. Circuit switching is characterised by the points in Table 1.8:

---

*Table 1.8 The characteristics of circuit switched networks*

Physical connection being set up on demand

Connection being maintained as long as required

All switches and lines are transparent to the communication

Terminal equipment (e.g. telephones) must be compatible

Generally suited to conversation, i.e. two-way traffic

Communications are in real time, i.e. no significant delays occur in signal transmission.

---

A slight amplification of the last point is probably required, as in today's world where many calls will be routed over satellite systems. Because of the vast distances involved the time for the signal to travel from earth to the satellite and back may be as along as half a second. This obviously is a delay but it is not significant in terms of conversational communications. This contrasts with the delays which do occur in message switched systems.

## Message Switching

Message switching is particularly suited to one-way communication, where long transmission delays can be tolerated. Often referred to as Store and Forward, these systems are based on networked computers. Messages will contain addressing information which is used by the computers to route the message to the correct destination. If the message can not be delivered for any reason, for example, the required terminal is currently receiving another message, or is out of service, the new message is stored in the computer and forwarded when the terminal becomes available.

Another major facet of message switching is that it may allow communications between incompatible terminals, such as teleprinters, (or teletypes, and VDUs) using different transmission parameters (Baud speed, Code, parity). The intelligence within the network is able to carry out speed and code conversion, and if required, transmit the same message to more than one recipient terminal. Examples of message switching systems include electronic mail systems.

## Packet Switching

Packet Switching is becoming a very popular system for public and private data networks. The system is an efficient method of data transmission, and has many of the advantages of circuit switching and message switching. British Telecom have been operating a public packet switched network in the UK since 1981.

BT's Packet SwitchStream provides communication between computers, terminals and other devices such as credit card readers, and is particularly suited to applications where there is a requirement for fast call set up with only small amounts of data to be transferred between the two parties to a call.

# 1.7 Other Uses of the Telephone Network

## 1.7.1 Digital Communications

In this section, we consider how the telephone network has come to be used for communications other than telephony. The two main items of concern are Facsimile (Fax) and Data Communications (Data Comms), both of which involve the transmission of digital signals.

The telephone network was originally designed to provide a communications system for voice purposes only. To use the telephony network for other forms of communications it is necessary to recognise the constraints imposed by the technology in use. Generally the constraints are due to the analogue nature of the network, and include:

❏ A limited bandwidth of 3KHz

❏ Interference in the form of noise and crosstalk

❏ Distortion due to uneven frequency and phase responses

## 1.7.2 Interface Requirements

The designers of Fax and Datacomms equipment must include interfaces within their equipment which convert the digital signals to suitable analogue signals capable of being transmitted through the network. This implies that the digital to analogue conversion process must produce a signal which can be transmitted through the network, suffering attenuation, distortion and interference and still be correctly re-converted to digital at the distant end.

The problem is compounded by the fact that the network is circuit switched. If a permanent circuit could be provided between Fax machines or computers and their terminals, this permanent circuit could be specially conditioned to optimise the quality of the circuit for the type of communications involved. In a circuit switched environment there are two reasons why this conditioning can not be applied:

❏ The trunk circuits in the network are used for all types of communications, and can not be conditioned and reserved for just one user, or one type of communications system.

❑ Because each individual call will be routed over different circuits, and maybe through different exchanges and different routes, it would be impossible to condition the subscribers' lines to match all the variations in distortion and so on that would occur from call to call.

## 1.7.3 Facsimile

Facsimile may be defined as the transmission of a replica of a document, picture, map and the like by radio, telephone, telegraph or cable. The basic principles of have been understood for over 100 years, and since 1968 recommended standards have been produced by the CCITT to ensure compatibility between fax machines of different manufacturers.

## 1.7.4 Facsimile Standards

Four fax groups have been defined, of which Groups 1 and 2 use analogue principles. Group 1 was introduced in 1968 and was based on frequency modulation (or frequency shift keying) techniques. An A4 document took six minutes to transmit. This group is now virtually obsolete.

Group 2 was established in 1976 based on Amplitude Modulation (or Tone On-Off Keying). The transmission speed and reproduction quality were superior to Group 1. Some Group 2 machines had the ability to work with Group 1 equipment.

Machines in groups 3 and 4 both use digital techniques. However group 3 machines are designed to work over an analogue PSTN, whereas those in group 4 are designed to operate over purely digital circuits. Group 4 machines are capable of transmitting an A4 document in four seconds over a 64Kbit/s data circuit. We will be discussing the use of group 4 machines in the chapters on facsimile and the Integrated Services Digital Network.

## 1.7.5 Datacomms

This of course is a huge subject, and many good books already exist in this area. Our main concern is how data is transmitted over the telephone network. Ever since the early days of computing, the computer engineers have sought ways of operating the computer terminals at a distance from the main computer. At the time the telephone network was already in place, and while it was not designed to handle data, for many datacomms applications it was a cost effective medium over which to transmit data signals.

## 1.7.6 Modems

The interface between the digital computer or terminal and the PSTN is called a Modem (from modulator-demodulator). A basic modem will simply carry out digital to analogue conversion (D/A) and digital to analogue conversion (A/D). A common D/A technique based on voice frequency telegraphy techniques involves Frequency Shift Keying (FSK). The modem contains an oscillator which is electronically switchable between two frequencies. When a binary 1 is to be transmitted the modem switches the oscillator to one frequency, for a binary 0, the oscillator is switched to the other. This is in fact a form of frequency modulation.

The maximum bit transmission rate of such a modem is low, and other modulation techniques involving phase and amplitude modulation together have been designed to allow bit rates of up to 9.6 KBits/s over circuit switched networks.

## 1.7.7 Modern Modems

The modems available today are designed around a microprocessor. This built in intelligence allows the modem to dial up the required number automatically after receiving instructions from the computer or terminal.

Some modems feature an automatic answer facility and would normally be located at the host computer site. As we have already mentioned each call through the PSTN will be subject to varying amounts of interference and distortion. Modern modems can automatically adjust themselves to transmit data at the highest possible speed over each circuit, and if necessary carry out limited error detection and correction.

## 1.7.8 Modem Standards

The V series of recommendations from the CCITT cover the transmission of data over analogue circuits. The most common standards in use at present include:

| | |
|---|---|
| V 21 | 300 Baud modem for use on the PSTN |
| V 23 | 600/1200 Baud modem for use on the PSTN |
| V 24 | Interface between computer and modem |
| V 25 | Automatic calling and answering equipments. |
| V 26 | 2400 Baud modem for use on 4 wire leased lines |
| V 26 bis | 2400 Baud modem for use on the PSTN. |
| V32 | A new standard for full duplex high speed modems for use on leased lines or over the PSTN. |

## 1.7.9 Digital Communications in the Future

This aim of this chapter was to provide an introduction to the subject of digital telecommunications networks by briefly describing the analogue telephone network. For some years now the telephone network has been undergoing a transition which has been unnoticeable to most users. Slowly the trunk network has been transformed from a three layer hierarchical structure based on analogue switches and FDM multiplex to a single layer digital network based on PCM transmission and digital exchanges. This network is the backbone of System X.

The System X programme involves replacing some 6,000 local exchanges with new digital exchanges, and eventually the whole telephone network will be digital and form the basis of a network capable of handling voice, data, telegraph and fax with equal ease. The remainder of this book is devoted to describing the components and principles of this integrated digital network.

# 2

# *Electronic Telephones*

The aim of this chapter is to provide the reader with an understanding of the operation of modern telephones. Although some of the material is quite detailed and some readers may wish to skip the detailed sections, these sections do provide an insight into the technology of the modern telephone.

## 2.1 An Introduction to Electronic Telephones

### 2.1.1 The Advantages of Electronic Telephones

Over the last decade telephone designers have incorporated silicon technology into the once simple basic telephone. The design of the telephone body is now the remit of artistic design consultants as the shape of the telephone is no longer constrained by the necessity to house physically large components such as the bell, rotary dial and transformer. Electronic components have replaced these large and expensive items. For example, modern telephones have a push button key pad rather than a rotary dial, and the magneto bell has been superseded by an electronic sounder device. The basic electronics which now include audio amplifiers obviate the need for a bulky and expensive transformer. Even the carbon granule microphone is in demise having been replaced by a sensitive dynamic microphone in most new telephones.

The advantages of electronic telephones from a supplier's and user's viewpoint are numerous. Some of the main advantages are listed in Table 2.1.

*Table 2.1 Some of the benefits of electronic telephones*

Faster and easier dialling
Greater reliability
Improved speech quality
Lower manufacturing costs
Lower maintenance costs
Lower power requirements

Most of these points are self explanatory, but some require amplification.

## 2.1.2 Post Dial Delay and Misdialling with Push Button Keypads

Using a rotary dial, a nine-digit number takes approximately nine seconds to dial. The call will not be completed until a few seconds after dialling is complete. The user will experience a short post-dial delay before hearing the busy tone or ring tone.

With push button dialling, reasonably dextrous users are able to dial a typical nine-digit number in under three seconds. A push button telephone connected to a conventional exchange will use Loop Disconnect signalling and must not transmit faster than 10 LD pulses per second, otherwise the exchange will not correctly connect the call. The dialled number is therefore stored, and subsequently transmitted by the electronics in the telephone. As far as the user is concerned, the time difference between keying the last digit, and the last digit actually being transmitted, significantly increases the post-dial delay.

Another point is that some users report more frequent misdialling with push button telephones. There may be two reasons for this. The first is that it is human nature to attempt to dial as fast as possible, and thus mistakes may be made. The second is that many telephone users are also computer and calculator users, and unfortunately the numeric keypad layout of computer keyboards and calculators is substantially different from that on a telephone.

## 2.1.3 Speech Quality

Although the new dynamic microphones and associated amplifiers greatly increase the intelligibility of telephone speech, it must be remembered that much of the noise and interference heard is actually picked up in the telephone network. While there is still a large proportion of analogue equipment in the telephone network, the new telephones will not greatly improve overall speech quality. However, in most offices there has been a noticeable improvement in speech quality when modern telephones and a new PABX has been installed.

## 2.1.4 Lower Power Requirements

The electronic telephone uses about one third of the power consumed by a conventional telephone. Even if some of this saved power is used to provide extra facilities such as memories and indicators, running costs will be reduced.

# 2.2 A Typical Electronic Telephone

## 2.2.1 The STC Elektron

One of the more popular telephones available for domestic and business users is the Elektron, manufactured by Standard Telephones and Cables (STC) and marketed by BT under the trade name of Viscount. This telephone, which is based on a single bipolar IC, provides the features listed in Table 2.2 which are typical of modern instruments.

*Table 2.2   The STC Elektron's major features*

Last number redial

Muting facility

Ringer volume control

Switching between LD and DTMF signalling

Exchange recall facility

10-number memory

## 2.2.2 Last Number Redial

This feature is inherent in the design of any integrated circuit (IC) produced for use in push button telephones. Since the IC must store the dialled number as it is keyed in, a small increase in circuit complexity and an extra key allow this number to be retransmitted without the need for the user to key in the digits again. Typically the number store will be large enough to hold a 24-digit telephone number which permits international dialling to a DDI extension on a PABX.

## 2.2.3 Muting Facility

When the mute facility is invoked, the microphone amplifier within the IC is disabled, thus preventing any sounds picked up by the microphone from being transmitted. This feature allows the user to talk to a colleague in the same room without being heard at the other end of the telephone line.

## 2.2.4 Ringer Volume Control

The magneto bell is replaced by an audio oscillator, operating at two frequencies around 1KHz. When a 17Hz incoming ring signal is received, the audio oscillator is operated and switches between the two tones to produce a warbling sound. The volume of this warbling sound may be controlled electronically or more simply by adjusting a knurled knob which moves a cover across the front of the output transducer to reduce the audio coupling.

## 2.2.5 Dual Tone Multi-Frequency (DTMF) Signalling

Almost without exception, modern digital PABXs are designed to operate with telephones using DTMF signalling. Users of BT's System X local exchanges will also use DTMF signalling if they wish to access some of the special services provided by System X. To avoid any possible confusion readers should be aware that in British Telecom's System X literature the DTMF system is referred to as the Touchtone system.

DTMF signalling provides a more flexible, and faster signalling system than the Loop Disconnect system. Figure 2.1 shows the layout of a DTMF keypad, with the optional four extra alphabetic keys labelled A to D.

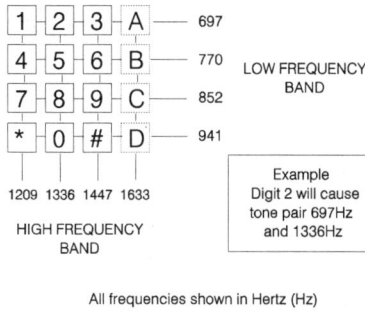

*Figure 2.1 The layout of a typical DTMF keypad*

The DTMF system uses eight-voice frequency tones, of which four are in a low band (<1KHz) and four are in a high band (> 1KHz). The lower frequency band is from 672Hz to 941Hz, while the high band runs from 1209Hz to 1633Hz. The depression of a key will thus cause one low band and one high band tone to be transmitted simultaneously by the telephone to the exchange.

There are 16 possible combinations in this 2 (1 out of 4) MF system, although most telephones in the UK will implement only 12, and not use the alphabetic keys. The keys marked # and * are used to access the special services of the exchange and do not normally invoke any function within the telephone.

Dual standard telephones are now becoming available. These are provided with a switch, often mounted on the underside of the instrument, which allows conversion for use on Loop Disconnect or DTMF systems.

## 2.2.6 Exchange Recall Facility

To access some of the special facilities of the exchange during a call, perhaps to transfer the call or set up a conference, the user has to dial an appropriate short code. In order for the exchange to reconnect digit receivers to the line to determine which code has been dialled, the exchange must be recalled. This can not be done by replacing the handset, as this is too clumsy and will result in the call being cleared. Another button normally labelled R (for Recall) is provided. Its operation will send a signal to the exchange informing the control system that a special facility is requested.

Two methods of recall signalling are found in practice, though generally only one will be implemented on any particular exchange.

## 2.2.7 Earth Recall (ER)

Earth Recall is the older method. Operation of the R button causes one wire of the telephone line to be connected to earth. A three-wire cabling system is required to implement earth recall, and this precludes the system from use in the PSTN where the cost of three-wire distribution is prohibitive. It is for this reason that System X local exchanges and most modern PABX implement a more modern recall system known as Timed Break Recall.

## 2.2.8 Timed Break Recall (TBR)

In the Timed Break Recall system, operation of the R button causes the DC loop to the exchange to be disconnected for a nominal period of 66 mSecs. This time is not sufficient for the exchange to detect this disconnection as a cleardown signal. The advantages of TBR are that it requires only two-wire distribution making it simpler and cheaper than ER to install, and thus is suitable for PSTN use. Another advantage, where TBR is implemented in electronic LD telephones, is that the 66 mSecs disconnect period is identical to the signal sent when digit 1 is dialled. Thus in this type of telephone TBR can be implemented by wiring the R button in parallel with the 1 key.

Many modern telephone instruments are switchable between loop disconnect and MF signalling, and between timed break and earth recall.

## 2.2.9 Ten-number Memory

Telephones which include a memory so that often used numbers may be stored and called up by single key press are called Repertory Telephones. Many of the ICs available today include this feature. Repertory phones normally have two extra buttons labelled P for Program, and T for Transmit.

The numbers are stored in the memory by depressing the Program button and a digit to select the store required. The number is then keyed just as for dialling, however the number is not dialled out, but stored in the selected memory location. When this number is required to be dialled, the Transmit button is pressed followed by a single digit to select the appropriate store. The IC then transmits this number as if it had just been dialled in.

The telephone IC will draw a minute current from the exchange even when idle to maintain the stores without the need for internal batteries. If the telephone is disconnected from the line for more than a short period, in most cases the stores will need to be reprogrammed.

# 2.3 Summary

This chapter has examined briefly, modern analogue electronic telephones. The result is that the costs of manufacturing the instrument have been reduced – a typical low cost telephone retails for under £10.

Analogue telephones will continue to be used at least into the next century, despite the emergence of digital telephone exchanges and networks. There will however be an increasing number of digital telephones available in the future as British Telecom introduces its Integrated Services Digital Network (ISDN), and the PABX manufacturers produce digital instruments for their exchanges.

*3*

# *Facsimile Systems*

## 3.1 Introduction to Facsimile

### 3.1.1 The Use of Facsimile

Facsimile (Fax) is the term given to the transmission, reception and subsequent reproduction of documents over a communications system. Most Fax machines are connected to public or private telephone networks, though there are some systems which can be used over radio links.

Fax permits the transmission of documents such as letters, orders and invoices at a rate of approximately 30 seconds per A4 page. However, unlike Telex which is significantly slower and only permits textual information to be transmitted, fax can also be used to transmit graphical images such as maps and line drawings. Some of the more modern machines have the ability to transmit photographs with a fair degree of quality.

Fax machines are becoming an increasingly more important communications asset to businesses of all sizes. Their use is likely to increase significantly in the near future when the introduction of ISDN permits fast digital fax machines to be used on the public network.

These digital fax machines will be able to transmit A4 documents in around four to five seconds. The reduction in transmission times is due to the much increased capacity of the ISDN network, making it possible to transmit a fax signal at 64KBit/s rather than the current maximum of 9.6KBit/s permitted over analogue circuits.

### 3.1.2 Basic Facsimile Principles

The basic principles of facsimile involve scanning a document in much the same way as a TV camera scans an image to be transmitted. Facsimile machines are however more similar to photocopiers, and in fact many of the photocopier manufacturers also produce fax machines by using technology which is common to both.

In a photocopy machine, the document to be copied is laid upon a transparent surface and then scanned by one of two techniques. Either the document is moved past a scanner, or the document remains stationary and the scanner moves. Generally fax machines use the first technique although newer machines will be employing movable scanners similar to those used in larger photocopiers.

In either case scanning the document effectively divides it into a narrow raster of horizontal lines as illustrated in Figure 3.1. Each scan line can then be considered as a sequence of picture elements (or pixels) which may be white or black.

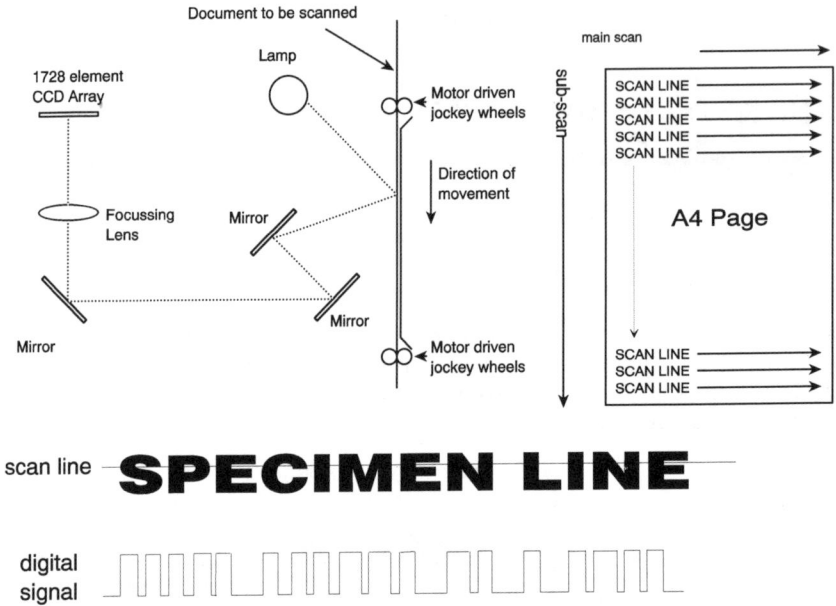

*Figure 3.1   Fax principles – scanning a document*

The horizontal resolution is eight pixels per mm, and thus for an A4 page there are 1728 pixels per scan line. For each of the 1728 pixels, a binary digital signal (1 bit) representing black or white is transmitted over the communications system to a fax receive terminal.

Once one scan line has been produced and is being transmitted the paper is moved up slightly so that the next scan line can be produced. The movement in the vertical direction is referred to as the sub-scan, and has a resolution of 3.85 lines per mm, although some machines offer higher resolution at 7.7 lines per mm.

The fax receiver consists of a decoder and a printer, which in many cases is a thermal printing unit. Once the receiving fax terminal is synchronised to the incoming signal, the signal is decoded into white and black picture elements, black elements causing the printer to mark the corresponding point on the page to be printed. In this way the fax receiver reconstitutes a copy of the original document, scan line by scan line as illustrated in Figure 3.2.

## 3.1.3 Other Functions of the Fax Machine

This simple explanation however hides some of the complexity of the fax machine. For a successful fax transmission to take place, it is first necessary to establish a connection through the telephone network to the called fax terminal. This can be done by using a telephone, though most modern fax machines include an integral dialler, based on the same type of integrated circuits found in conventional push-button telephones. It is also therefore essential to ensure that the dialler is compatible with the local telephone exchange (i.e. 10pps or DTMF dialling).

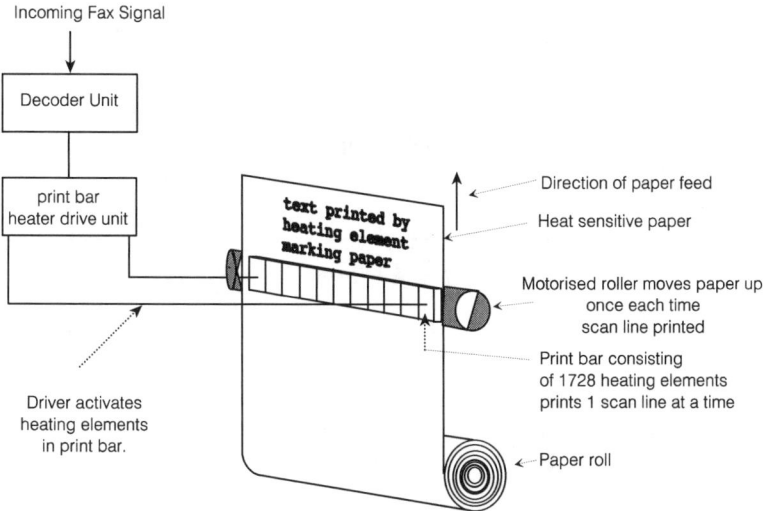

*Figure 3.2 The basic principle of facsimile reception*

The local exchange to which the called fax terminal is connected will send the same type of ringing signal to the fax machine as it would to a normal telephone. This call can of course be answered by an operator using a telephone handset, and then the fax machine connected to the telephone line by operating a switch on the fax machine.

To save on operator time and transmission costs most modern machines have automatic diallers which can be programmed to place calls during off peak periods. It this therefore necessary for the called machine to have an automatic answer facility.

Fax machines are actually digital in nature, but machines in groups 1, 2 and 3 are designed to be connected over an analogue telephone network, or at least over a conventional analogue telephone line to the local exchange which can of course be analogue or digital.

The fax machine must therefore also include some form of modem, to permit the transmission of the digital fax signal over the analogue line. It should come as no surprise to those with experience of datacomms that the transmission speeds of such modems range from 2.4KBits/s to 9.6KBit/s, and use the same phase modulation and quadrature amplitude modulation techniques as modems employed for datacomms applications.

## 3.1.4 The Evolution of CCITT Facsimile Standards

Fax machines are a good example of communications standards in operation. As it is now possible to purchase a fax machine from many vendors it is reassuring to know that a newly purchased machine will be compatible with almost all other machines connected to the network, irrespective of manufacturer.

Facsimile standards are to be found in the CCITT T series of recommendations, and have evolved since the late 1960s as technology has improved the quality of reproduction and increased transmission speeds.

Table 3.1 gives outline details of the four basic fax standards, referred to as groups G1 to G4. Readers unfamiliar with the abbreviations used to describe the transmission type should note that these terms are explained later in the chapter.

*Table 3.1   Basic Facsimile Data*

| Group | Date | CCITT Rec'n | Transmission Type | A4 Transmission Speed |
|-------|------|-------------|-------------------|----------------------|
| 1 | 1968 | T2 | Analogue FM | 6 mins |
| 2 | 1976 | T3 | Analogue AM PM VSB | 3 mins |
| 3 | 1980 | T4 | Analogue QAM or PM | 30 secs |
| 4 | 1984 | T | Digital up to 64KBit/s | < 10 secs |

# 3.2 A Typical Fax Machine

## 3.2.1 Block Diagram of a Typical Fax Machine

Figure 3.3 is a simplified block diagram of a typical group 3 fax machine incorporating a dual standard dialler (i.e. switchable between 10pps and DTMF), and an automatic answer function. This machine is also capable of making local copies. This is of use when a document of questionable readability is to be transmitted, as the local copy will give an indication of the readability of the document as it will be received at the distant terminal.

The main units of the machine are controlled by a microprocessor, with a memory element which may be extended to hold pages for retransmission without the necessity of re-scanning.

## 3.2.2 Fax Transmission

The document is illuminated by a lamp. Light reflected from the document is focused by a mirror and lens arrangement on to a photosensitive detector, such as a charge coupled detector (CCD) array with 1728 elements, each element in array corresponding to one pixel on the scan line. The voltage level stored in each of the CCD elements will be dependent upon the amount of light reflected from the corresponding point on the scan line. The amount of light reflected from each point will itself be dependent upon the image to be transmitted and will vary from a low level of reflection for black (i.e. printed) areas, and a high level of reflection from white, or unprinted areas. Note that all elements in the CCD are activated at the same time, equivalent to loading the CCD in parallel.

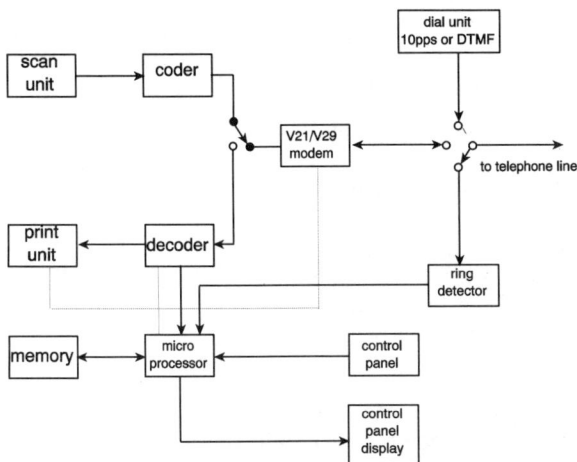

*Figure 3.3   Simplified block diagram of a typical fax machine*

The signal from the CCD in the scan unit is clocked out in series to form a waveform similar that of a pulse amplitude modulated signal which is then fed to the coder.

The main function of the coder is to make a decision about each incoming pulse from the CCD. A threshold circuit is employed to decide whether each pulse should be considered as black or white, so that the output from the coder is a binary sequence with each pulse of equal amplitude. However the coder has another function which concerns reducing the number of bits actually transmitted in order to reduce transmission time.

Again using the A4 page as a typical document, at a sub-scan density of 3.85 lines per mm, there are 1145 scan lines per page. The total number of pixels per page is the number of pixels per line (1728) multiplied by the number of lines per page (1145) giving a total of nearly 2 million pixels. Since each pixel is represented by one bit, it is necessary to transmit 2MBit of fax data to the distant fax terminal.

The nominal transmission speed is 9.6KBit/s, and thus it would take approximately 3.5 minutes to transmit an A4 page unless some digital processing is carried out to reduce the total number of bits that must actually be transmitted.

## 3.2.3 Data Compression

The processing required is known as data compression. Consider that a page of text is to be transmitted. Much of the page will contain *white* which when scanned will produce variable length strings of binary zeroes. Similarly the actual letters of the text will produce shorter variable length strings of binary 1s. An underline in the text will produce a long string of binary 1s, although the probability of this occurrence is far lower than that of a string of zeroes.

The length of each string of 1s or zeroes is called its run length. Data compression involves transmitting a new code which represents the run length of the string, rather than the string itself.

Two common data compression techniques are used in practice, and are known as:

❏ Modified Huffman Coding (MH or MHC)

❏ Modified READ Coding (MR or MRC)

## 3.2.4 Modified Huffman Coding

MH coding is used to reduce the number of bits required to transmit each horizontal scan line, and is based on the mathematical probability analysis of the strings of bits to be transmitted. Strings with a high probability of occurrence are replaced by a short binary code, while strings with low occurrence probability are encoded with longer binary codes.

This system involves two types of code word. For strings with run lengths of up to 63 bits, a direct replacement known as a terminating codeword (for reasons which will become obvious later) is used. A few examples of the terminating codewords used as replacement strings are given in Table 3.2. The complete table can be found in Table 1 of CCITT recommendation T.4.

*Table 3.2   Examples of terminating codewords in the modified Huffman coding system for data compression*

| White run length | Terminating Codeword | Black Run length | Terminating Codeword |
|---|---|---|---|
| 1 | 000111 | 1 | 010 |
| 2 | 0111 | 2 | 11 |
| 4 | 1011 | 4 | 011 |
| 10 | 00111 | 10 | 0000111 |
| 30 | 00000011 | 30 | 000001101000 |
| 63 | 00110100 | 63 | 000001100111 |

Line 1 of this table may seem a little strange in that the minimum length string of one white bit is replaced by a longer string six bits in length. However when you consider that the probability of a single white pixel occurring in a real document is very small this replacement will very rarely be necessary. On the other hand a run of 63 consecutive white elements will be replaced by a string of only eight bits in length. This particular replacement represents a compression ratio of 8:1.

Run lengths of two and three are the most probable for black data. As these are both already short strings, it is impossible to replace them with much shorter codes, although a run length of three blacks (i.e. 111) is replaced by the two bit code 10. As the string 111 will very frequently appear, a compression ratio of 3:2 is achieved in these cases.

The second type of codewords are known as makeup codewords. Replacement for strings of up to 63 bits in length takes place as described above. For strings with run lengths of between 63 bits and 1728, the coding system is modified to provide even greater compression using makeup codewords. The string is divided in to multiples of 64 bits, with each 64 multiple assigned its own makeup codeword as shown in Table 3.3. The run length of the remaining bits in the string is then coded using terminating codewords. If the run length is an exact multiple of 64, there will be no remaining bits, and thus to comply with the coding rule that a makeup codeword must be followed by a terminating codeword, a special terminating codeword representing a run length of 0 is used.

*Table 3.3  Examples of makeup codewords in the modified Huffman coding system for data compression*

| White Run Length | Makeup Codeword | Black Run Length | Makeup Codeword |
|---|---|---|---|
| 64 | 11011 | 64 | 0000001111 |
| 128 | 10010 | 128 | 000011001000 |
| 1024 | 011010101 | 1024 | 0000001110100 |
| 1600 | 010011010 | 1600 | 0000001011011 |
| 1728 | 010011011 | 1728 | 0000000000001 |

The examples below show clearly how run lengths of between 64 and 1728 bits are encoded using a makeup codeword followed by a terminating codeword.

**Example A**

A white run length (RL) of 1054 is coded as follows:

```
Coding   =  Makeup     +   Terminating
1054        codeword       codeword
            (1024)         (30)

RL 1054  =  RL 1024    +   RL 30
         =  16 x 64    +   RL 30
Coding   =  "11010101" &"00000011"
```

This produces a total of only 17 bits to be transmitted rather than the original string length of 1054 bits representing a compression ratio of approximately 62:1.

**Example B**

A Black Run Length of 1600 is coded as

```
Coding   =  Makeup     +   Terminating
1600        codeword       codeword
            (1600)         (0)

Coding   ="0000001011011" &"000000111"
```

Note the coding rules permit one terminating codeword to be followed by another, but a makeup codeword must also be followed by a terminating codeword, hence the need for a terminating codeword for a zero run length.

To enable manufacturers to build machines which are capable of accepting documents larger than A4 size, the CCITT recommendation also includes makeup codewords for run lengths greater than 1728.

As the number of bits transmitted per scan line will not be constant, it is also necessary to transmit a special codeword, EOL which represents the End Of (the scan) Line.

Intuitively you can appreciate that the amount of data compression that occurs will depend upon the nature of the document to be transmitted, e.g. a typed letter will be compressed far more than a photograph. Of course the processing required will take some time, however the savings to be gained in terms of transmission time far outweigh the disadvantage of the extra complexity involved.

At the receive fax terminal, the incoming signal is decoded according to the same rules, however the system has a built-in mechanism which detects some transmission errors. When the signal is decoded, an error will be detected if:

❑ The number of bits in the scan line becomes less than 1728

❑ A makeup codeword is followed by a makeup codeword, or EOL instead of a terminating codeword

❑ A code sequence not in the coding table is detected.

## 3.2.5 Modified READ Coding

The MH coding system described in the previous section operates in one dimension only, i.e. on each scan line. By contrast MR coding operates in two dimensions, and is based on the fact that in many instances the bit sequence representing one scan line is the same or very similar to the previous line.

The term READ refers to a technique known as Relative Element Address Deviation which is used to reduce the number of bits required to transmit a scan line which is similar to the scan line previously transmitted.

The system is too complex to describe here, other than to note that it compares each scan line with the previous line and then sends a code which represents only the difference between the scan lines.

## 3.2.6 Modulation Systems

After data compression the fax signal to be transmitted is fed still in digital form to the data transmit side of the modem. Here the signal is converted into an analogue form which is suitable for transmission over a telephony circuit. Readers with a background in data communications may be interested to note that the techniques employed in this area are specified in the CCITT V series of recommendations for digital communications over analogue telephony networks.

### 3.2.7 Modulation in Group 1 Machines

Various modulation systems have been used. For example Group 1 machines used a form of frequency modulation (FM), also known as frequency shift keying (FSK), whereby a carrier of 1700Hz was shifted up in frequency by 400Hz to represent a black signal. A white signal caused a decrease in frequency from 1700Hz to 1300Hz.

### 3.2.8 Modulation in Group 2 Machines

Group 2 machines used a mixture of amplitude modulation (AM) and phase modulation (PM) techniques. The amplitude modulation element was based on On-Off keying of a 2100Hz carrier. The carrier signal was switched off to indicate black data, and switched on to represent white. In order to reduce the effects of noise in the transmission system, the 2100Hz carrier was shifted in phase by 180 degrees for each consecutive *white* bit.

To make the most use of the 3.1KHz bandwidth available this system also employed a form of single sideband (SSB) transmission known as vestigial sideband (VSB). The concept behind this system is based upon the fact that all the information to be transmitted is contained in one sideband, the other sideband being a duplication. It is therefore only necessary to transmit and receive one sideband. However reception of one sideband requires more complex circuits than those required to demodulate a full AM signal. VSB is a compromise which permits economy of bandwidth and simple demodulation.

### 3.2.9 Modulation Systems in Modern Group 3 Machines

Today's Group 3 machines are able to transmit fax data at 9.6KBit/s and 7.2KBit/s by using the Quadrature Amplitude Modulation (QAM) techniques of CCITT recommendation V.29 over communications links where the circuit quality is good enough to support this type of (relatively) high speed transmission.

On poorer quality circuits these machines are able to automatically switch down in speed and use the phase modulation systems specified in recommendation V.27 ter for transmission at 4.8Kbit/s and 2.4Kbit/s.

The two types of modulation technique used in Group 3 permit significant increases in transmission speed over simple AM and FM techniques. In a simple modulation system the carrier is altered in some predetermined manner by each bit in the signal to be transmitted. Figure 3.4 illustrates this concept for AM, FM and PM systems. Note that for each bit in the data signal there is a single change in the transmitted carrier, i.e. a change in amplitude, frequency or phase. In these systems, the rate at which the carrier is modulated is numerically equal to the data rate.

The term Baud (after Baudot, a Frenchman who invented the first telex) is used to indicate the rate at which a carrier signal is modulated (the Modulation rate). In the simple AM and FM systems described above there is a direct 1 to 1 relationship between the data rate and the modulation rate, and thus the baud rate of the carrier signal is numerically equal to the bit rate of fax data signal.

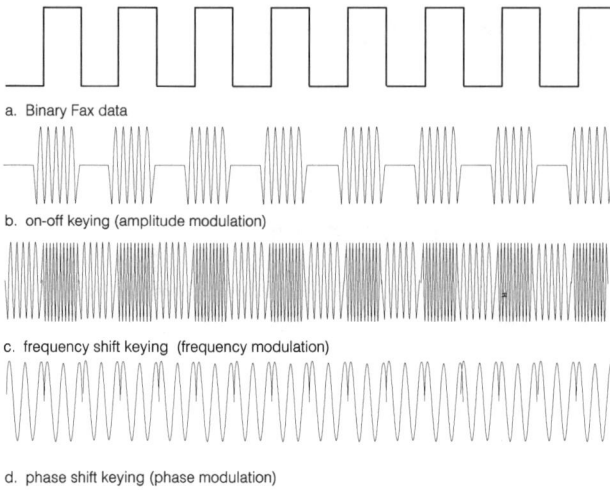

a. Binary Fax data

b. on-off keying (amplitude modulation)

c. frequency shift keying (frequency modulation)

d. phase shift keying (phase modulation)

*Figure 3.4   AM, FM and PM techniques for fax transmission*

The more complex modulation systems used in Group 3 fax systems provide the opportunity of faster data transmission speed at a given modulation rate, or permitting the modulation rate to be reduced for a given data transmission rate. The advantage here is that systems with low modulation rates are more resilient to amplitude and phase distortion.

Basic QAM techniques improve data transmission speed by providing a mechanism in which each modulation of the carrier signal contains more than one bit of fax data. As an example a single change in the carrier could be made to represent two bits of data, then a carrier modulated at a rate of 2.4 KBaud can be used to transmit data at 4.8Kbit/s. In this case the data rate would be numerically equal to twice the baud rate.

## 3.2.10 V27 ter Phase.Modulation Used in Group 3 Fax Machines

The V27 ter system uses a carrier of 1800Hz. This carrier is phase shifted to one of four possible phases relative to the reference. Each one of the four possible phases represents two bits of data.

The actual baud rate employed is 1200 baud, but as two bits of fax data are transmitted in every single element, this modulation system permits a data rate of 2.4Kbit/s.

As Figure 3.5 illustrates, V27 ter includes an option for transmission of 4800Bits/s using an eight state phase modulated carrier. In this case, the baud rate is increased from 1200 baud to 1600 baud, and each possible signal represents three bits of fax data rather than two.

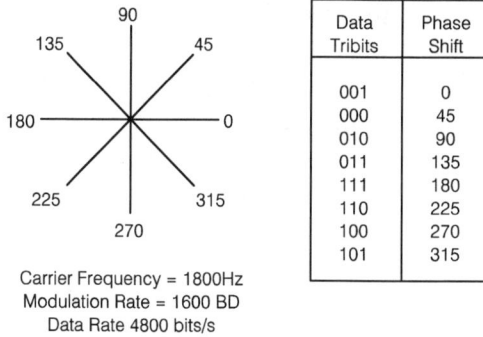

Carrier Frequency = 1800Hz
Modulation Rate = 1600 BD
Data Rate 4800 bits/s

| Data Tribits | Phase Shift |
|---|---|
| 001 | 0 |
| 000 | 45 |
| 010 | 90 |
| 011 | 135 |
| 111 | 180 |
| 110 | 225 |
| 100 | 270 |
| 101 | 315 |

*Figure 3.5  V27 ter Fax Data Transmission at 4800Bits/s using a phase modulated carrier*

## 3.2.11 V29 Quadrature Amplitude Modulation.

The V29 system used to transmit data at a rate of 9.6KBits/s takes this technique further in two respects. Firstly, there are eight possible phase shifts, rather than four, and secondly the carrier can also be amplitude modulated between one of two possible levels for each phase shift, thus giving a total of 16 possible signal elements.

This concept is illustrated in Figure 3.6 which shows 16-point QAM, in which each signal element represents four fax data bits. The carrier, in this case 1700Hz, is modulated at a baud rate of 2.4KBaud, and so the actual data transmission rate is 9.6KBits/s.

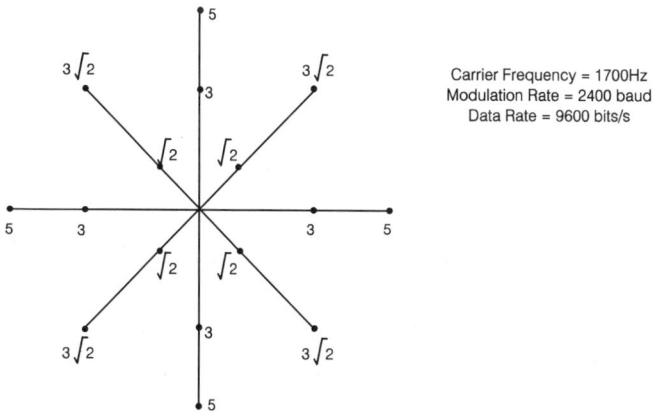

Carrier Frequency = 1700Hz
Modulation Rate = 2400 baud
Data Rate = 9600 bits/s

*Figure 3.6  Fax Transmission at 9.6KBit/s using 16 point Quadrature Amplitude Modulation*

Although this system is still digital in nature, it is not a two state binary system. In a binary system the receiver must simply distinguish between the presence or absence of a pulse, a task which is relatively easy to perform even the presence of quite severe noise. The receiver in a 16-point QAM system must be able to distinguish between eight possible phase shifts, separated by only 45 degrees. The receiver must also be able to distinguish the two levels of amplitude modulation, and for each signal element arriving correctly reproduce the four bits originally transmitted.

For these reasons this system can only be used on good quality connections, as the receiver would find it too difficult to distinguish correctly one signal element from another in the presence of anything other than minor noise disturbances. For example, a single incorrectly detected signal element will result in four data bits being received in error.

When the connection is provided by the network is not good enough to support this system, the fax machine can automatically fall back to a lower speed which has only eight possible signal elements. This system uses the same carrier and baud rate but as in this case each signal element represents only 3 bits, a data rate of 7.2KBits/s results. This fall back system is illustrated in Figure 3.7.

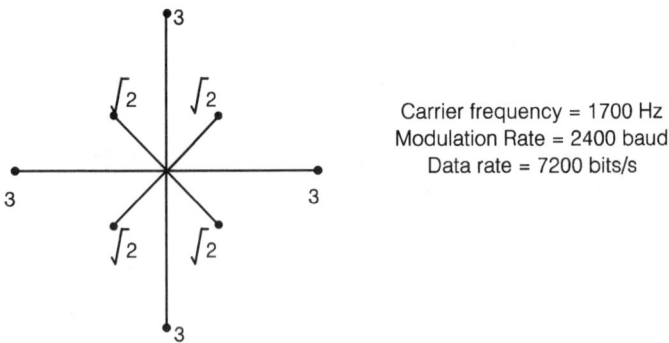

*Figure 3.7  Fax Transmission at the fall back rate of 7.2KBit/s using QAM*

Note that the phase techniques remain, but there is only one possible amplitude level for each phase element, although the amplitude level of the system is not constant. In the presence of circuit impairment, it is less difficult for the receiver to distinguish individual signal elements, as elements in the system which are adjacent and thus most likely to be confused are different from one another in two respects.

For example, the signal with a phase shift of 45 degrees is most likely to be mistaken for 90 degrees or 0 degrees. By making these signals of significantly different amplitude, the chance of error is significantly reduced

The circuit quality required for 9.6KBits/s and 7.2KBits/s can not be guaranteed on every dial up connection in the analogue network, so provision is made for fall back to 4.8KBit/s, or even 2.4KBits/s using the V27 techniques if necessary.

# 3.3 Facsimile Protocols

## 3.3.1 Introduction to Fax Protocols

Before the transmitting fax terminal can begin to send the data representing the document it is essential that there is an agreement between the transmit and receive terminals regarding compatibility and transmission speed.

In the early days of fax, this agreement was reached by the two fax operators talking to each other. Now of course it is carried out automatically by the machines themselves.

Protocols involving the exchange of a series of messages containing various commands and responses have been defined by the CCITT and are contained in recommendation T.30.

Basically the procedure involves a series of commands and responses prior to the transmission of the document. Following transmission of the document another series of responses and commands are exchanged to indicate whether or not the document was received correctly prior to the connection being released.

Although our block diagram shows only one modem in the fax machine, there are in fact normally two. Document transmission and reception uses a V29 QAM modem, while the transmission of the command and response messages is at 300Bits/s using a V21 frequency shift keying modem.

## 3.3.2 Basic Group 3 Procedure

The basic procedure can be divided into the following five phases:

❏ Call establishment, which may be manual or automatic

❏ Pre-document procedures (V21 and V29 training)

❏ Document transmission (V29)

❏ Post-document procedures (V21)

❏ Call release

To ensure that all commands and responses are correctly received, they are sent within High Level Data Link Control (HDLC) frames. (Readers unfamiliar with the operation of HDLC should refer to *The NCC Handbook of Datacommunications p371*)

As an example of the procedure we will briefly describe a fax transmission of a single page between auto-transmit and auto-receive machines. This procedure is shown diagramatically in Figure 3.8.

| Transmit Fax Terminal | Telephone Network | Receive Fax Terminal |
|---|---|---|
| 1. Dials number of required fax terminal, switches to V21 modem. | | 1. Incoming ring detected. |
| | | 2. Sends CED as 2100Hz tone |
| | | 3. Sends NSF,CSI,DIS using 300Bits/s V21 modem |
| 2. Detects NSF,CSI, DIS. Selects facilities, sends NSS, TSI using V21 Modem | | |
| 3. 3 Switches to V29 modem, sends TCF ie ZEROES for 1.5 secs (see fig 3.9) | | 4. Training signal correctly received with errors < 4. Sends CFR |
| 4. CFR received. | | |
| 5. Send fax data | | 5. Fax data received and printed, counts any errors detected |
| 6. Send RTC | | 6. RTC received |
| 7. Switch to V21 modem, send EOP | | 7. EOP received |
| 8. MCF received | | 8. Send MCF as confirmation that document received OK |
| 9. Send DCN | | 9. DCN received |
| 10. Disconnect | | 10. Disconnect |

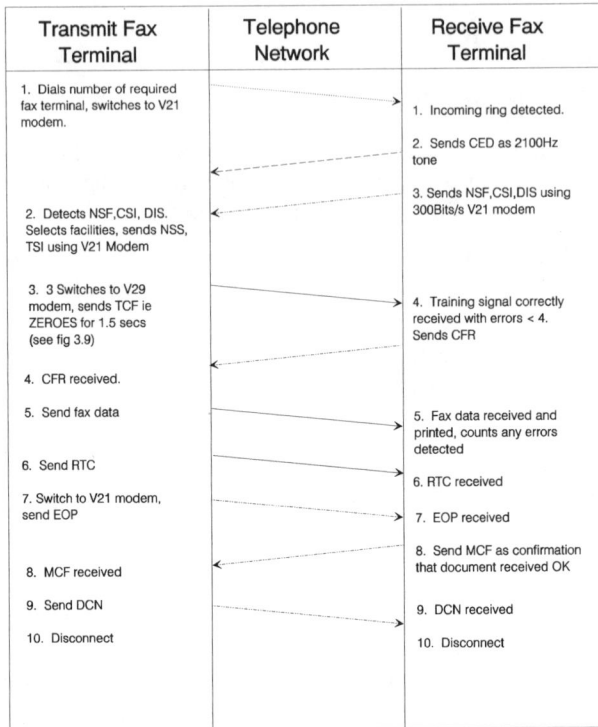

*Figure 3.8  Protocols used in the transmission of a single page fax*

### 3.3.3 Fax Call Establishment

The fax call begins with a call set up over the telephone network, the transmitting machine dialling the telephone number of the fax terminal which is to receive the transmission. At the receive terminal this call is detected as an incoming ring signal. The terminal goes into its auto answer routine which involves going off hook and sending a call received response known as CED. The CED response is in the form of a 2100Hz tone which has two purposes. Firstly it identifies the terminal as a fax machine (rather than a computer terminal, or other device), and secondly the tone is used to Cancel Echo Detectors (CED) which may be in circuit on the route, especially if it is over a satellite system.

### 3.3.4 Pre-document Procedures

Following the CED, the receive fax terminal uses the slow speed V21 modem to transmit three HDLC frames containing the following information:

❑ Frame 1: Non Standard Facilities (NSF)

❑ Frame 2: Called Station Identification (CSI)

❑ Frame 3: Digital Information Signal (DSI)

The NSF frame is included to permit fax manufacturers who have incorporated features which are additional to the basic system as defined by the CCITT to indicate to the transmitting machine which facilities are available on the receiving machine. Such features include the ability to transmit in a mode known as half tone, which is suitable for photographs.

These extra facilities are included in all machines. If a transmitting fax terminal wishes to use the facility indicated in the NSF frame, the frame is answered by an Non Standard Set Up (NSS) frame to indicate to the receive terminal which facility is to be used. This information can then be displayed on the front panel of both machines as an indication to the operators.

The CSI frame includes the telephone number and the name of the terminal being called. This information is normally displayed on the front panel, and saved to be printed on the transmission confirmation slip at the end of the transmission.

The DIS frame contains information regarding the type of fax machine. In this case it will identify the receiving machine as a Group 3 terminal, and identify which Group 3 facilities are available.

The transmit machine responds to these three frames with two frames, as follows:

❏ Frame 1: Non Standard Set Up (NSS)

❏ Frame 2: Transmit Station Identification (TSI)

The TSI contains the same information as the CSI but for the transmitting terminal.

If the machines are of different manufacture then a Digital Command Signal (DCS) frame may also be transmitted to indicate to the receive terminal that the transmit machine is also a Group 3 machine.

## 3.3.5 Modem Training

Modem training is necessary to ensure that the document can be transmitted without errors at the highest possible speed, over the circuit which has been provided by the telephone network. Although two machines may regularly be in communication, this training will be necessary on every fax call as the switched connection between the two machines will rarely be routed over the same circuits as the previous connection.

Following the transmission of the TSI and DCS frames, V29 modem training takes place in order to establish the speed of transmission of the fax data signal. The transmitter commences training by sending a Training Check Frame (TCF). The TCF signal is a series of binary zeros which is sent for 1.5 seconds, initially at the highest possible modem speed i.e. 9.6KBit/s.

As shown in the illustration of the modem training procedure in Figure 3.9, the receiver checks this signal for errors. The receiver does not check for the whole 1.5 seconds transmission of the TCF, checking only for 900 mS, neglecting the first 400 mS and the last 200. If less than a preset number of errors (normally four) occur during this period, the receive terminal sends a Confirmation to Receive (CFR) to the transmit terminal to indicate that the circuit quality is satisfactory for transmission at this data rate.

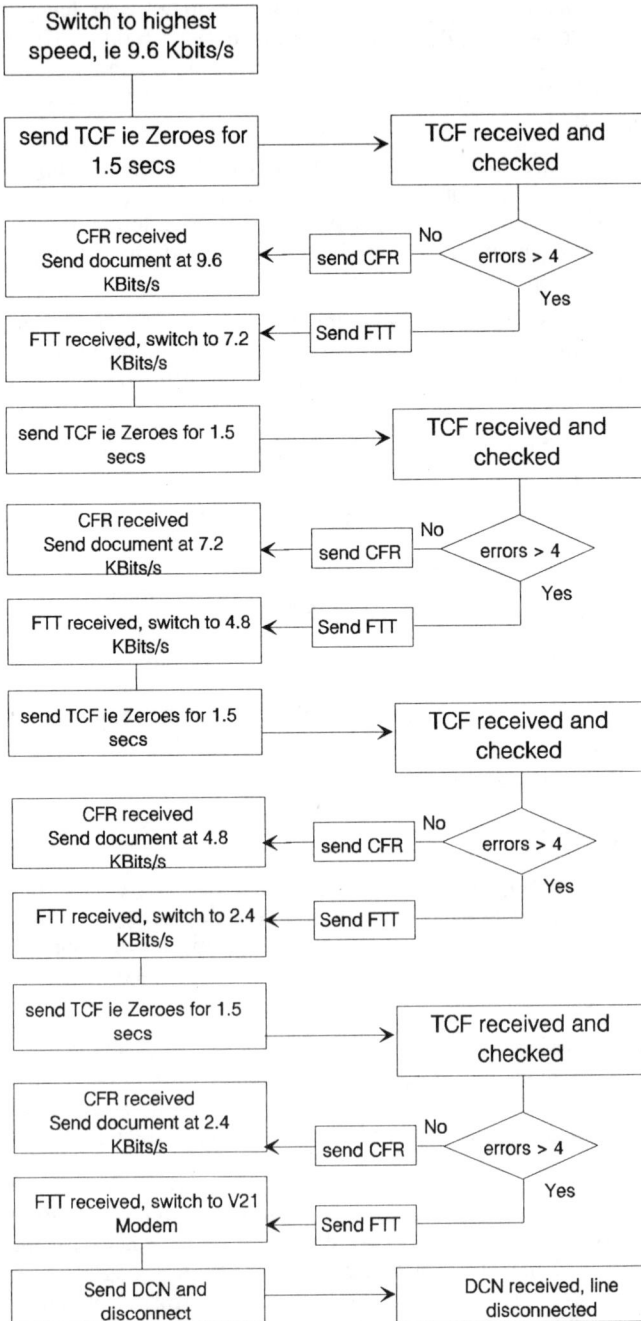

*Figure 3.9 Modem trianing protocols*

If however the circuit quality is not good enough to permit the TCF to be received with less than four errors, the receiver transmits a Fail To Train (FTT) signal.

At the transmitter the reception of the FTT causes the modem to switch down in speed to 7.2Kbit/s and to retrain with another TCF. The procedure is the same as above. If it is necessary for the receiver to return a further FTT, the transmitter will attempt TCF at 4.8Kbit/s, and if necessary at 2.4Kbit/s. If the circuit quality is not good enough to support error free transmission at 2.4KBit/s the transmitter will reply to the last FTT with a Disconnect (DCN) message, which initiates a cleardown of the fax call, and both machines disconnect from the telephone line.

## 3.3.6 Document Transmission

Document transmission, as described in Section 3.2, takes place following reception of the CFR by the transmit terminal. Most machines use the Modified Huffman and Modified READ data compression codes mentioned. However the more sophisticated modern machines are able to use to use even more powerful data compression codes to achieve even quicker document transmission times.

As an example the Siemens HF2305 has MHC and MRC as standard, but this machine also has an express mode which uses compression coding systems called Special Modified READ code (SMRC), and Modified Modified READ Code (MMRC). These codes can be used to transmit A4 documents between two similar machines at a rate of 10 seconds per page.

The use of data compression codes means that some form of synchronisation system must be established to ensure that the receiver can correctly identify the end of each line. Under noise and disturbance free conditions the decoding process will always produce a line of exactly 1728 pixels. But if an error occurs during transmission the decoding process cannot be relied upon to produce the correct number of pixels on each line.

Line synchronisation is maintained by the transmission of an End of Line (EOL) signal, which consists of 11 zeroes followed by a single 1. As in either the run length or the make up encoding tables there is no other code that has more than six consecutive zeros, should an error occur in the EOL signal, this should not prevent it being correctly interpreted.

As documents are often of different lengths, it is not possible to rely on counting the number of lines received, and so a different system has to be used to indicate the end of a page. The end of page signal is actually called Return to Control (RTC), i.e. switch from fax transmission to control signal transmission.

The RTC signal consists of six consecutive EOLs followed by a 1. This signal indicates that the fax is returning to the V21 mode to transmit the end of document procedures.

## 3.3.7 Post-document Procedures and Call Release

Having switched to V21 mode to handle the end of page handshake, if there are no more pages to be transmitted the transmitter will follow the RTC with an End of Procedure (EOP) Signal.

When the receive fax detects the EOP signal, it will respond with a Message Confirmation (MCF) signal, which in turn causes the transmitter to issue a Disconnect (DCN) signal. At this point both machines disconnect themselves from the telephone line and go into auto-answer mode.

The transmitting fax machine will also normally print a transmission confirmation slip which includes the number of the called fax and the called terminal identity (both of which are taken from the CSI, rather than simply using the number that was actually dialled), the number of pages transmitted, which mode was used and the date-time stamp for the transmission.

## 3.3.8 Multi-page Transmission

Multi-page transmission is also handled by the post-document handshake. If the transmitter has more pages to send, it does not follow the RTC with an EOP, but instead sends a Multi-page (MPS) signal. The receiver still sends the MCF signal but instead of waiting for a DCN, now prepares itself to receive the next page.

## 3.3.9 Document Reception

This section is rather brief as the description of the receive side of the fax machine is basically the opposite of that of the transmit side.

In its idle state the fax machine will be set to receive incoming fax messages, either automatically, or with manual intervention. Where the machine is connected to a telephone line which has been designated solely for fax use, the usual practice is to set the machine to automatic receive. If, however the fax machine has to share a line with a normal telephone, it should be set to manual receive.

We will consider automatic operation. The incoming ring from the local exchange activates the internal ring detector, which then signals to the microprocessor that a new fax call has been received. The modem is switched to the line, thus supplying the necessary loop to seize the line and cause the ringing signal from the exchange to be stopped.

After the exchange of the necessary protocols with the transmitting fax terminal (see Section 3.3) the incoming fax message is received as an analogue signal by the modem. After demodulation the fax signal now in digital form, but still in a compressed data format, is passed to the decoder.

The decoder carries out the inverse function of the coder in the transmit terminal, by decoding the run length and make up codes. Each code is replaced by the relevant strings of 1s and 0s to reproduce the original fax signal, line by line, as was produced by the scan unit of the transmit machine. All 1728 bits of each scan line are passed, in parallel format, to the print unit.

In most cases the print unit will be a thermal device, which requires the use of special paper. The device consists of 1728 heaters, each of which can reproduce one pixel of the scan line. The presence of a 1 causes the heater to be switched on, thus marking the heat sensitive paper, while a 0 causes no mark.

Not all machines use the traditional heat sensitive paper technology. The latest fax machines make use of the technology of the photocopier to permit the use of plain paper.

Another useful facility available on newer machines designed for small offices, where the fax terminal has to share a line with the telephone, allows the machine to be set to delayed auto-answer, with automatic switch over from voice to fax.

In this mode, a delay of typically 10 – 20 seconds is programmed into the fax machine. When an incoming ring is detected, the machine will wait for the specified time, and, if the call has not been answered by the telephone at the end of this time, the fax terminal will go off hook and *listen* for a further 10 – 20 seconds to ensure that the call has not subsequently been answered.

After this second delay, the receive fax will automatically switch to fax mode and send the fax answer tone, if the call is in fact from a fax machine, rather than another telephone user, the fax call will proceed as described in the previous section.

## 3.3.10 Summary of Fax Protocols

The protocols to be used to ensure reliable fax transmission and reception are contained in CCITT recommendation T.30. These protocols include a series of handshakes to permit transmit and receive machines to identify each other, agree on the fax facilities to be used and the speed at which transmission is to take place.

Following transmission there is a further handshaking to confirm delivery, and indicate whether the transmission is complete or that there are more pages to be sent.

# 3.4 Group 4 Facsimile

Although facsimile is by a nature a digital system, the facsimile systems described in the previous sections are designed to be connected to an analogue telephone network. Although much of the telephone network is now digital, transmission on the local line to which the fax machine is connected is still analogue, and will remain so until the introduction of an ISDN service (e.g. ISDN2 in the UK).

Group 4 fax is designed to use ISDN and other public data services capable of operating up to 64 Kbits/s and will permit very high speed document transmission with high quality reproduction. Using the 64Kbit/s capacity of an ISDN circuit it will be possible to transmit documents with much greater definition than at present without increasing the actual transmission time.

A typical Group 4 fax machine incorporates the CCITT defined error correction mode and a grey rather than simple black or white transmission system. The error correction system allows the receive machine to detect, correct and recover from transmission errors which would otherwise cause a distorted image to be printed at the receive terminal. The grey scale, in which each pixel is defined by several bits rather than one, permits up to 64 shades of grey including all black and all white. This facility is often referred to as half tone and is ideal for high quality transmission of photographs. Machines capable of colour transmission and reception are beginning to emerge.

Most Group 3 machines permit two levels of resolution: standard with a sub-scan density of 3.85 lines/mm, and detail with a sub-scan density of 7 lines/mm. The new Group 4 machine offers an even higher resolution, called superfine, which has a sub-scan density of over 15 lines/mm.

Obviously the higher quality system produces far more than the 2MBits of data per A4 page that is used in Group 3. However by using the same types of data compression codes and some new proprietary techniques, coupled with the much increased data transmission rate the actual time to transmit an A4 document is reduced to around five seconds. It is speculated that a colour fax will soon be available with a transmission rate of 30 seconds per A4 page.

One further advantage of Group 4 fax will come when ISDN services are available to the small office user. With an ISDN connection the fax terminal can be connected to the same line as the telephone, but both will be treated independently by the local telephone exchange. In fact the telephone and the fax machine will be allocated a completely different network identity numbers, and thus it will be possible to have an auto-answer fax terminal without losing the use of the telephone which is sharing the line.

Although Group 4 machines are designed mainly for ISDN circuits, it is also possible to purchase Group 4 machines for connection to private data circuits, e.g. point to point Kilostream circuits with an X.21 interface. An X.25 interface is also available to permit Group 4 fax machines to be connected to a private packet switched network, and thus get the benefit of a switched system.

## 3.5 Chapter Summary

The current range of fax machines conform to the CCITT Group 3 recommendations, though many of them offer functionality and features above those specified for this group. These machines are designed for connection to analogue telephone lines, which may be connected to public or private telephone networks. Although these networks may use 64KBits/s digital switching and transmission, the maximum data rate of the fax machine is limited to 9.6KBit/s due to the nature of the local line.

The future range of machines will offer far greater resolution, and even colour transmission. These new Group 4 machines will connect to an ISDN line, and make use of a 64KBit/s ISDN circuit giving high speed transmission, and high quality reproduction as a result of error correction systems, and higher transmission speeds.

*4*

# An Introduction to Digital Networks

## 4.1 The Evolution of Digital Networks

This chapter provides an introduction to the remainder of the book by describing how the analogue networks, briefly discussed in chapter 1, have evolved into digital systems. Certainly digitisation has not been an overnight affair; the process has been going on since the 1960s, and is planned to go on well into the 1990s and the next century. Although it would be impossible to state the dates at which various events in the digitisation process occurred, some significant milestones have been reached. These milestones are listed in Table 4.1

*Table 4.1   Major milestones in the evolution of digital networking*

Digital Transmission
Exchange Control by Digital Computer
Digital Switching in the Exchanges
Integration of Switching and Transmission
Introduction of Common Channel Signalling
Digital Transmission in the Subscriber's Loop

These topics form the basis of this book. Before each topic is discussed in detail, a brief introduction is given in the remainder of this chapter.

## 4.2 Digital Transmission

The first step towards a digital network was the introduction of digital transmission systems, in place of the Frequency Division Multiplex (FDM) systems linking telephone exchanges on trunk and junction routes. In the 1960s, 24-channel digital multiplex systems were first introduced. These systems used Pulse Code Modulation (PCM) and Time Division Multiplex (TDM) techniques, and many systems based on this 24-channel concept are in use in areas such as North America, Japan and Asia. Within Europe, including the UK, the 24-channel systems have largely been replaced by a 30-channel digital multiplex system. PCM/TDM techniques are used, but there are several differences between the European and American systems which make them incompatible with each other. This has provoked the CCITT to recommend that

30-channel systems should be used across international borders, where the countries involved use different systems within their own networks.

Figure 4.1 illustrates the deployment of digital multiplexers within a telephone network consisting of analogue exchanges. The multiplexers are used on inter-exchange links such as those between local exchanges and trunk exchanges, and of course on links between trunk exchanges.

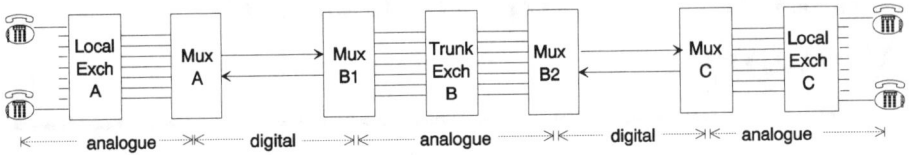

*Figure 4.1   The introduction of digital transmission into an analogue network*

In Figure 4.1, consider a call originating from a subscriber at local exchange A, to be routed through trunk exchange B to a subscriber at local exchange C. The subscriber's line, local exchange A and the trunk connections to multiplexer A are all analogue. The multiplexer converts the analogue signal on each trunk channel to digital form using Pulse Code Modulation (PCM) techniques. All 30 trunk channels are then multiplexed together using Time Division Multiplex (TDM) techniques to produce a single digital signal running at approximately 2MBits/sec. This digital signal is then transmitted to the receive multiplexer B using cable or microwave radio relay.

Multiplexer B1 separates each channel from the composite digital signal and then carries out digital to analogue conversion, so that an analogue signal is presented to trunk exchange B. A similar arrangement is used between trunk exchange B and local exchange C. Thus the original voice signal from the subscriber at local exchange A is subjected to two analogue to digital conversions, and vice versa before being received by the subscriber at local exchange C.

During the digital to analogue conversion process at multiplexer B1 a small amount of noise is added to the original signal. This noise cannot be removed from the analogue signal and so it passes through Trunk Exchange B to Multiplexer B2 where the noise and voice signal are converted to digital once again. At multiplexer C, the digital to analogue conversion process also produces a small quantity of noise, but will also reproduce the noise that was introduced into multiplexer B1. The noise from these two successive digital to analogue conversions is thus additive and could cause problems unless steps are taken to keep noise levels to a minimum.

In this description, only the transmission of voice signals using Digital Multiplex has been mentioned so far. It is important to remember that the multiplex equipment must also transmit the signalling information associated with each trunk channel from one exchange to another so that the call can be routed correctly. This is achieved by converting the analogue signalling conditions on each channel to a digital form and transmitting this digital signalling on a multiplexer channel specifically reserved for this purpose. The essential aspects of digital multiplexing and signalling are introduced in Chapters 5 and 8.

## 4.3 The Introduction of Exchange Control by Digital Computer

The Common Control techniques applied to crossbar and electronic exchanges have evolved into control by digital computer. Usually referred to as Stored Program Control (SPC), this step in the process of network digitisation has continued to evolve by taking advantage of the progress made in computer technology.

The introduction of SPC was an essential prerequisite for digital switching, since digital switching itself relies heavily on computer control techniques. Many different approaches to SPC have been developed, each trying to improve the reliability of the control system upon which the operation of the exchange depends. Other work in this area includes the introduction of specialised computer languages designed to make the task of producing the necessary software more efficient.

As well as improving the actual operation of the exchange, SPC also offers several other advantages. These include the ability to offer new telephony services, such as Call Transfer, to the subscriber.

SPC also provides the ability to maintain network information and make this available to the network management so that faults can be quickly isolated and rectified.

The essential aspects of Stored Program Control are introduced in Chapter 7.

## 4.4 The Introduction of Digital Switching

The term digital switching implies that within the telephone exchange, analogue voice signals from subscribers' lines are converted to digital form, and then switched from one line to another before being converted back to analogue.

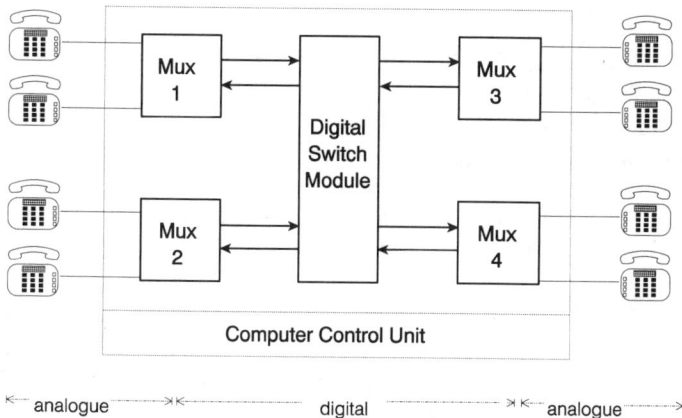

*Figure 4.2 The basic components of a digital exchange*

Digital switching is based heavily on PCM techniques. Figure 4.2 shows the basic components of a digital switching exchange, complete with an SPC computer control unit. Subscribers' lines are connected to 30 channel multiplexers which are essentially similar to the multiplexers used for digital transmission. Switching between one circuit and another is accomplished within the Digital Switch Module by storing the incoming digital voice signals from the multiplexers in memory devices similar to those found in digital computers. The digital signals can then be transmitted to any multiplex channel, converted back to analogue and passed on to the receiving subscriber's line.

To show how digital switching may be incorporated into a telephone network, the diagram in Figure 4.1 has been revised slightly. In Figure 4.3 the analogue Trunk Exchange has been replaced by a Stored Program Control Digital Switching Exchange. Since this new exchange must interface with existing equipment within the network, the analogue trunk circuits must be connected to the digital multiplexers B1 and B2 as before.

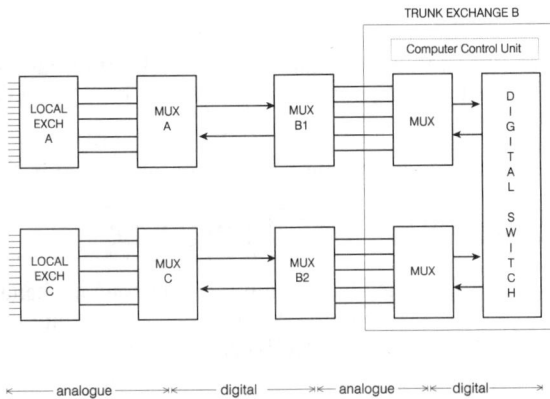

*Figure 4.3  The introduction of SPC and digital switching*

Consider the call that was described in Section 4.2, originating from a subscriber at local exchange A. This call now undergoes three successive analogue to digital, and digital to analogue conversions and thus more noise is added to the voice signal.

In small private telephone networks where calls will be routed through a maximum of three or four digital exchanges, the noise due to the conversion process will be no more than would have occurred in an all analogue system and so can be considered acceptable, as other benefits accrue from the introduction of such technology.

# 4.5 The Integration of Digital Switching and Transmission

A network that consisted of a large number of digital switching exchanges and digital multiplexed transmission links would prove to be rather noisy, especially for calls that were routed through several trunk exchanges. Fortunately this is not the case, as modern networks are based on exchanges in which the functions of switching and multiplexing have integrated into one unit.

Figure 4.2 showed the essential parts of a digital exchange, namely the digital switch module at the centre, surrounded by digital multiplexers. Consider now the situation where a number of this type of digital exchange are connected together over digital multiplexed links as shown in Figure 4.4.

In this case the link between local exchange A and trunk exchange B actually contains four multiplexers. Two of these are located within the exchanges to provide part of the switching function, and two, external to the exchanges, providing the transmission function. This would be clearly uneconomic in large public networks, and would have other unsatisfactory side effects. The point about noise has already been mentioned. Another point worth noting is that the reliability of the network as a whole is lower than need be, because of the number of component modules in series in any link.

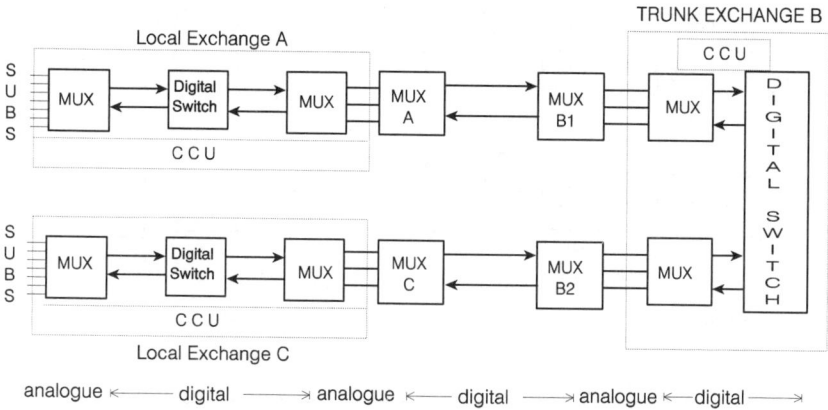

*Figure 4.4  Digital transmission and switching*

A sensible solution to these, and other, problems is to remove all four multiplexers from the link, and simply connect the digital switch module of local exchange A to the digital switch module of trunk exchange B, and likewise the link between trunk exchange B and local exchange C. A Digital Line Termination (DLT) is required at each end of the link, but as this was part of the multiplexer anyway, it is not a significant disadvantage. This now produces an Integrated Digital Network (IDN), so called because the transmission functions and switching functions have been integrated into one unit, the exchange. Figure 4.5 shows how the simple telephone network of Figure 4.1 has now evolved into an IDN.

Now reconsider the call described previously. Commencing at local exchange A, the voice signal is converted from analogue to digital in the subscribers' line multiplexer and the digital signal is switched on to a digital trunk within the digital switch module. It remains in digital form until it arrives at the digital switch module of trunk exchange B, where once again it is switched in digital form to a digital trunk terminating on the digital switch module of local exchange C.

Conversion back to analogue does not occur until the signal reaches the subscribers' line multiplexer at C. The call has thus travelled through the telephone network entirely in digital form, except for the small portion of line between the subscribers' telephones and the subscribers' line multiplexers in the local exchanges. Moreover the voice signal has only undergone one analogue to digital, and one digital to analogue conversion and so the noise added to the signal becomes minimal, and certainly will be far less than would have occurred in an analogue network.

British Telecom's System X Network, which is gradually being introduced to replace ageing analogue systems in the PSTN, is a typical example of an Integrated Digital Network.

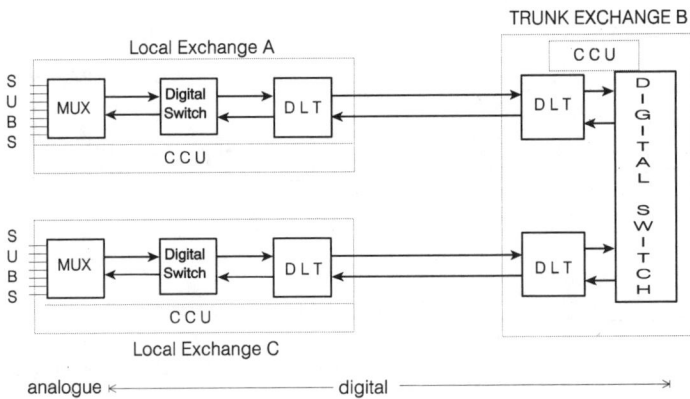

*Figure 4.5  The integration of switching and transmission*

# 4.6 The Introduction of Common Channel Signalling

This aspect of the evolution process stems from digital transmission, SPC and integration. The introduction of digital transmission included a Channel Associated Signalling (CAS) system, which was a means of transmitting signalling information from one exchange to another by converting the analogue signalling to digital form, sending it in a dedicated signalling channel over the multiplexed link, and converting back to analogue prior to finally passing it to the distant exchange.

Figure 4.6 shows the concept of channel associated signalling for just two trunk channels. Two wires (a and b) carry speech in both directions between the exchange and the multiplex. Signalling is also carried on these wires, and the third wire (p). A dedicated multiplex channel, normally channel 16 in 30-channel systems, is used to carry the signalling information extracted from the a, b and p wires of all speech channels. This is achieved by further multiplexing the channel into a number of sub-channels. Each sub-channel is then used to carry the signalling for one voice channel.

It is important to note that although a single multiplex channel is used to carry signalling for all voice channels using sub-multiplexing techniques, this system is not defined as Common Channel Signalling.

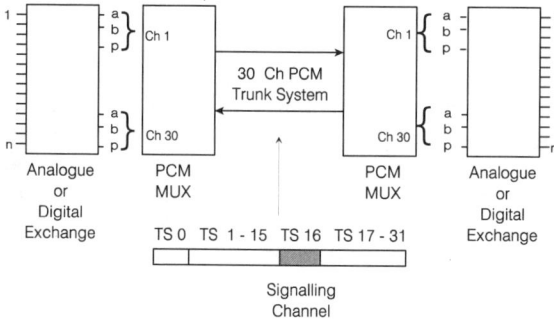

*Figure 4.6   Channel associated signalling*

The introduction of SPC involved using a digital computer in the exchange to handle call processing. Figure 4.2 illustrates this by showing the computer controlling the call processing function as a separate unit from the other exchange functions which can be considered as included in the switching module (which in this case is digital). A situation now arises in which signalling information is generated from within the computer, obviously in digital form. This is converted to an analogue form to be compatible with existing exchanges and passed to a digital multiplex where once again it is converted to digital, albeit nothing like the original digital form within the computer. There was clearly room for the improvements which came with Common Channel Signalling systems (CCS).

Three factors are worth considering:

❏ Firstly signalling is now produced in digital form within the computer

❏ Secondly, a digital transmission system exists between the exchanges

❏ Thirdly, on inter-exchange links the multiplex equipment has in fact disappeared with integration of switching and transmission.

Common Channel Signalling involves using a dedicated channel on the digital link between the exchanges as a data transmission path to enable the computers controlling the exchanges to communicate with each other and co-operate directly with the setting up of inter exchange calls.

The basic concepts of CCS are shown in Figure 4.7, in which the two exchanges must be SPC Digital Switches.

In this diagram, two options for providing a dedicated common signalling channel are shown. The signalling channel may be entirely separate from the channels carrying voice traffic as shown at the top of the diagram. Although this was the method used in the early CCS systems, the preferred method nowadays is to use a traffic channel within the multiplex system to provide a digital signalling pathway between the two exchange computers.

Each computer has access to its own database of real time information about circuits in use on trunk links. If a new call is to be set up between the exchanges, say from A to B, computer A can select a free channel on the link for the call by referring to the database. Once a channel has been selected, computer A sends a message in the digital signalling channel to the computer at B, informing B that a given channel has been selected. Computer B then stores this information in its own database and sends an acknowledgement back to A, accepting the call. Computer A can then transmit the routing information, the dialled digits, in another message over the digital signalling channel to B, which can process the call. As the call proceeds through the stages of dialling, testing, ringing, called subscriber answer and clearing, the two computers keep each other informed by sending appropriate signalling messages over the signalling channel.

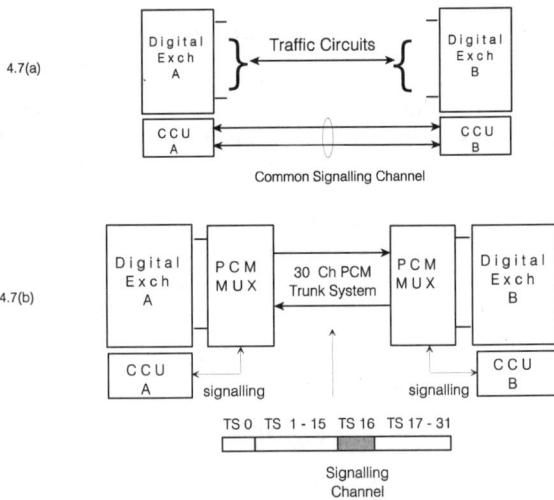

*Figure 4.7  Common channel signalling*

System X uses CCS for all its links between digital exchanges. Most new private telephone networks use an adaptation of the CCS system used in System X to provide their users with all the features you would expect of a modern telephony network.

Chapter 5 includes some information on Channel Associated Signalling in the discussion of PCM transmission. The principles of Common Channel Signalling are introduced in Chapter 8.

# 4.7 Digital Transmission over the Subscriber's Line

An IDN comprises a number of SPC digital exchanges connected by digital transmission links utilising Common Channel Signalling. The subscribers' lines are still analogue, and thus use analogue telephones of the type described in Chapter 2. There is probably little to commend extra expenditure on the part of the user to provide a digital telephone and digital transmission over the subscribers' line purely for voice communications. But this is not the case for those users who transmit data and facsimile over the telephone network. These users stand to gain considerably from digitising the subscribers' line, which is now the only analogue section of the whole network.

Currently users wishing to use the network for data communications must purchase a modem to convert the digital output of their terminals, or PCs, to an analogue form suitable for transmission over the subscribers' line.

At the exchange, the subscribers' line multiplex will convert this signal into digital form so that it can be switched in the same way as digital voice signals.

One of the problems associated with this approach is that while the subscribers' line has a bandwidth far in excess of 3KHz, this limit is imposed by the voice frequency filters within the subscribers' line multiplex. The effect of limiting the bandwidth to 3khz is that the data rates achievable are limited to about 9.6KBd, and then only on good connections. Modern modems are capable of automatic data rate reduction if a noisy circuit is being used. Data rates are then reduced in steps from 9.6KBd, through 4.8KBd, 2.4KBd and so on until an acceptable performance is achieved.

The irony of this situation in an IDN is that at the subscribers' line multiplex the PCM analogue to digital conversion produces a digital signal that is actually transmitting voice *data* at 64KBd.

Many data communications applications do not require data transfer at speeds in excess of about 9.6KBd. In interactive systems where databases are interrogated from remote terminals connected over the telephone network, quick response times in human terms are required. Response times of such interactive systems are governed by:

❏ Dial up and call set up time

❏ Time to transmit the query

❏ Access time to find the required information on disk

❏ Time to transmit the reply.

Since in most of these cases, the actual quantity of data to be transmitted is small, response times cannot be significantly improved by increasing data rates. In many cases call set up time far exceeds the total of the other three.

There are however several instances when high speed data transfer is a distinct advantage. Consider a remote PC accessing files on a central mainframe computer. The PC could be used simply as a terminal with all processing taking place in the central computer. This approach introduces a high communications cost penalty if a switched telephone circuit is used for the entire session. Other approaches would involve transferring the whole file, or large parts of it, to the PC, and processing could take place within the PC itself. This approach, using the telephone network, has not been advocated previously due to the time it would take to transmit the file, without errors, to the PC before processing. If the process involved updating the file, the new version of the file must be transmitted back to the mainframe computer after processing.

Although packet switching can be used for high speed data transfer it is generally not suited to applications in which large amounts of data are to be regularly transferred.

As no circuit switched data service existed (in the UK) until the introduction of BT's IDA service (see later) the only viable alternative was to lease high capacity digital lines from carriers such as BT and Mercury. The X-stream services such as Kilostream and Megastream are typical of the type of service available.

By making the subscribers' line a digital, rather than analogue, transmission medium, several benefits accrue to the data user:

❏ Circuit Switched Data Service available

❏ Extremely fast call set up using digital signalling, typically <1 sec.

❏ High speed error free data transmission, typically 64Kbit/s

❏ High capacity – more than one call can be transmitted over the same line

❏ Versatility – the line can be used for voice, data, fax and so on.

British Telecom, in common with several other national PTTs, is offering its business customers digital transmission over the subscribers' line. Of the six milestones listed in Section 4.1, this is the latest to have been reached.

Digital transmission in the local loop has been introduced under a variety of names. An IDN which also includes digital transmission to subscribers' premises is defined by the CCITT as an Integrated Services Digital Network (ISDN). British Telecom's System X is an IDN, and business customers connected to System X local exchanges are now offered ISDN facilities.

Chapter 11 describes the technology of ISDNs, including transmission and signalling.

## 4.8 Summary

Integrated digital networks have evolved from the old analogue telephone system in stages and the term IDN is synonymous with *Digital Telephone Network.*

Digital Transmission in the trunk and junction network was followed by the introduction of computer controlled exchanges using SPC techniques. Analogue switching in its various forms is being superseded by digital switching systems in which computer memory devices are used to store incoming digital signals before they are retransmitted to the required line.

By about 1980 the switching functions and transmission functions, both of which are based on PCM techniques, could be integrated in the exchange, saving costs, reducing analogue to digital conversion noise and increasing the overall reliability of the system.

The introduction of computer control and digital transmission opened up the opportunity for high speed data circuits to be used to connect exchange computers together for signalling purposes. Since the computers pass signalling information about all speech channels over a common data channel, this technique is known as Common Channel Signalling. The introduction of all these aspects into the telephone network creates an Integrated Digital Network.

The last part of the telephone network to be digitised is the subscribers' line. In cases where only voice transmission is required, there is little benefit to be gained from transforming this section of the network. But businesses which require instant access to large amounts of information may benefit from the advantages of fast call set up, high speed data transmission, the ability to connect more than one type of equipment to the line and the possibility of more than one call being in progress on the same line.

ISDNs are still very much in their infancy. Research is ongoing to produce solutions to problems such as inter-connectivity between public and private ISDNs, and between the ISDNs of different countries.

# 5

# *Digital Transmission and PCM Systems*

## 5.1 Introduction to Digital Transmission Systems

### 5.1.1 Bits and Transmission Speed

Digital transmission involves sending information as a series of simple electrical signal elements each of which is well defined in terms of voltage level and duration. These signal elements are often known as bits, and the rate at which information is transmitted is measured in bits per second (bits/s).

In some early digital systems information was transmitted at the rather slow speed of 50 bits/s. Today we are sending information much faster, and we tend to measure transmission rates in terms of thousands of bits per second (Kilobits/s, Kbits/s) and even millions of bits per second (Megabits/s, Mbits/s). Any information that can be represented in a digital form can be transmitted, although we are more familiar with systems that transmit voice, computer data, facsimile and video.

### 5.1.2 How Information is Carried in a Digital Signal

In most of the digital systems in use today the duration of each signal element, or bit, is constant. For example in a 300 bits/s system each bit is approximately 3.33 mS long, while in a 64 Kbit/s system the duration of each bit is much shorter, at around 15.6 uS.

As the duration of each bit in a particular system is constant, information can only be carried by varying the voltage level of the signal elements in accordance with a encoding rule which is relevant to the system in use.

In many systems only two voltage levels are permitted: 0 Volts and 6 Volts, representing binary 1 and binary 0. Such systems, for example TTL logic, are often known as binary systems, due to the two state nature of the transmission. In actual fact a pure binary system is based upon there being equal probabilities of each bit being either a binary 1 or a binary 0, but this exact definition need not concern us here.

In other digital systems there may be three or more possible voltage levels. These systems are still digital although they are not described as binary. A three level system in which the possible voltage levels may be +6 Volts, -6 Volts or zero Volts would be classed as a ternary, or three-state system.

Note that using a large number of different levels in a digital system would not be particularly useful as the electronic circuits required to distinguish one level from another in the presence of noise would be too complex and expensive to justify. Figure 5.1 illustrates various common forms of digital transmission.

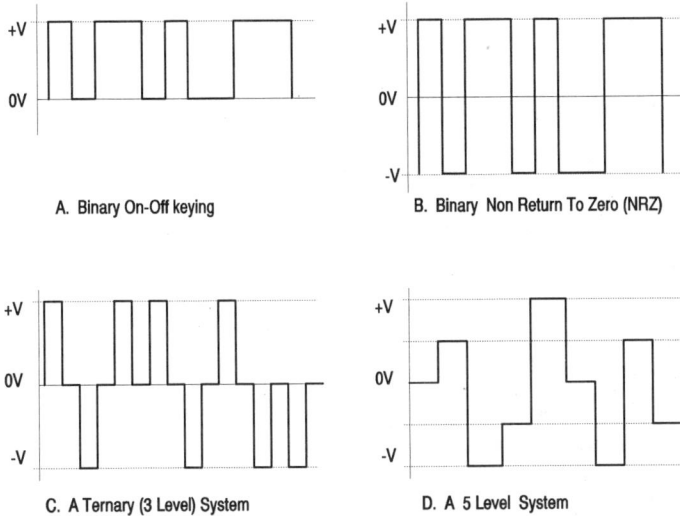

*Figure 5.1   Various forms of digital transmission system*

Note that one of the most common forms of digital transmission used in telecommunications systems is a three-level, or ternary system known as High Density Bipolar 3, or HDB3 for short. We will be examining this system later in the chapter.

## 5.1.3 The Requirement of Repeaters in Digital Systems

In all digital systems the number of possible signal elements is limited. As the diagram of Figure 5.1 shows, binary and ternary systems consist of two or three electrical signal elements, each of which differs substantially in voltage level from all other possible signals in the system.

During transmission, distortion and electrical noise from various possible sources will change the original signal. The receiver has the task of determining what signal element was originally transmitted despite the presence of this distortion in the received signal. Since there are a only few limited possibilities, fairly simple decision circuits can be used to discriminate one digital signal from another.

Due to the effects of signal attenuation and increased noise, on very long links the total amount of distortion would make it impossible for the receiver to detect the correct signal, and so the link must be divided into a number of shorter digital sections.

Figure 5.2 shows how, at the point where the digital sections are joined together, an electronic device known as a repeater or regenerator is used to receive a distorted signal from one section and transmit the recovered digital signal, with the distortion removed over the next section. Very simply, a repeater is a receiver followed by a transmitter.

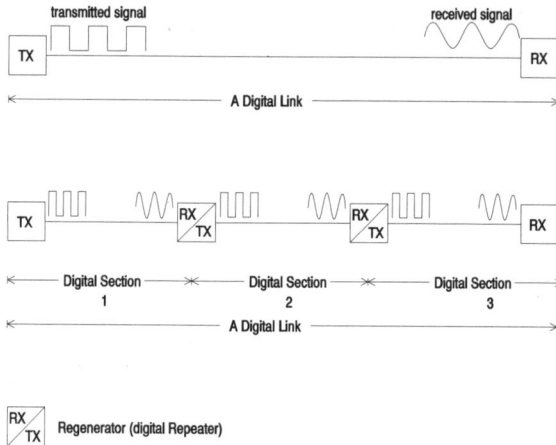

*Figure 5.2  Digital links with and without regenerators*

The diagram shows a typical digital link comprising three digital sections, with the associated regenerators. The number of digital sections required on any particular link will depend largely upon the length of the link, and the speed of transmission.

The maximum length of a digital section is determined by the following main factors:

❏ Type and output voltage of transmitter

❏ Sensitivity of the receiver

❏ Transmission speed

❏ Characteristics of the cable, e.g. attenuation/Km

❏ Maximum permissible bit error ratio.

# 5.2 Analogue to Digital Conversion

## 5.2.1 Conversion Techniques

A digital telephone network must incorporate some form of analogue to digital (A/D) conversion. There are several techniques available although in the main only two systems are used. One of them, Pulse Code Modulation (PCM) provides high quality speech , and is used extensively in telephone networks, both public and private, throughout the world. Almost all public telephone systems employing digital transmission use one of two types of PCM, both of which are transmitted at 64Kbits/s. We will be dealing with both types later in this chapter.

The other main A/D technique is known as Delta Modulation (DM). This system does not give such high quality speech reproduction as PCM, but it is used in mobile military telephone networks where constraints of bandwidth make PCM an unattractive proposition.

Table 5.1 lists several types of analogue to digital conversion techniques which are currently in use, or being developed for commercial use.

*Table 5.1  Analogue to digital conversion techniques*

| | |
|---|---|
| Pulse Amplitude Modulation | Delta Modulation |
| Pulse Width Modulation | Delta Sigma Modulation |
| Pulse Position Modulation | Adaptive Differential PCM |
| Pulse Code Modulation | Linear Predictive Coding |
| Digital Speech Interpolation | |

## 5.2.2 Pulse Modulation Systems

The first three techniques on this list, pulse amplitude modulation, pulse width modulation and pulse position modulation are not found in commercial use, except to note that they often form part of the process of changing an analogue signal to a PCM signal. Several types of PCM system exist and will be discussed later in this chapter.

## 5.2.3 Delta Modulation Techniques

The basic principle of Delta Modulation (DM) is illustrated in Figure 5.3. The analogue signal is sampled at a sufficiently high rate. The amplitude of the signal at the sampling time is compared with the amplitude of the previous sample. If the analogue signal has increased in amplitude since the last sample, the transmitted bit is set to 1. If the analogue signal has decreased in amplitude, the transmitted bit is set to 0. Thus only one bit per sample is transmitted, and the output bit rate equals the sampling rate. Basic DM has several limitations, one of which is that DM is unable to cope with signals which include a DC content which must be preserved.

Another limitation of simple Delta Modulation is known as slope overload. Slope overload is the inability of the system to cope with analogue signals that have a large amplitude high frequency content.

Reconstituting the original analogue signal i.e. D to A conversion from a Delta modulated signal is a simple matter: the incoming pulse train is passed via an integrator circuit to a low pass filter which removes any components of the sampling frequency from the output. The simplicity of the D to A conversion is one of the merits of delta modulation.

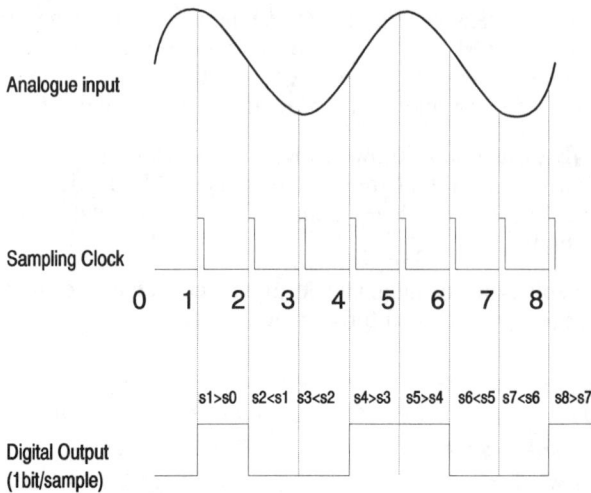

*Figure 5.3  Delta modulation and demodulation*

## 5.2.4 Delta Sigma Modulation (DSM)

Delta Sigma Modulation (DSM) is an adaptation of DM which overcomes slope overload without the requirement for very high sampling rates. Typical military systems employing DSM use a 16 KHz sampling rate which thus produces a 16 Kbits/s DSM signal. The achievable quality of these systems does not quite match that obtainable from PCM, but this is acceptable when you consider that 16 Kbits/s DSM occupies only a quarter of the bandwidth of 64 Kbits/s PCM.

## 5.2.5 Other Analogue to Digital Coding Systems

The other A/D coding systems mentioned in Table 5.1 are generally attempts to provide reasonable quality speech reproduction while utilising the lowest possible bit rate. For example Linear Predictive Coding is a technique which can encode speech signals at bit rates of under 5 Kbits/s, which would make for highly efficient use of bandwidth. The reproduced speech is rather synthetic, and although this would not be suitable for commercial networks, certain applications are found in satellite communications systems where bandwidth is at a premium.

# 5.3 Pulse Code Modulation

## 5.3.1 The History of PCM

The essential concepts of PCM were developed by A H Reeves, and patented in France as long ago as 1938. It was not until the development of the transistor that PCM could become a practical reality. In fact it was 1962 when the first PCM system was introduced by the American Telephone and Telegraph Corporation (AT&T). Since that time PCM systems have been further developed for use in transmission and switching systems. Research has been carried out to produce PCM variants which use less bandwidth than the traditional 64 Kbits/s standard system. Differential PCM (DPCM) and Adaptive DPCM (ADPCM) now offer as good quality speech as PCM, but at bit rates of 16 Kbits/s or 32 Kbits/s. Although later developments have produced A/D conversion techniques which are more efficient, PCM remains the most common system in use, and will remain so for time due to the requirement that any new equipment must be compatible with existing systems.

A thorough understanding of PCM is required by anyone working in a digital communications environment today. The term PCM has come to mean far more than just analogue to digital conversion – it is often used, perhaps wrongly, to describe multiplexing and switching techniques. The next section considers basic PCM as a conversion technique, subsequent sections describe PCM multiplexing.

## 5.3.2 PCM Fundamentals

The concept of PCM is to produce a digital representation of an analogue voice signal. This section describes the traditional method of producing a PCM signal while a later section in the chapter will describe the operation of a modern integrated circuit which produces PCM via a rather different process.

Conventionally a PCM signal is produced in three stages, known as sampling, quantisation, and encoding.

## 5.3.3 Sampling an Analogue Signal

The purpose of sampling is to capture sufficient information about the original analogue signal so that after encoding and transmission it can be faithfully reproduced. The sampling process produces a regularly spaced series of pulses which have been effectively amplitude modulated by the analogue signal.

The rate at which an analogue signal must be sampled has been shown (by Shannon) to be dependent upon its highest frequency component. Shannon showed that for a signal to be sampled and then reproduced it be must be sampled at a rate that is at least twice as high as the frequency of the signal itself.

The basic sampler consists of a switching circuit (or gate) which is operated by a series of clock pulses. The voice signal is passed to the gating circuit which opens during the clock pulse to pass a small portion of the analogue signal. This gating produces a narrow sample pulse equal in amplitude to the analogue signal. This is in fact Pulse Amplitude Modulation (PAM) and is illustrated in Figure 5.4.

For telephony purposes a speech bandwidth up to 3.4 KHz is considered sufficient. According to Shannon the minimum rate at which a signal containing components of 3.4 Khz could be sampled and successfully reproduced would be 6.8 KHz, or twice the highest frequency present.

In the FDM systems that existed when PCM was developed, the bandwidth of each channel slot was 4 KHz, the extra bandwidth often being used for voice frequency signalling. Since PCM was being developed to replace FDM, it made sense to choose 4 KHz as the highest modulating frequency, and thus a sampling rate of 8 KHz was required. Typical PCM equipment will include a pre-sampling filter to limit the bandwidth of the audio signal to 3.4 KHz.

Analogue Input                                                Digital Output

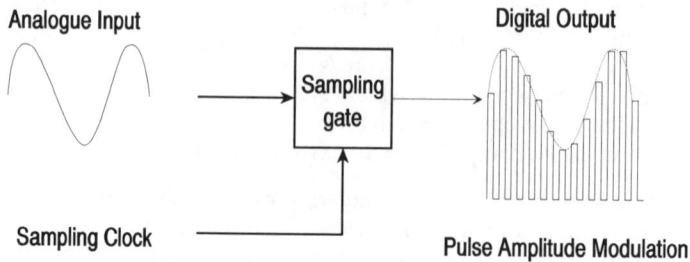

Sampling Clock                                                Pulse Amplitude Modulation

*Figure 5.4  Sampling and pulse amplitude modulation*

## 5.3.4 Quantising the Amplitude of the Sample Pulse

The second stage of the process is to quantify the amplitude of the PAM sample pulse. This stage allocates a numeric value to each sample pulse in accordance with a set graduated scale. In digital systems binary numbering schemes are preferred to the decimal system because they are easier to implement in hardware. For this reason the graduated scale used to quantify the sample amplitude will have p levels, where $p = 2$ raised to the power n. For reasons of simplicity at present, this example will use $n = 8$, thus the scale has 256 levels. The sample pulse may have any amplitude ranging from some maximum negative voltage to some maximum positive voltage, both of which must be defined. Any signal which exceeds these limits will not be correctly processed.

The 256 levels are divided into 128 levels to *measure* from 0 volts to the maximum permissible positive voltage, and 128 levels to *measure* from 0 volts to the maximum permissible negative voltage.

Each level can be considered to represent a small voltage range. Consider that the permissible input voltage was +/- 1.28 Volts. Each level would represent a range of one hundredth of a volt, as illustrated in Figure 5.5. For example :

        level 1  =  0.00 volts to 0.01 volts
        level 2  =  0.01 volts to 0.02 volts
        level 3  =  0.02 volts to 0.03 volts

etc. until:

        level128 =  1.27 volts to 1.28 volts

The same procedure is adopted for negative voltages, so that for example:

        level 0  =  0.00 volts to -0.01 volts
        level-1  =  -0.01 volts to -0.02 volts

etc. until:

        level-127=  -1.27 volts to -1.28 volts

The slight disparity between positive and negative ranges is because the binary number 0 must be used to represent a level.

*Figure 5.5   Quantisation of PAM signals*

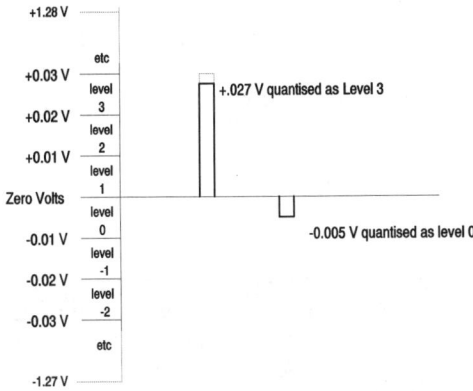

From Figure 5.5 it can be seen that any sample pulse between +/- 1.28 volts can be quantified by being allocated the binary number which represents the amplitude of the sample. A sample of 0.027 volts will be allocated level 3, while a sample of -0.005 volts will be allocated level 0. Input voltages greater than 1.28 volts can only be allocated level 128.

Although this description has used the term *quantified*, a word with which the reader may be familiar, the technical term that should have been used to describe this stage is quantised.

## 5.3.5 Encoding the Quantised Sample

The third stage in the process involves encoding the numeric value allocated to each sample in the quantisation stage into a suitable binary code that can be transmitted to line. In this example 256 levels are used to quantify the sampled signal, thus eight binary digits (or bits) are required to uniquely encode each sample. The terms octet and PCM code word are often used to refer to the 8 bit encoded sample. The most significant bit of the code word is used to denote the polarity of the sample, 0 indicating positive samples and 1 indicating negative samples. The remaining seven bits are used to indicate the amplitude of the sample. For example:

| PAM Sample | Quantisation level | Encoded as |
|---|---|---|
| 0.027v | 3 | 00 00 00 11 |
| -0.005v | 0 | 00 00 00 00 |
| 1.278v | 128 | 01 11 11 11 |
| -1.278v | -127 | 11 11 11 11 |

## 5.3.6 Summary of Basic PCM Analogue to Digital Conversion

PCM analogue to digital conversion consists of three basic steps, sampling, quantisation and encoding. In telephony PCM systems, the analogue signal must be filtered to remove components above 3.4 KHz before sampling. The sampling interval is 125 uS, i.e. the sampling rate is 8 KHz, and for each sample an 8-bit PCM code word is produced. All eight bits of each code word must be transmitted in the sampling interval, i.e. before the next sample has been taken, quantised and encoded.

This process produces an overall digital transmission rate of 8 bits x 8 KHz or 64 Kbit/s. You should note that in terms of bandwidth a 64 Kbit/s signal will have a far higher frequency content than the original voice signal.

# 5.4 Multiplexing

## 5.4.1 Multiplexing Concepts

Several PCM channels may be transmitted over a single coaxial cable, optical fibre, or radio link using a technique known as Time Division Multiplexing (TDM). Unlike Frequency Division Multiplexing (FDM), in which the bandwidth of the transmission medium is divided into frequency slots, TDM involves allocating the available bandwidth on a time sharing basis. Each channel is allocated a discrete period known as a Timeslot (TS) during which information relating to the channel is transmitted.

In order to present the fundamental concepts of TDM, the principles of two standard TDM systems will be described. This approach has the merit that you will be aware that several different TDM systems exist in networks world-wide, and that it brings out certain points that would not be dealt with if only one system was covered in detail.

The CCITT considers two basic types of PCM multiplexing system. One, a 24-channel system, is adopted for use in North America, Japan and parts of Asia. The other, a 30-channel system, is used extensively in Europe, including the UK. Both these basic PCM systems are known as primary multiplex, as they are the basic building blocks from which large capacity digital networks are designed.

## 5.4.2 24-Channel PCM Multiplex

This system is described in CCITT recommendation G733, and is also known as the T1 system. Each of the 24 channels is sampled, quantised and encoded once every 125 uS (i.e. 8000 times/second). During this 125 uS period the code word produced by each channel is transmitted as a sequence of bits commencing with the eight bits for channel 1, followed immediately by eight bits for channel 2 and so on until the eight bits for channel 24 have been transmitted. Figure 5.6 shows how this produces a digital structure known as a multiplex frame which consists of 192 bits (24 channel x 8 bits/channel).

The subsequent code words from each channel are similarly transmitted in the next frame.

## 5.4.3 Bit 193 in the 24-Channel System

To permit the receiving equipment to correctly demultiplex the incoming bit stream, it must correctly identify the first bit of each frame. A further bit (bit 193) is added to the 24-channel frame structure for this purpose.

The CCITT recommendation states that this extra bit should be the first bit (i.e. bit 1) of the frame to be transmitted, and that bits 2 to 193 are to used for the 24 voice channels. Despite this, the term *Bit 193* is used in many texts to refer to this particular bit, even though it may be transmitted as bit 1.

The use of bit 193 to provide the following signals is described in the sections which follow:

❑ Frame Alignment Signal (FAS)

❑ Multiframe Alignment Signal (MFAS)

❑ Alarm Indication Signal (AIS)

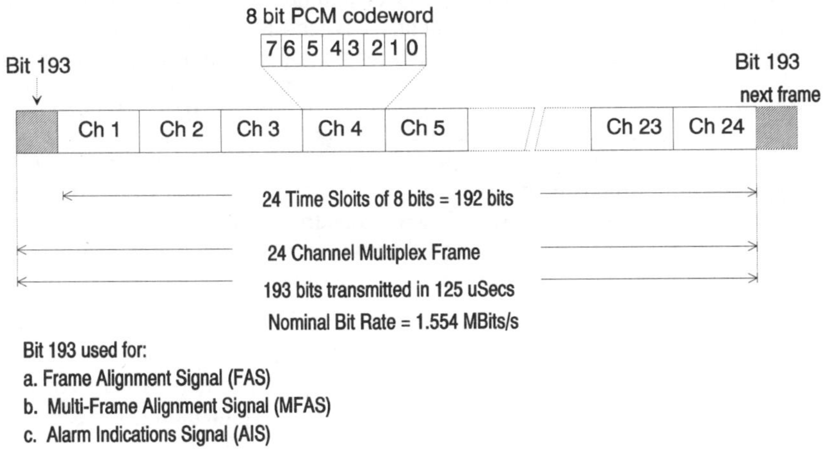

*Figure 5.6   24-channel frame structure*

## 5.4.4 The Use of Bit 193 as a Frame Alignment Signal

The Frame Alignment Signal (FAS) is transmitted in bit 193 of odd numbered frames, and is simply a string of alternating 1s and 0s. Bit 193 = 1 in frame 1, 0 in frame 3, 1 in frame 5 and so on.

The presence of a FAS in bit 193 is detected by the receiving equipment and the relevance of the remaining 192 bits in the frame is determined by their position relative to the frame alignment bit. Immediately following the frame alignment bit, the eight bit octet for channel 1 is received, followed by eight bits for channel two etc, until after the eight bits for channel 24 have been received, bit 193 is again received.

## 5.4.5 Channel Associated Signalling in the 24-Channel System

In most PCM systems provision is made to convey signalling information between the telephone exchanges which are connected by the PCM link. In the T1 system Channel Associated Signalling is achieved by a technique known as bit stealing. This involves using one of the eight bits normally associated with a PCM channel, to transmit signalling information for that channel. During every sixth frame (i.e. every sixth PCM sample) the least significant bit of the PCM code word is ignored and this bit position is used for signalling purposes.

The capacity of the signalling channel provided in this way can be determined by considering:

❑ That the sampling rate = 8000 samples/sec

❑ That one signalling bit is transmitted every six samples therefore, Signalling rate = 8000/6 or 1333 bits/s

❑ Signalling capacity is approximately 1.3 Kbits/s

Some versions of the T1 system split this capacity to provide two signalling channels, known as the A and B channels. The A channel is transmitted in Frame 6 and the B channel in Frame 12. This gives each channel a capacity of 666 bits/s (from 8000/12).

To allow the receiving equipment to determine in which frame the signalling information is being transmitted a digital structure known as a multiframe is used. In this case, as illustrated in Figure 5.7, a multiframe consists of 12 consecutive frames, signalling being transmitted in bit 0 of each channel during the sixth and twelfth frames. A Multiframe Alignment Signal (MFAS) permits the receive demultiplex to extract signalling bits from the channels in the correct frames.

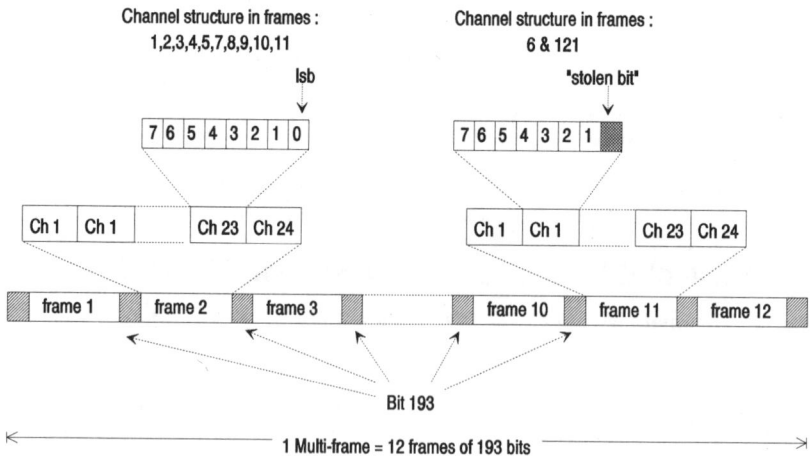

*Figure 5.7 Multiframe structure of the CCITT/Bell system for channel associated signalling*

The FAS and MFAS are both predetermined bit patterns which are sent in alternate frames in Bit 193. The MFAS is transmitted in even numbered frames. The bit sequence of the MFAS is slightly more complex than that of the FAS, being 0 0 1 1 1 0. Both the FAS and the MFAS are shown in Table 5.1.

Table 5.1  Use of bit 193 for FAS, MFAS and AIS indications

| Frame No. | FAS | MFAS | AIS |
|---|---|---|---|
| 1 | 1 | | |
| 2 | | 0 | |
| 3 | 0 | | |
| 4 | | 0 | |
| 5 | 1 | | |
| 6 | | 1 | |
| 7 | 0 | | |
| 8 | | 1 | |
| 9 | 1 | | |
| 10 | | 1 | |
| 11 | 0 | | |
| 12 | | 0 | 1 |

## 5.4.6 Use of Bit 193 as an Alarm Indication Signal

The third use of bit 193 is to provide an Alarm Indication Signal (AIS). The presence of a fault or some other alarm condition at one end of link can be indicated to the distant end by setting the MFAS bit in frame 12 to 1 rather than 0.

## 5.4.7 Common Channel Signalling

There is provision for a 24-channel multiplexing scheme in which Common Channel Signalling is used. This does not use bit stealing and hence all voice samples are transmitted in full 8-bit format, the signalling being carried in Bit 193. The 12-frame multiframe is replaced by one containing just four frames. The FAS signal is sent in bit 193 of odd numbered frames, while bit 193 in frames two and four provides a 4 Kbit/s common digital pathway to carry signalling between exchange processors for all 24 channels.

## 5.4.8 Bit Rate of 24-Channel Systems

The overall bit rate of this 24-channel system can be determined as follows:

> Output Bit Rate = No of Frames/s x No of Bits/frame
> where, No of Frames/s = 8000 i.e. the sampling rate
> and,   No of Bits/frame = 24 channels x 8 bit/ch + 1 (FAS etc) = 193
> thus,  Output Bit Rate = 8000 x 193 bits/s = 1,544,000 bits/s or 1.544 Mbits/s.

Often referred to as 1.5 Megabit systems, these 24-channel systems are used in many areas outside Europe. The 30-channel system adopted in Europe differs in several respects from the T1 system and will be described in detail later.

# 5.5 Demultiplexing and Digital to Analogue Conversion

## 5.5.1 Demultiplexing of the 24-Channel System

Using the 24 Channel system as an example, the demultiplexing and conversion processes involve the reception of a stream of bits transmitted at a rate of 1.544 Mbits/s from a PCM multiplex via a digital communications channel. The incoming stream of bits must be separated into a sequence of frames, each containing 24 PCM code words, each of which must be decoded into a positive or negative voltage, in order to reproduce the original voice signal output for each channel. Additionally signalling must be extracted from the relevant frames, converted back to its original form and passed onto the exchange. Several problems must be overcome for these processes to be successful.

## 5.5.2 Overcoming Timing Problems

The first problem to be overcome is that of timing or synchronisation. The PCM multiplex will derive its timing from some master oscillator. In some designs the master oscillator is built into the PCM multiplex equipment, in others timing is derived from some external source, for example a master oscillator supplying several items of equipment in one location.

The PCM demultiplex must operate at exactly the same clock speed as the PCM multiplex otherwise the incoming stream of bits will not be correctly received, and a problem known as bit slip will occur.

When both transmit and receive clocks are running at exactly the same rate, the receive circuits are designed so as to sample every incoming bit once, and only once. The purpose of this sampling is to measure the actual received voltage of the digital signal against some threshold to decide whether the signal received was a binary 1 or a binary 0.

As shown in Figure 5.8(a) timing circuits in the receive equipment ensure that each incoming bit is sampled at its midpoint to allow the greatest probability of the bit being correctly detected in the presence of noise and distortion.

Figure 5.8(b) shows, in an exaggerated form, how bit slip may occur when the master oscillator, or system clock, in the demultiplex is running slightly faster than that in the multiplex. Since the period of the demultiplex clock is shorter than that of the transmit multiplex, some incoming bits will be sampled, and thus detected, twice. This leads to a corruption of the received bit sequence by effectively inserting a bit that was not present in the digital signal as it was originally transmitted.

The converse of this situation is shown in Figure 5.8(c), where the demultiplex clock is running slower than that of the multiplex. Here, now that the demultiplex clock period is greater than that of the multiplex, some of the bits arriving at the demultiplex will not be sampled at all. The corruption of the received sequence is now due to an effective extraction of a bit that was originally transmitted.

Clearly the multiplex and demultiplex must run at the same speed if bit slip is to be avoided. This can not be achieved simply by allowing each equipment to use its own independent master oscillator. Even with very tight frequency control, some tolerance must exist and eventually this will lead to bit slip, making it impossible to correctly decode the incoming sequence.

The most common way of ensuring that the equipment at each end of a digital link run at the same speed, is to transmit a timing signal along with the transmitted PCM signal. This is normally achieved by converting the pure binary PCM signal into a pseudo-ternary signal such as HDB3. Among other features an HDB3 signal contains both the PCM information and timing information.

All bits sampled at mid point of pulse duration

Transmitted Bit Sequence
as received

Receive clock synchronised to
transmit clock

Recovered Bit Sequence

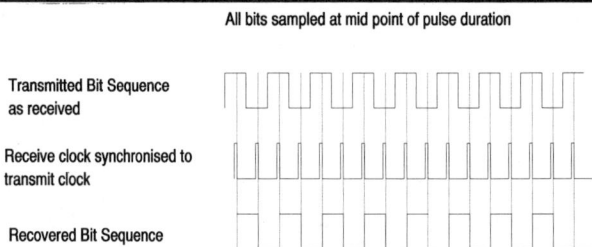

Figure 5.8a  Correct sampling and detection of incoming bit
sequence due to synchronisation of transmit and receive clocks

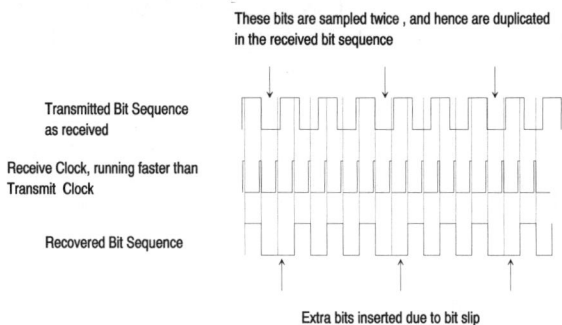

These bits are sampled twice , and hence are duplicated
in the received bit sequence

Transmitted Bit Sequence
as received

Receive Clock, running faster than
Transmit  Clock

Recovered Bit Sequence

Extra bits inserted due to bit slip

Figure 5.8b  Bit slip causing insertion of "extra" due to receive clock running
faster than transmit clock.

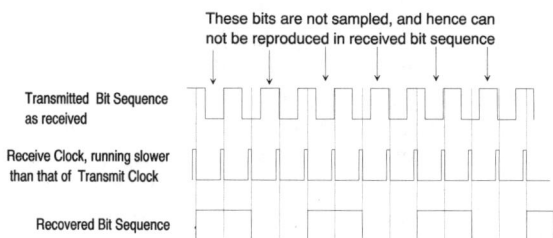

These bits are not sampled, and hence can
not be reproduced in received bit sequence

Transmitted  Bit Sequence
as received

Receive Clock, running slower
than that of  Transmit Clock

Recovered Bit Sequence

Figure 5.8c  Bit Slip causing loss of bits  due to receive clock running slower than transmit clock.

*Figure 5.8  Causes and effects of bit slip*

At the demultiplex equipment the timing information is extracted from the received
HDB3 signal and is used to control the master oscillator in the demultiplex so that it
runs at exactly the same speed as the master oscillator in the distant end transmit

multiplex. In this way, transmit and receive multiplex equipment are kept in synchronisation.

### 5.5.3 Frame Alignment

The second problem is be overcome is that of frame alignment. The incoming bit stream is organised as a frame structure in which the start of each frame is indicated by the presence of bit 193 containing either a FAS or a MFAS. Since bit 193 is actually no different from any other received bit, the receive multiplex must search all incoming bits for the presence of this FAS. This will involve checking all incoming bits looking for the FAS pattern of 1 0 1 0 1 0 in the first bit of every alternate frames. While this search is being carried out the equipment is not capable of transmitting intelligible information over any of the channels, as any digital signal received in a channel during this process will appear as unwanted noise. To prevent this interfering with the exchange to which the multiplex is connected, the traffic channels are automatically disabled. Once frame alignment has been achieved, the traffic channels can be enabled and it is then only necessary for the receive equipment to check those bits in which it expects to receive a FAS.

If the FAS is missing, or incorrect for any reason, some tolerance must be permitted before it can be assumed that frame alignment has been actually lost. For example it may be the case that noise has caused a single FAS bit to become corrupted, while frame alignment still exists.

If the FAS can not be found in the checked bit of several frames, loss of frame alignment must be assumed, and a new search begun by the receive equipment. It is also necessary for the receive multiplex to inform the distant end equipment that it is not in frame alignment. This is done by setting the AIS bit on the associated transmit multiplex. The basic sequence of events involved in regaining frame alignment upon loss in one direction is shown in Figure 5.9.

With frame alignment established in both directions, the AIS signals are reset indicating an alarm clear condition. During this period both items of multiplex equipment monitor the FAS by checking only bit 193 in the relevant frames. If frame alignment is lost in the direction B to A, A indicates this condition by setting the AIS, and then proceeds to search all incoming bits for an appearance of the FAS in the correct relative bit positions (i.e. 193 bits apart). As frame alignment still exists in the direction A to B, the AIS sent by B to A remains reset. After a period of time, A will have recognised the FAS from B and re-established frame alignment, A resets the AIS to inform B that an alarm condition no longer exists at A.

Once frame alignment in both directions is established a similar process is required to establish multiframe alignment, in order that the signalling channel transmitted in bit 0 of each octet during frames 6 and 12 can be extracted.

The process of decoding the traffic channels involves dividing the remaining 192 bits in each frame into the relevant 24 code words and directing them to correct channel decoder. Each channel decoder receives one code word every 125 uS and decodes these to produce a pulse amplitude modulated signal, which is essentially a replica of the PAM signal originally produced during the sampling process in the transmitter The decoded PAM signal is then passed through a low pass filter which removes unwanted high frequencies associated with a pulse signal and reproduces the analogue signal.

It is at decoding stage that unwanted noise is introduced on to the analogue signal. As long as no bits have become corrupted during transmission, most of the distortion picked up during transmission will have no effect on the recovery of the original signal. But noise will be present in the analogue output of the decoder due to the nature of the PCM encoding process.

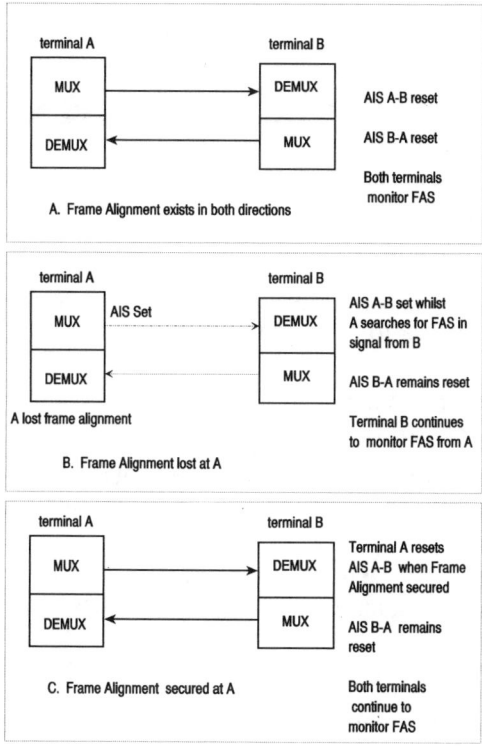

*Figure 5.9  Frame alignment procedures*

## 5.5.4 Quantisation Noise

In the example on quantising, it was suggested that a voltage step, or quantising level of .01 volts be used, and that all 256 levels in the coding system would be of equal amplitude. Thus the transmitted PCM signal can only be encoded with an accuracy of +/- .005 volt. At the decoder then, it is only possible to recreate pulses whose amplitudes are multiples of the quantisation level of 0.01 volt. This introduces a small error voltage on each sample, which may be heard as a background mush during periods of silence, but in most cases it is not noticeable during speech. This quantisation noise as it is called can, however, be unpleasant especially when the volume of transmitted speech is low. This is because the signal to noise ratio (S/N) caused by quantisation noise is not constant over the range of possible input levels.

The problem is compounded when a PCM signal is converted to analogue to pass through an analogue exchange and then retransmitted over another PCM link when more quantisation noise is added.

To illustrate the problem consider two samples:

Sample A    actual voltage = 0.706v
Sample B    actual voltage = 1.116v

Both are 0.006 volts above a quantising level, and in this example will be quantised as:

Sample A    quantised voltage = 0.71v
Sample B    quantised voltage = 1.12v

At the decoder amplitude modulated pulses will be recreated as follows:

Sample A    PAM pulse = 0.71v
Sample B    PAM pulse = 1.12v

In both cases an error, or noise voltage, of 0.004 volts has been added to each sample. However in terms of signal to noise ratio the effect is worse on sample A than on sample B.

S/N (A) = 0.706/0.004 = 176
S/N (B) = 1.116/0.004 = 279

If these samples represent the peak amplitudes generated by two different talkers it is evident that the total signal generated by A will suffer more from quantisation noise than B.

It should be noted that the maximum error voltage is equal to half the amplitude of the quantising level, and that this error voltage may be positive or negative:

$$\text{V Error(max)} \quad = \frac{Vq}{2}$$

where Vq = Amplitude of one quantising level.

One way of overcoming this problem to a significant extent would be to reduce the amplitude of each level used in the quantising process. This would have the effect of reducing the quantisation noise by reducing the maximum possible error voltage that could be added to the signal. If for instance in our example the level of each step was reduced from 0.01v to 0.005v, the maximum error voltage would be reduced from +/- .005v to +/- .0025v. This approach would require that the number of quantising steps to encode the same range of input voltage (-1.28v to + 1.28v) to be increased to 512, and this in turn would require a PCM code word of nine bits in length. The major disadvantage of this approach can be seen when the output bit rate of a multiplexer employing such a scheme is calculated.

Output Bit Rate = 24 chs x 9 bits/sample x 8000 Hz sampling rate + 1 bit x 8000

This would produce an overall bit rate of 1,736,000 bits/s or 1.736 Mbits/s, the effective bit rate for each channel having been increased from 64 Kbits/s to 72 Kbits/s. This represents a 12.5% increase in bit rate for just a small increase in circuit quality, which in fact would be undetected during most conversations. Also this solution does not address the problem outlined before in which low amplitude signals suffer a poorer signal/ratio than high-level signals. This solution merely reduces signal/noise ratios for all levels of signal.

The type of quantising and encoding process described is known as linear quantising because each quantising level is of equal amplitude and thus there is a linear relationship between the voltage represented by each level, and the number of the level as shown in Figure 5.10.

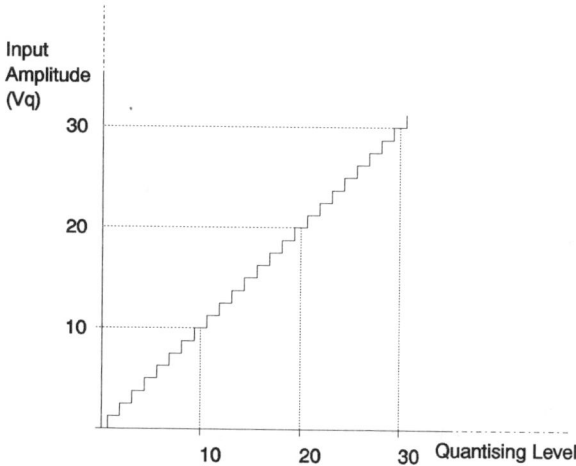

*Figure 5.10  Linear quantising system*

## 5.5.5 Non Linear Quantising

The use of Non linear Quantising does not eliminate quantisation noise, as this is an inherent function of PCM. Non linear quantising does however provide a reasonably constant signal/noise ratio for a wide range of input levels. There are two non linear quantising systems described by the CCITT, One of them, μ law encoding, is particularly associated with the 24-channel systems described earlier in this chapter. The other, A law encoding is associated with the European 30-channel systems. The two systems are incompatible in that a signal encoded using a μ law equipment can not be decoded using equipment designed to use the A law standard.

## 5.5.6 μ law Encoding

The μ law system can be explained in terms of a mathematical formula. However the concept may be more readily understood if presented graphically. Figure 5.11 shows the first few μ law quantising levels in the positive direction only. Taking the smallest level from 0 to level 1 as a reference (Vq), the diagram shows that each quantising level between level 1 and level 15 is twice the amplitude of the reference level (i.e. 2Vq). Thus at level 16 the voltage represented by this level is 31 Vq (from (15 x 2Vq) + Vq). These first 16 levels make up the first coding segment. The second segment comprises another 16 levels, but in this segment the voltage step of each level = 4Vq. The voltage represented by level 32 = 95 Vq (from (16 x 4Vq) + (15 x 2Vq) + Vq). In total there are 16 segments, eight in the positive direction and a further eight in the negative direction, each level in the eighth segment representing a voltage step of 256 Vq. Since all segments contain 16 levels, there is a total of 256 quantising levels, just as before in the linear system, so only eight bits are required in each PCM code word.

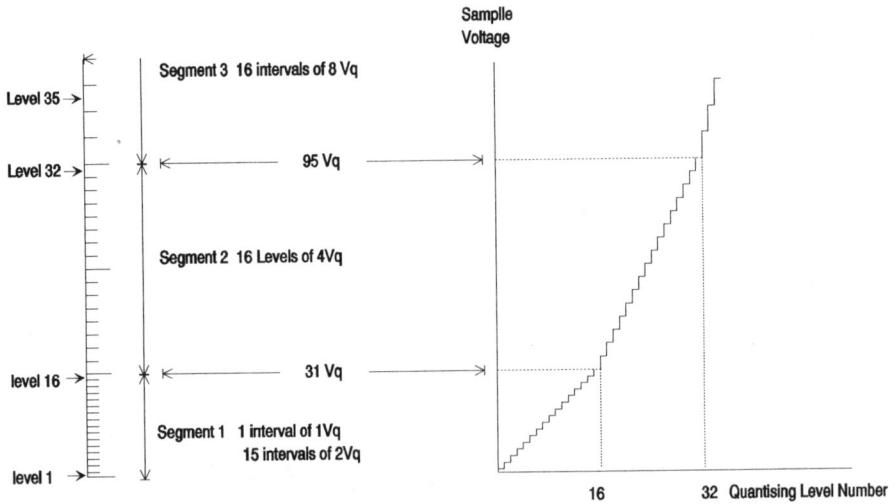

*Figure 5.11  μ law quantising*

The μ law 8-bit PCM code word is made up of three parts:

❏ One bit to indicate polarity

❏ Three bits to indicate in which segment the sample lies

❏ Four bits to indicate the level within the segment.

During decoding quantisation noise will be produced, the maximum amount for any sample being dependent upon the segment within which the sample has been coded. For example a sample coded in segment 2 will be subject to a maximum quantising error voltage of Vq, where as a sample being coded in segment 8 will have a maximum quantising error voltage of 128Vq. Now the maximum amount of quantising noise produced for any sample is almost proportional to the amplitude of the sample.

The theoretical performance of μ law encoding is to provide a signal power to quantising noise power ratio of 1000 (30dBs) for a 48db range of input level.

The dynamic range of 48dbs refers to the minimum and maximum input levels can be encoded in this system, where:

$$10 \log \frac{\text{Power In max}}{\text{Power In min}} = 48 \text{ dB}$$

The amount of improvement in performance offered by this non linear quantising law may be appreciated by considering that to achieve an equivalent performance using a linear quantising scheme would require 8159 quantising levels, and thus a 13-bit PCM code word. The effect of such a scheme in terms of multiplex output bit rates would be to almost double the amount of bandwidth required for a given number of channels.

The μ law system is defined by CCITT in recommendation G711 Table 2.

## 5.5.7 24-Channel TDM/PCM System Summary

| | |
|---|---|
| Nominal bit Rate | 1.544 Mbits/s |
| No of Channels | 24 |
| No of bits/sample | 8 |
| Encoding law | μ law |
| Input Level (max) | 3.17 dBm0 |
| Sampling Rate | 8000Hz |
| Channel Bit Rate | 64 Kbits/s |
| No of bits/frame | 193 |
| Frame alignment | by FAS in bit 193 of odd numbered frames |
| Multiframe | 12 frames |
| Multiframe Alignment | by MFAS in Bit 193 of even numbered frames |
| Signalling | |
| Channel Associated | by bit stealing in 6th and 12th frames of multiframe |
| Common Channel | by using bit 193 of even numbered frames (No multiframe required) |

To close this section it should be mentioned that early 24-channel systems in the USA were somewhat simpler than the T1 system. Their main drawbacks included a signal/noise performance that was not good enough to be used on switched circuits and the lack of provision for common channel signalling.

## 5.5.8 Basic Summary of PCM/TDM

PCM/TDM allows several digital voice channels to be transmitted over the same media on a time divided basis. Additional signals must be transmitted along with the digital voice channels to allow for functions such as frame alignment, multiframe alignment and alarm indications. These extra bits are sometimes referred to as the housekeeping bits. Provision for signalling must also be incorporated. Channel associated signalling systems require use of a multiframe structure which may not be required when common channel signalling systems are employed.

The actual bit rate for any multiplex scheme may be calculated using the formula:

Bit Rate = No of Channels x No of bits/sample x Sampling rate + No of additional bits transmitted per second

# 5.6 CCITT 30-Channel PCM System

This system, often referred as the CEPT 30-Channel system, is described in CCITT recommendation G703.

## 5.6.1 Voice Channels

Each of the 30 voice channels is sampled 8,000 times per second. Each sample is quantised and encoded using a non linear quantising scheme known as A law. The concept of A law quantising is not unlike that described for u law and is illustrated in Figure 5.12. The significant difference is that the first four segments (two positive, two negative) are co-linear, i.e. they have the same quantising interval. The first four segments together consist of 64 quantising intervals of size 2Vq extending from -64 Vq to + 64 Vq.

In the CCITT recommendation G711 these first four segments are together labelled segment 1.

| Segment | No of steps x step size | Input Amplitude Range | Level Numbers |
|---------|-------------------------|------------------------|---------------|
| +7 | 16 x 128 | 2048 to 4096 | 113 to 128 |
| +6 | 16 x 64 | 1024 to 2048 | 97 to 112 |
| +5 | 16 x 32 | 512 to 1024 | 81 to 96 |
| +4 | 16 x 16 | 256 to 512 | 65 to 80 |
| +3 | 16 x 8 | 128 to 256 | 49 to 64 |
| +2 | 16 x 4 | 64 to 128 | 33 to 48 |
| 1 | 64 x 2 | 64 to -64 | 32 to -31 |
| -2 | 16 x 4 | -64 to -128 | -32 to -47 |
| -3 | 16 x 8 | -128 to -256 | -48 to -63 |
| -4 | 16 x 16 | -256 to -512 | -64 to -79 |
| -5 | 16 x 32 | -512 to -1024 | -80 to -96 |
| -6 | 16 x 64 | -1024 to -2048 | -97 to -111 |
| -7 | 16 x 128 | -2048 to -4096 | -112 to -127 |

The remaining 12 segments, six in the positive direction, six negative, each have 16 levels. The size of the quantising interval doubling from one segment to the next, so that in segment 2, for example, the quantising interval is 4Vq, and in segment 7 it is 128Vq.

The A law PCM code word consists of eight bits, organised as:

❏ One bit to indicate sample polarity

❏ Three bits to indicate in which segment the sample is coded

❏ Four bits to indicate the level within the segment.

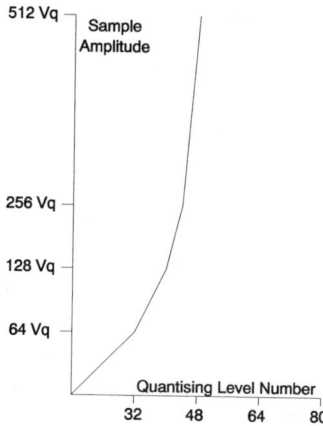

*Figure 5.12 Diagram to illustrate A law quantising values*
*Table 1a*

The maximum input level accommodated at +/- level 128 is 3.14 dBm0. This scheme is equivalent to using 4,096 quantising levels in a linear scheme, which would require a PCM code word of 12 bits in length.

Since the sampling rate is 8 KHz, and an 8-bit PCM code word is produced for each sample, this system produces a 64 Kbits/s PCM signal just as in the T1 system.

As has already been stated this type of PCM signal is not compatible with a 64 Kbits/s PCM signal produced in the T1 system as a different quantising process has been used to encode the analogue signal.

A full table showing the A law scheme can be found in Table 1b of recommendation G711. Interesting points to note are that the bits in each code word are numbered so that the most significant bit (i.e. the sign bit) is bit number 1, and the least significant bit is bit number 8. Secondly, the even bits of the 8-bit code word are inverted before transmission.

## 5.6.2 30-Channel Frame Structure

In the section on 24-channel systems it was shown how time division multiplex systems employ a multiplex frame containing sufficient timeslots to cater for the transmission of all channels plus some housekeeping information. Figure 5.13 shows that in the 30-channel systems timeslots are organised into a multiplex frame structure which actually comprises 32 timeslots (TS) identified as TS0 to TS31. The voice channels are allocated as follows:

| | |
|---|---|
| Voice Channels 1 – 15 | TS 1 – 15 |
| Voice Channels 16 – 30 | TS 17 – 31 |

The two remaining timeslots are numbered TS0, and TS16. These are used for housekeeping, such as frame alignment and alarm indications, and signalling respectively. Note that because TS16 is used for signalling, voice channel number 16 is transmitted in TS17.

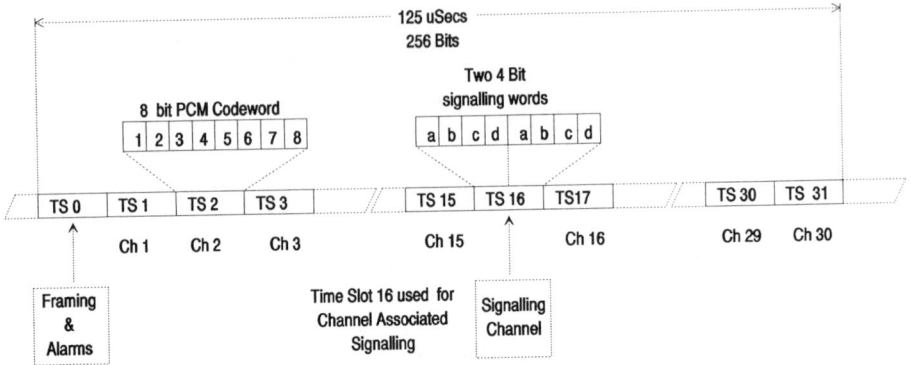

*Figure 5.13   30-channel frame structure*

## 5.6.3 Calculation of the Bit Rate of 30 Channel Systems

The sampling rate, and thus the rate at which frames are transmitted is 8KHz. For each of the 30 voice channels, an 8-bit TS is allocated in which an 8-bit code word is transmitted. There are also two other TS in each frame, both of which are also eight bits long. Thus the output bit rate may be calculated from:

Output Bit Rate   = ((No of Chs x No of bits/sample) + No
                    of extra bits per frame) x Sampling rate

In this case:

Output Bit Rate   = ((30 x 8) + 8 (TS0) + 8 (TS16) ) x 8000
                  = 2.048 Mbits/s.

Another way of considering this system is as a frame of 32 8-bit TS, which is repeated 8,000 times per second. Thus the output bit rate = 32 x 8 x 8000 = 2.048 Mbits/s.

In later chapters we will meet some 32-channel PCM systems which are based on this system but which do not have frame alignment or signalling timeslots. However for the time being you should consider that the 30-channel system is the standard.

## 5.6.4 Frame Alignment Signal (FAS)

The concept of a frame alignment signal has been introduced in the section on the T1 24-channel system. The CCITT defines a Frame Alignment Signal, or FAS, as a distinctive signal used by the receive multiplex to secure frame alignment. Frame alignment is required to ensure correct demultiplexing of the multiplex frame into its constituent channels.

In the CEPT 30-channel system, the FAS is a 7-bit sequence chosen so as to keep the probability of it being mimicked in traffic signals to a minimum. The actual sequence used in the CEPT system is :

0 0 1 1 0 1 1

and is transmitted in bits 2 – 8 of TS0 in even numbered frames only. The remaining bit, bit 1, is reserved for other purposes.

The frame alignment strategy adopted is based upon the fact that although occasionally some bits will be corrupted during transmission, frame alignment is not necessarily lost. Once frame alignment has been established on a link, the receiving multiplex monitors the seven bits in which FAS signal is transmitted to check that frame alignment is being maintained. Should a bit from the FAS be corrupted, an incorrect FAS will be detected at the receive multiplex. A single error such as this should not be taken as an indication of loss of frame alignment, as traffic circuits will be unnecessarily disabled during a search for new alignment.

Frame alignment is considered to be lost when three consecutive FAS are received with errors. This means that a total of five frames must have been received, and that in the three even numbered frames, the correct FAS has not been received.

Once loss of frame alignment is detected the traffic channels are disabled and a FAS search involving all incoming bits is started. The search scans all bits for the 7-bit FAS pattern 0 0 1 1 0 1 1. Once this has been found, the seven corresponding bits of the next frame are checked to ensure that they do NOT contain a valid FAS (see the next section on Not Frame Alignment Signal). If the corresponding bits of the third frame contain the correct FAS pattern, then frame alignment is considered to have been established, and voice channels can once again be enabled for traffic.

## 5.6.5 Not Frame Alignment Signal (NFAS)

This oddly named signal is transmitted during TS0 of odd numbered frames, i.e. when the FAS is not being transmitted. The NFAS has eight bits, bit 1 being reserved for international use. Bit 2 is always set to 1, thereby ensuring that whatever the combination of the other seven bits, there is no possibility of the NFAS being the same as the FAS. Note that bit 2 of TS0 is always set to 0 when a FAS is being transmitted.

Bit 3 is used to transmit an Alarm Indication Signal (AIS). Normally this bit is set to 0 to represent an Alarm Clear condition. When set to 1 by the transmit multiplex an alarm condition e.g. loss of incoming signal, or frame alignment is indicated.

The remaining five bits, bits 4 – 8 are reserved by CCITT for national use, to be used by national telecommunications administration for their own purposes.

## 5.6.6 Channel Associated Signalling (CAS)

Although CAS has been introduced in the section on 24-channel systems, this section demonstrates that the implementation of CAS in 30-channel systems is somewhat different.

Channel Associated Signalling is defined by the CCITT as a method in which the signals necessary for the traffic carried by a single channel are transmitted in the channel itself, or in a channel permanently associated with it. This definition holds true for analogue and digital systems. For example FDM voice frequency signalling whether in-band or out-of-band is true CAS, and similar terms are used to describe digital CAS systems as In-slot signalling, or Out-slot signalling.

In-slot signalling is defined as signalling associated with a channel and transmitted in the same TS as the channel itself. In the 24-channel T1 system for example CAS is provided In-slot, by bit stealing.

Out-slot signalling is defined as signalling associated with a channel but transmitted in a TS (or TSs) other than the channel TS. CAS in the 32-channel system involves Out-slot signalling, by using a submultiplexing technique, in which TS16 is used to transmit the signalling associated with all 30 voice channels. This obviates the need for the bit stealing technique adopted in T1, and provides a system capable of easy modification to common channel signalling.

Each voice channel is allocated four signalling bits labelled a – d (note lower case). These four bits can used to carry signalling information by relaying the DC conditions present at the exchange/multiplex interface as changes in status of individual bits. Several different versions of this type of signalling exist to cater for the large variety of analogue exchange presentation possible. For example there are 2-wire and 4-wire loop disconnect systems, 2-wire and 4-wire E and M systems. Irrespective of the analogue signalling system in use the digital transmission of these four signalling bits in TS16 is carried out as described below.

TS16 is submultiplexed to provide 30 signalling channels by transmitting signalling for each channel in different frames. This involves the use of a 16-frame multiframe structure. TS16 of each frame contains eight bits, and thus can be used to transmit four signalling bits for two voice channels. This would require a multiframe of 15 frames if it were not for the requirement to send a Multiframe Alignment Signal (MFAS). The MFAS is transmitted as a 4-bit pattern of all zeros in the first frame of the multiframe. Signalling is transmitted in the other frames as shown in Table 5.2.

Note that throughout, this system is based on binary multiples, rather than duo-decimal multiples. This leads to a slight confusion when referring to timeslots and frame numbers since the first in each case is numbered zero, rather than 1 as is the case in T1.

*Table 5.2   Allocation of TS16 for channel associated signalling*

| Frame No | 1 2 3 4 | 5 6 7 8 | Remarks |
|---|---|---|---|
| | | TS16 Bit No | |
| 0 | 0 0 0 0 | X X X X | 0000 is the MFAS |
| 1 | a b c d | a b c d | ch 1 and ch 16 Signalling |
| 2 | a b c d | a b c d | ch 2 and ch 17 Signalling |
| 3 | a b c d | a b c d | ch 3 and ch 18 Signalling |
| ..... | | | |
| ..... | | | |
| 15 | a b c d | a b c d | ch 15 and ch 30 Signalling |

The capacity of each signalling subchannel may be calculated as follows:
> Frame Time = 125 uS (due to 8000Hz sampling)
> No of frames in multiframe = 16
> Thus Multiframe Time  = 16 x 125 uS
> = 2000 uS or 2 mS
> Thus number of multiframes per second = 500
> Each channel is allocated four signalling bits/multiframe
> Thus subchannel signalling capacity = 2 Kbits/s

When it is considered that most of the time the signalling subchannel is idle, this seems an unnecessarily large amount of capacity especially when you realise how much data could be moved over such a channel.

The total capacity of TS16 is 64 Kbits/s, 60 Kbits/s of which is used for 30 channels' worth of CAS while 4 Kbits/s is used to transmit the MFAS. The remaining 4 Kbits/s is used for other housekeeping tasks.

## 5.6.7 Common Channel Signalling (CCS) within PCM/TDM Multiplex

Common Channel Signalling is defined as a method in which signalling relating to a multiplicity of circuits is conveyed over a single channel.

CCS is used between modern computer controlled exchanges. A data link is required to pass signalling information about each voice channel between the exchanges. This data link could be provided over an entirely separate digital path. However it is usually the case that the data link is provided by using a single TS in the TDM multiplexing scheme. For historical reasons the chosen TS is TS16, although in truth any TS, other than TS0 could be used.

Although the main parameters of CAS and CCS PCM/TDM systems are identical, the multiplex frame and multiframe structures used for CCS systems differ in several respects from those of CAS systems. The concepts and principles of Common Channel Signalling are described in a later chapter. In this chapter, only the effects that the use of CCS has on 30-channel frame structure etc are discussed.

At this stage it should be stressed that CCS is a function of the two exchanges at either end of a link, and that the involvement of the PCM multiplex is purely as a bearer for the CCS messages between the exchanges. The multiplex equipment is completely transparent to the signalling messages, and takes no part in the signalling protocols.

Analogue exchanges such as Strowger and Crossbar units with loop disconnect, or E&M signalling are by their very nature channel associated systems, since the signalling for each channel is carried in the very wires of the channel. Even some digital exchanges, in which PCM switching is used, interface to other exchanges using Loop Disconnect or E&M signalling. In all these cases a Channel Associated Signalling PCM/TDM multiplex, entirely separate from the exchanges, will be used if digital transmission is to be employed between the two exchanges.

If the exchanges are both Stored Program Control (SPC) digital units, and suitable control software is installed in the processor unit, Common Channel Signalling may be used. However when this is the case the function of the PCM/TDM multiplex is often built into the exchange and will not be a standalone external unit as is the case for Channel Associated systems.

## 5.6.8 Refinements to the 30-Channel System

This section deals with refinements to the 30-channel system which provide two features. The first is that the refined system is designed with common channel signalling in mind, and is generally not used with out-slot channel associated signalling. Secondly, a form of error detection mechanism is built into the system. These refinements are described in CCITT recommendation G704.

In a PCM system carrying CCS there still exists a requirement for the 7-bit Frame Alignment Signal (FAS) transmitted in TS0. But, as no TS16 submultiplexing takes place, there is strictly no need for a multiframe and its associated Multiframe Alignment Signal. In fact since TS16 is used to provide the CCS data link, the MFAS cannot be sent in TS16. There are, however, other aspects of these systems which do require the use of a multiframe structure.

## 5.6.9 Detection of Errors

The CCITT defines digital error as a single digit (bit) inconsistency between the transmitted and received signals. Although digital systems are generally very resilient to noise, high noise levels and irregularities in timing will cause some errors to be introduced. These errors may have adverse effects on speech and will almost certainly corrupt signalling circuits. It is therefore important to be able to detect that errors are occurring.

An important parameter in this respect is the rate at which errors occur. This is known as the Bit Error Ratio (BER) and is defined as the proportion of the number of digital errors received to the total number of bits received. The BER is normally expressed as a normalised number. For example if two bits were received in error, for every million bits received:

$$\text{BER} = \frac{\text{No of Bits received in error}}{\text{Total No of Bits received}} = \frac{2}{10^6}$$

This figure would normally be quoted as a BER of $2 \times 10^{-6}$. If two bits were received in error every 1,000 bits, the BER would be $2 \times 10^{-3}$, and obviously the second case is the worse of the two.

When transmission quality specifications are quoted, they generally include an error performance criteria which will be based upon the error ratio that can be tolerated by the application. For voice grade circuits over PCM equipment, error rates of less than $1.10^{-6}$ are preferred, and will be specified as BER < $1.10^{-6}$. PCM voice circuits start to get noticeably noisy when the error rate is around $1.10^{-5}$, and are unworkable at error rates of $1.10^{-3}$, even if frame alignment could be maintained at such high error ratios.

Although test equipment is available to measure the error performance of a digital link, this equipment can not generally be used when the multiplexing equipment at either end is in use. A method of measuring error performance without interfering with the transmitted traffic is therefore built into the PCM equipment.

Basic 30-channel PCM/TDM multiplexers, particularly those used with CAS systems use a simple method for determining the error performance of the interconnecting digital link. This method suffers from several problems which make it unsatisfactory for many of today's applications.

To detect errors in any system it is necessary for the receiving equipment to either know the status of every transmitted bit in advance, or be given sufficient information for it to be able to determine whether bits have been corrupted.

The first case is obviously not entirely practicable, as there is no way that the receive multiplex could know what was actually transmitted in any TS, other than TS0.

The simple error detection system operates only on the FAS transmitted in TS0. Since the frame alignment strategy allows for three consecutive FAS to be received in error before loss of frame alignment is assumed, errors detected in non consecutive FAS can be used as a guide to the total number of errors being received. The concept is that if the number of errors occurring in the FAS over a given period of time is known, the total number of errors arriving in the same period can be extrapolated from this figure.

## 5.6.10 Requirement for Better Error Detection

A multiplex frame consists of 256 bits, of which only seven carry the FAS, and then in odd numbered frames only. Thus only seven bits every 512 are checked. The effects of this low checking ratio on the ability of this system to detect errors are:

❏ Many errors will go undetected if errors in TS 1-31 occur while there are no corresponding errors in the FAS. There is a high probability of this occurring when actual error rates are low.

❏ Fall off in performance may take a long time to detect. For example should the actual error rate fall from better than $1 \times 10^{-7}$ to $1 \times 10^{-5}$, it may take up to 200 seconds to detect the deterioration. During this time users will have poor quality circuits, or worse, signalling circuits will have been corrupted and calls incorrectly routed.

A further disadvantage of this system is that the FAS was designed so that there is a low probability of it being mimicked by PCM encoded speech signals. Since many channels will now be carrying Fax or data traffic, the probability of FAS mimicking is considerably increased.

## 5.6.11 Cyclic Redundancy Coding for Error Detection

The CCITT have recommended an error detection system which is based on the concept of providing the receive multiplex with sufficient information for it to detect whether errors have occurred in any part of the multiplex frame, not just in the FAS. The system uses a powerful technique known as cyclic redundancy coding (CRC), which has also found applications in the data communications field in such systems as the High Level Data Link Control (HDLC) protocol used in X 25 Packet switching systems.

The CRC technique involves transmitting extra bits, known as check bits, along with the actual PCM traffic. The actual status of these CRC check bits, i.e. whether they are 1 or 0, will be determined by the content of all the bits in the multiplex frame using a simple binary division algorithm in the transmit multiplex. The check bits are set so as to equal the value of the binary remainder after the division has been carried out.

At the receive equipment exactly the same binary division algorithm is used on all received bits. If no errors have occurred, the division will yield a remainder of zero. However, if an error has occurred in the actual traffic, or in the check bits, the remainder will no longer be zero indicating that an error has been detected in the received bit sequence. This form of the CRC technique is not powerful enough to identify which bits have been corrupted.

## 5.6.12 Illustration of CRC Techniques

To illustrate the concept of CRC coding using all 256 bits of a multiplex frame would be very time consuming and probably very confusing. However the technique is very simple in concept, and the CRC coding process actually takes place in hardware which is very simple to implement. The simple example shown below will suffice to illustrate the mechanism involved.

Consider that the following digital signal is to be transmitted:

1 1 0 0 1 1 1 0 1 1 1 (Total 11 Bits)

Unless given extra information, the receiver is unable to detect if an error occurs during transmission and the signal is received as:

1 1 0 0 0 1 1 0 1 1 1
!

i.e. bit 5 has been corrupted.

In this example the extra information required by the receiver is two check bits, C1 and C2, transmitted as two extra bits at the end of the original data as shown below.

1 1 0 0 1 1 1 0 1 1 1 C1 C2 (Total 13 bits)

The status of the two extra bits will be determined by initially setting C1 and C2 to 0 and carrying out binary division of all 13 bits by a smaller binary number, known as a generator polynomial. In this case the generator polynomial will be the three digit binary number 1 0 1. The result of the binary division will produce a two digit remainder which will then be placed into the 13-bit sequence to replace the two C bits previously set at 0. The binary division is shown overleaf. It simply involves binary subtraction with no carry. This process is identical to modulo 2 addition, i.e. addition with no carry and can be implemented in hardware using Exclusive OR gates.

```
                    1 1 1 1 0 0 1 0 0 : 1 1        <=  RESULT
        1 0 1 | 1 1 0 0 1 1 1 0 1 1 1 : 0 0
        1 0 1                             C1 C2
GENERATOR
POLYNOMIAL    1 1 0
              1 0 1

                1 1 1
                1 0 1
                  1 0 1
                  1 0 1
                  0 0 0 1
                    0 0 0

                      0 1 0
                      0 0 0

                        1 0 1
                        1 0 1

                        0 0 0 1
                          0 0 0

                            0 1 1
                            0 0 0

                              1 1   0
                              1 0   1

                                1   1 0
                                1   0 1

            Remainder  => 1 1
```

The actual value of the result of this division process is not relevant, only the remainder is significant.

The remainder 1 1 is now placed after the 11 bits of original data in place of the two C bits, and the whole sequence of 13 bits transmitted.

The receiver will carry out the division process dividing the whole 13-bit sequence by the same generator polynomial i.e. 1 0 1, which will result in a remainder zero if no errors have occurred. This can be appreciated by considering that the 13-bit number originally divided by the generator polynomial ended in 0 0. The two digit remainder was substituted for the two zeros. This has the same effect as subtracting the remainder using no carry arithmetic. Since the remainder has been subtracted there should be no remainder when the division process is carried out.

Should an error have occurred during transmission the probability of there being a remainder is extremely high (typically well over 95%) and the occurrence of at least 95% of these can be detected. It is important to note that this system can not determine which bit or bits were actually received in error.

## 5.6.13 CRC Error Detection in 30-Channel PCM systems

The CCITT CRC system will detect errors which have occurred in any part of the multiplex frame, not just the FAS. The system is based on a 16-frame multiframe, but for the purposes of error detection the multiframe is divided into two submultiframes, each of eight frames.

Error checking on all bits in each submultiframe is carried out by transmitting a CRC check bit in bit position 1 of TS0 of the four even numbered frames in the next submultiframe, i.e. the first bit before the FAS. These four CRC bits, shown as C1 – C4, make up a four bit CRC check word and are initially set to zero. All 2,048 bits in the submultiframe, which includes the CRC check word initially set to its all-zero state, form the original digital signal to be divided by the 5-bit generator polynomial 1 0 0 1 1. A 4-bit remainder is produced which then becomes the CRC check word to be transmitted in bits C1, C2, C3 and C4 of the next submultiframe. Exactly the same process happens in subsequent submultiframes.

Figure 5.14 illustrates the multiframe structure of 30-channel systems using the CRC techniques, TS 1 – 31 are shown as bits 9 – 256 and are designated the payload area. This payload area is the part of the frame that is used to carry user traffic, which may be digital voice and signalling, or it may be 256 bits of data if the link is part of a data communications network.

Bit 1 of even numbered frames in the submultiframe carries a CRC check bit, while bits 2 to 7 carry the FAS. There is still a requirement for a MFAS to be transmitted, to ensure correct CRC decoding of the two submultiframes, but as this can not be transmitted in TS16, other arrangements are made. Bit 1 of odd numbered frames 1 to 11 carry the MFAS 0 0 1 0 1 1.

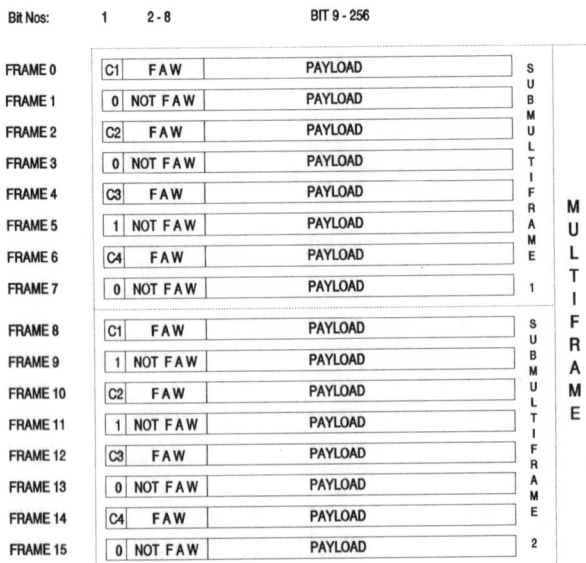

*Figure 5.14  Multiframe structure of CCITT 30-channel PCM system with CRC*

Bit 1 of frames 13 and 15 are still reserved for international use. These two bits may be used within national networks, but should be set to 1 if the link crosses international borders.

## 5.6.14 Advantages of the CRC System

One of main benefits to be gained from using this CRC technique is that error rate detection times are substantially shorter than in the simple FAS monitoring system. Note that there is no improvement in actual error performance, it is the ability to detect the presence of errors which has been improved.

Typical detection times of just under two seconds to detect a BER of $1.10^{-5}$ can be expected. This time reduces to just under half a second when the error performance degrades to $1.10^{-3}$.

A second benefit is that errors in the payload are detected by this system. In the FAS monitoring system, errors may have occurred in the payload without being detected.

The power of this CRC techniques is illustrated by the figures in Table 5.3.

*Table 5.3  Error detection performance of CCITT CRC check*

| Type of error | Probability of detection | |
|---|---|---|
| Single bit | 100% (1) | |
| | | as a 5 bit generator |
| 2-bit burst | 100% (1) | |
| | | polynomial has been |
| 3-bit burst | 100% (1) | |
| | | used for CRC coding |
| 4-bit burst | 100% (1) | |
| 5-bit burst | 87% (.87) | |
| All bursts | 97% (.97) | |

In many cases the 64 Kbits/s capacity of a TDM channel is used as the physical circuit for high speed data rather voice. These applications are less resilient to errors and further steps need to be taken to protect these circuits against the errors which inevitably will occur, even at error ratios of less than $1 \times 10^{-6}$. Error protection for these data circuits will be provided as part of the software controlling the flow of data over the physical circuit, and will not be a function of the multiplex itself. Since CCS systems use TS16 as a data circuit, they must also include such an error protection mechanism.

## 5.6.15 30-Channel TDM/PCM System Summary

| | |
|---|---|
| Nominal Bit Rate | 2.048 Mbits/s |
| No of Voice Channels | 30 |
| Sampling Rate | 8000 Hz |
| No of bits/sample | 8 |
| Encoding Law | A law (with A = 100) |
| Input Level Max | 3.14 dbm0. |
| Channel Bit Rate | 64 Kbits/s |
| No of Bits/frame | 256 |
| Frame Alignment | by 7-bit FAS in TS0 of even numbered frames |
| Multiframe | 16 Frames |
| Multiframe Alignment (In CAS systems) | 4-bit MFAS in bits 1 – 4 of TS16 in frame 0 |
| Multiframe Alignment (In CCS systems) | 6-bit MFAS in bit of frames 1,3,5,7,9 and 11 |
| Signalling | |
| Channel Associated | Four bits per channel per multiframe, by submultiplexing TS16 |
| Common Channel | by addressed messages in TS16, between processor controlled exchanges |

# 5.7 Typical 30-Channel PCM/TDM Multiplex Equipment

## 5.7.1 Block Diagram

Figure 5.15 is a block diagram of an early 30-channel multiplex equipment, which illustrates well the concepts discussed thus far. The equipment consists of four basic areas.

❏ Individual voice channel cards (for transmit and receive)

❏ Individual signalling cards (for transmit and receive)

❏ Transmit common card (common to all channels)

❏ Receive common card (common to all channels)

Each channel is equipped with analogue filtering, pulse amplitude modulation circuits and a channel gate in the transmit direction, plus a channel gate and filtering circuits which are used in the receive direction. In the example equipment a channel card contains the circuitry listed above for six channels, and therefore five such channel cards are required per 30-channel equipment.

This multiplex is specifically designed to handle Channel Associated Signalling as described in Section 5.6.6. A signalling card contains the transformer hybrid that is necessary for 2-wire to 4-wire conversion, plus signalling extraction circuits which convert the DC signalling present on the a, b and p wires from the exchange equipment to the appropriate four signalling bits (a – d). In the example equipment the relatively large size of the transformer hybrid means that only two circuits can be accommodated on a single signalling card, and thus 15 signalling cards are required per equipment.

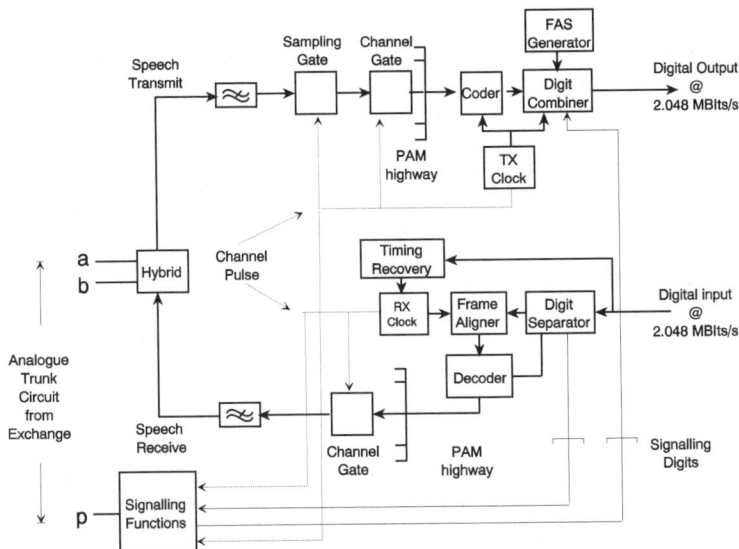

*Figure 5.15 Block diagram of a 30-channel PCM multiplex equipment (Reproduced by permission of Marconi Instruments Ltd)*

The transmit common card contains the single PAM to PCM encoder, transmit clock, the FAS generator, and a data selector used as a digit combiner. The output stage of the transmit common card includes a line code converter which transforms the pure binary output of the digit combiner to the pseudo ternary code, HDB3. Only one transmit common card is required per equipment.

The receive common card contains and HDB3 to binary converter, a digit separator, timing recovery circuit, frame aligner and PCM to PAM decoder. Again only one receive common card is required per equipment.

Additionally, although not shown on this diagram, there is a DC power conditioning card which produces the positive and negative DC voltages required by the equipment from either AC mains input or the 50 Volt DC exchange supply.

There will also be an alarm card to decode the alarm signals, display alarm conditions on LED indicators and if necessary operate alarm relays for remote indications of multiplex status.

## 5.7.2 Equipment Operation

In this case the exchange/multiplex interface is a 3-wire system. Voice signals to and from the exchange are carried solely on the a and b wires of the interface. DC signalling is carried on the a, b and p wires – generally the p wire is used to indicate that the circuit is use, while the a and b wires are used for the transmission of the seize signal, dialled digits and clear signals. These three signals are usually in the form of loops and/or loop disconnections.

Voice signals on the a and b wires which form the input to the hybrid from the exchange are fed to the input of a 3.4 KHz low pass filter. The band limited signal is then passed to a sample and hold circuit which produces a new pulse amplitude modulation sample when triggered by the channel pulse which arrives every 125 microseconds. Each individual channel pulse is timed so that the PAM signal produced by each channel is fed on to the transmit PAM highway in its own timeslot. The transmit PAM highway takes these channel pulses to the single PCM codec which produces an 8-bit A law encoded PCM code word for each PAM pulse. The output of the PCM codec is connected to one of the inputs of the digit combiner. The digit combiner is a selector which will switch one of the following signals to the digital output:

❑ Output of the FAS generator during TS0

❑ Output of the PCM codec during TS 1-15, and TS 17-31

❑ Output of the signalling converter during TS16.

On the receive side the incoming digital signal is fed from the digit separator to the timing recovery circuit. This circuit extracts timing information from the digital signal and uses it to control the receive clock so that it runs at exactly the same speed as the transmit clock in the multiplex equipment at the other end of the link (see Section 5.5.1).

The digital input is also switched by the digit separator to the frame aligner until frame alignment, and multiframe alignment have been established. Once alignment has been established the digit separator will switch the incoming signal to :

❑ The frame aligner during TS0 to check FAS for errors

❑ The PCM to PAM decoder during TS 1-15, and TS 17-31

❑ The Signalling converter during TS16.

The output of the PCM to PAM decoder is a series of PAM pulses, one for each channel, which are fed via the receive PAM highway to all channel gates. Only the appropriate channel gate will receive a channel pulse which opens the gate allowing the PAM pulse to pass through to the 3.4 KHz filter, thus reproducing an analogue signal which is fed to the exchange 2-wire line via the receive side of the hybrid.

Signalling bits are transmitted in the appropriate TS16 of the multiframe structure by the digit combiner, and similarly routed to the correct signalling converter in the receive direction by the digit separator.

This equipment may be easily modified to cater for other channel associated signalling systems such as 2-wire, or 4-wire E&M by replacing the signalling and hybrid card.

The equipment may also be modified to transmit data at up to 64Kb/s over certain channels by replacing the voice channel cards with data cards, and the transmit common card with one in which the required data channels are connected directly to the digit combiner and digit separator, thereby by-passing the PAM to PCM conversion circuitry.

# 5.8 Typical PCM Integrated Circuits

## 5.8.1 Introduction to PCM Codec Design

Traditional PCM codec (CODer-DECoder) designs have required a hybrid of analogue and digital circuitry. This makes them difficult and expensive to implement in LSI circuits, with the result that multiplex equipment such as that described in the previous section includes only one codec, and that this codec is used by all voice channels as part of the time division multiplexing operation. This section describes a single channel PCM codec based on two LSI integrated circuits produced by Plessey Semiconductors.

The codec uses an interesting design in which the sampling, quantising and encoding processes required to produce PCM from an analogue signal are rather different from those described in Section 5.3. The design is inexpensive to implement, and thus each channel in the multiplex may have its own dedicated codec. There are other advantages from this approach that will be discussed later.

The codec converts analogue signals to 64 Kbits/s A law encoded speech by a three stage process as shown in Figure 5.16.

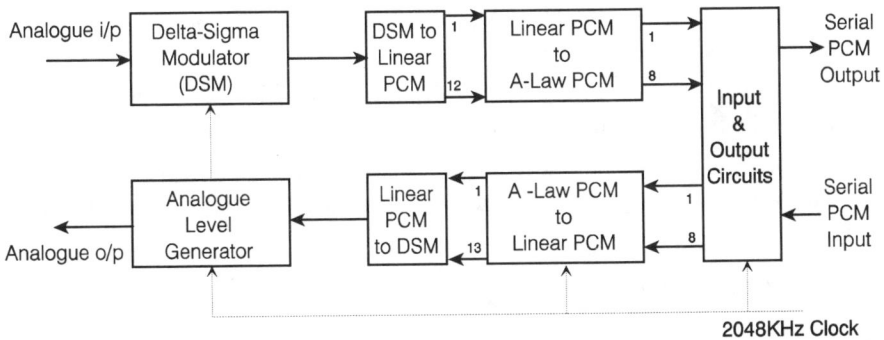

*Figure 5.16 Single channel PCM Codec using an intermediate code conversion*
*(Reproduced by permission of Plessey Semiconductors)*

❏ The Analogue Signal is converted to 2.048 Mbits/s Delta Sigma Modulation

❏ The Delta Sigma Modulation is converted to 12-bit Linear PCM at a rate of 8,000 PCM words per second, equivalent to 96 Kbits/s.

❏ The 12-bit Linear PCM is converted to 8-bit A law PCM at 64 Kbits/s

## 5.8.2 Simplified Operation of the Single Channel Codec

The band limited analogue signal is fed to a delta sigma modulator IC which is triggered by a 2.048 MHz clock signal. The digital output of the modulator is a 2.048 Mbits/s DSM signal which is the intermediate code from which the final PCM signal is produced.

The DSM signal is fed to the input of the PCM codec where it is first converted to 12-bit linear PCM.

Since the PCM sampling rate will be 8KHz, 256 bits from the DSM are used to produce each PCM code word. During each 125uS period the DSM to PCM converter produces a 12-bit linear PCM code word by multiplying each input pulse by a weighting factor which depends upon the time position of the pulse in the 125uS sampling period, and adding the result to the sum of all previous pulses in the same sampling period.

The 12-bit linear PCM code word is then converted to an 8-bit A law PCM code word using a look up table implemented in read only memory or combinational logic.

The final stage is the conversion from 8-bit parallel format to 8-bit serial format. The 2.048 MHz clock is used to time the transmission of each 8-bit PCM code word as a burst at 2.048 Mbits/s, each eight bit burst being separated from the next by just under 125 uS.

Three advantages of this design are:

❑ It is sufficiently cheap to be used on a one per subscriber line basis

❑ It has a fast start circuit which allows the components to be powered down when not in use, thereby saving power

❑ It can be easily modified to become a timeslot assignable codec. This is a codec which can be programmed to transmit the 8-bit PCM word in any one of 32 TS.

In Chapter 10 which deals with private exchanges, we will show that the line card of a typical modern exchange is a practical application of this type of single channel codec.

# 5.9 Transmission Line Codes

## 5.9.1 Introduction to Transmission Line Codes

In Section 5.1 we introduced the idea that there were various forms of digital transmission, and Figure 5.1 showed some simple examples with no further explanation. In this section we discuss the reasons for employing techniques which are slightly more complex than the simple on-off keying that was used in the early days of telegraph transmission.

As PCM transmission at 2.048 Mbit/s is one of the most common forms of digital transmission in use today. This section will describe the basic parameters for this type of transmission as given by the CCITT in recommendation G703.

Although it is possible to transmit digital signals as pure binary sequences of on-off pulses as shown in Figure 5.1(a), there are in fact several aspects of this form of transmission which make it unsuitable for use in a practical system.

## 5.9.2 The Requirements of a Practical Transmission Code

In a practical system in which two items of PCM equipment are connected by a digital link consisting of several repeatered digital sections, it will be necessary to feed electrical power to the repeaters. By far the most economical way of delivering this power to the repeaters is by connecting a DC source to the actual cable used to transmit the digital signals. Figure 5.17 illustrates the principle of DC power feeding and illustrates how the digital signal and the DC power are separated at each repeater by a transformer bridge.

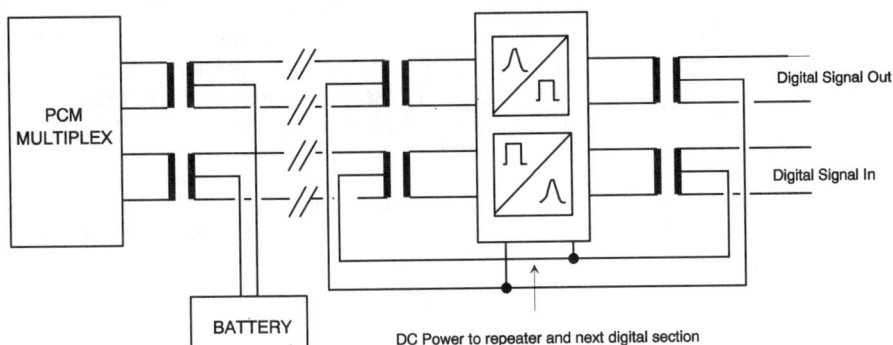

*Figure 5.17  The principle of DC power feeding by use of a transformer bridge*

One of the major problems with transmitting a binary signal is that this type of signal has a DC component, which would make it impossible to separate the digital signal from the DC power feed using a transformer bridge.

The most common form of cable used at these transmission speeds is coaxial cable with an impedance of 75 ohms. If separate power feeding pairs had to be included in the cable, its cost would be considerably increased. This leads us to the conclusion that if economic power feeding using simple transformer bridges is required, a practical line code must have no DC component.

The second problem with pure binary transmission is that the basic frequency content of the signal is continuously varying, from a minimum of zero Hertz to a maximum of half the transmission rate. Long strings of marks or spaces produce a continuous DC level and hence account for the zero frequency component, while strings of alternating marks and spaces account for the highest frequency component. Most of the time, of course, the instantaneous frequency component is varying anywhere between the two extremes as the bit sequence is generally random.

This varying frequency would give rise to two further problems in a practical system. The first is that a practical cable route will include multi-pair cables some of which may be used for digital transmission, while others are used for audio and FDM transmission. Despite some pair screening in the cable there will be sufficient low frequency power radiated from the digital cable to cause crosstalk interference on the other pairs in the cable. Although this cross talk is unlikely to have a serious effect on the digital circuits, its effect on the audio and FDM circuits is likely to be considerable. This discussion leads us to conclude that a practical digital line code should not cause crosstalk on to audio and FDM circuits which are carried on adjacent pairs in the same cable.

We have already stated that it is necessary for PCM transmit and receive equipment to operate synchronisation to avoid the problems of bit slip. For exactly the same reason it is necessary for the repeaters in the digital link to be synchronised to the transmit equipment. Some data transmission systems provide the synchronisation by transmitting the reference clock signal to the receive equipment along a completely separate pair to that used for the transmission of the actual data. This is perfectly satisfactory when the distance involved is short, typically less than 25 metres or so. However at longer distances the extra cost of cable and repeaters purely to carry clock signals, is prohibitive. Nevertheless it is essential that the receive equipment is provided with information concerning the transmit clock, otherwise it will not be able to gain synchronisation.

We now conclude that a practical line code must contain sufficient clock information to enable the receive equipment to recover timing and operate in synchronisation with the transmit equipment.

Summarising so far, a practical line code must:

❏ Have no DC component

❏ Not cause crosstalk in adjacent audio and FDM cables

❏ Contain timing information.

In addition to these three main requirements, it would be advantageous if the line code incorporated some degree of error detection, and was relatively easy to implement in hardware.

## 5.9.3 The 2.048 Mbit/s Interface

Section 6 of CCITT recommendation G703 contains the preferred parameters for the 2.048 Mbit/s interface. The first point to note is that although the nominal bit rate specified is 2.048 Mbit/s, the allowed tolerance is as high as 50 parts per million. In reality this allows for an actual transmission of between 2,047,898 bit/s (i.e. 2.048 Mbit/s – 102 bit/s) and 2,048,102 bit/s (i.e. 2.048Mbit/s + 102 bit/s). With tolerances such as these the requirement for effective synchronisation becomes abundantly clear.

The preferred cable systems are 75ohm coaxial, as already mentioned, or 120ohm balanced. The recommendation specifies that a mark should be represented by a nominal voltage level of 2.7 volts in the case of coaxial systems, and 3 volts on balanced cables. In each case zero volts is to be used to represent a space, although a tolerance equal to one tenth the level of a mark voltage is permitted.

The line code to be used on both coaxial and balanced cables is specified as High Density Bipolar 3 (HDB3). This is one of a number of codes which conform to the requirements outlined in the previous section.

It is becoming rare for new digital systems to use traditional metallic cables. Increasingly the requirement is for long distance optical fibre systems as described later in Chapter 15. It should be noted that in the interests of conformity the electrical interface between PCM equipment and the optical fibre transmission system is almost always an HDB3 interface.

## 5.9.4 The High Density Bipolar 3 (HDB3) Line Code

The HDB3 line code is described as a pseudo ternary code. By this we mean that it is a 3-level code in which individual signal elements may be one of three discrete voltage levels. In a coaxial system, these three levels would nominally be + 2.7 volts, zero volts and − 2.7 volts. In a pure ternary system, each level would have equal probability of being assigned to a particular element.

Figure 5.18 shows the form of an HDB3 signal. We can see from this diagram how the HDB3 code conforms to the requirements of a practical line code, in that:

❑ The average DC level is zero volts, due to the cancelling effect of approximately equal numbers of positive and negative voltage pulses

❑ There are no continuous strings of marks and spaces, and thus the main frequency content is concentrated within a small high frequency band, thus avoiding crosstalk problems

❑ There is a change in signal level at every element, thus permitting a relatively simple circuit to recover all the original timing information.

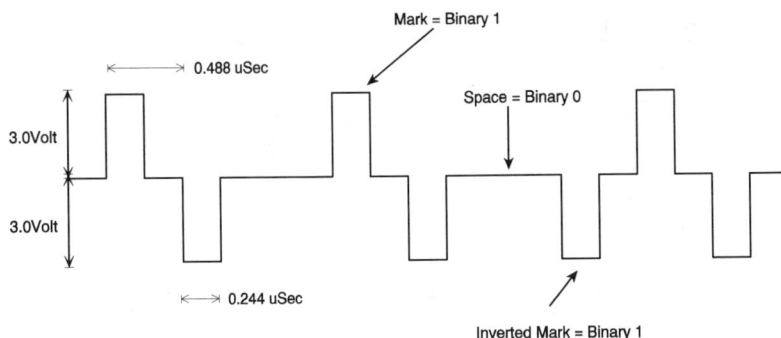

*Figure 5.18  Example of a 2.048 Mbit/s HDB3 encoded signal*

The process by which a unipolar binary code, such that derived from the output of a TTL logic gate, is transformed into HDB3 coded form is governed by a set of encoding rules which can be implemented within a single LSI integrated circuit.

The process can be summarised as a series of steps, which are illustrated in Figure 5.19. An example unipolar binary signal, which is to be HDB3 encoded, is shown in Figure 5.19(a).

## Step 1: Alternate Digit Inversion (ADI)

Every alternate bit to be transmitted by the PCM multiplex is inverted. This is actually carried out within the multiplex itself, immediately after the PCM encoding process (see Section 5.6.1). As shown in Figure 5.19(b) this step still produces a unipolar signal which has a DC component, and although strings of eight zeros, which would be transmitted on idle channel timeslots, are converted to a series of alternating marks and spaces, the same process can also produce unwanted strings of marks or spaces.

## Step 2: Alternate Mark Inversion

In the unipolar ADI signal each mark is represented by a pulse of positive voltage. This second step, illustrated in Figure 5.19(c), produces a bipolar signal by reversing the polarity of every alternate mark in the ADI signal. This results in a ternary signal with an average DC level of zero volts. Although this signal can not now contain continuous strings of marks, there is a possibility that strings of zeros, which contain no timing information, may be present in the output.

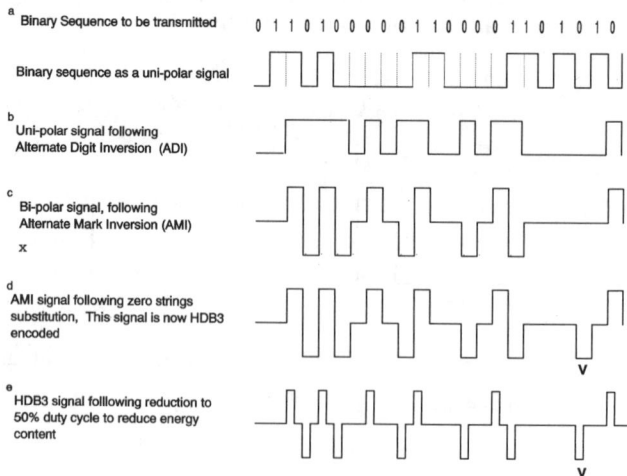

*Figure 5.19  Conversion of a unipolar binary signal into an HDB3 signal*

## Step 3: Zero string substitution

This step ensures that the output signal will not contain any strings of zeros longer than three bits duration. The timing circuits used to recover the clock signal can free-wheel for this relatively short period of time without loss of synchronisation. Figure 5.19(d) shows that this step involves substituting any string of four consecutive spaces, with three spaces followed by a mark. The total number of bits to be transmitted remains unaltered.

This process would result in a corruption of the received data after decoding if the alternate mark inversion rule is followed, as the receiver would not be able to discriminate between a true mark from the actual signal and an inserted mark. To enable the receiver to make this discrimination, and thus correctly reconstitute the original signal, the mark substituted for the fourth zero is set to the same polarity as the last previously transmitted mark. Because this inserted mark violates the alternate mark inversion rule, it is referred to as a violation pulse, and is indicated by a V in the diagram.

The HDB3 encoding rule has one further complexity. Consecutive violation pulses of the same polarity would introduce a DC level into the signal. This situation would occur if there were an even number of marks between violations. To prevent any DC level being added to the signal, successive violation pulses must be of opposite polarity. Where it is necessary to insert two successive violation pulses which would have the same polarity, the string of four zeroes is replaced by mark, zero, zero, mark. In this case the first mark follows the AMI rule, and the second mark is a violation pulse, thus maintaining an average DC level of zero volts.

## Step 4. Return to zero coding

The digital signal which results from the previous three stages is termed a non return to zero (NRZ) code. The term NRZ refers to the fact that each mark stays at its positive or negative potential for the full duration of the pulse interval and does not return to zero. The fact that zeros are transmitted as zero volts is not relevant here, a fact which sometimes leads to confusion for students meeting this term for the first time.

By reducing the pulse duration of each mark pulse to 50% of its original duration, the signal becomes a Return to Zero (RZ) code, as shown in Figure 5.19(e). Note that this step may not always be the final step to be implemented – RZ coding is possible while the signal is still in its unipolar form. RZ coding significantly reduces the energy content of the signal, resulting in crosstalk reduction, and reduction in a form of distortion known as inter-symbol interference.

The effect of inter-symbol interference is dependent upon distance, and occurs as a result of the individual pulses gradually becoming broader during transmission. This can happen to such an extent that the end of one pulse overlaps the beginning of the next by the time the signal is actually received. If the overlap becomes excessive it will be impossible for the receive equipment to distinguish one pulse from another. By reducing this form of distortion, the length of digital sections can be increased, thus reducing the number of repeaters required on any particular link.

# 5.10 Typical HDB3 Encoder/Decoder LSI IC

This short section introduces another LSI IC that can be found in typical equipment. The Plessey Semiconductors MV1441 HDB3 encoder/decoder is one of a series of ICs specifically produced for use in PCM systems. Other ICs in the same family include:

❑ Timeslot zero (FAS) transmitter and receiver

❑ A law and u law PCM Codecs

❑ Timeslot 16 transmitter and receiver

A simplified block diagram of the MV1441 IC is shown in Figure 5.20.

This IC includes the following features:

❑ An on chip digital clock regenerator capable of operating at 1.544 Mbit/s as well as 2.048 Mbit/s

❑ HDB3 encoding/decoding to CCITT G703 recommendations

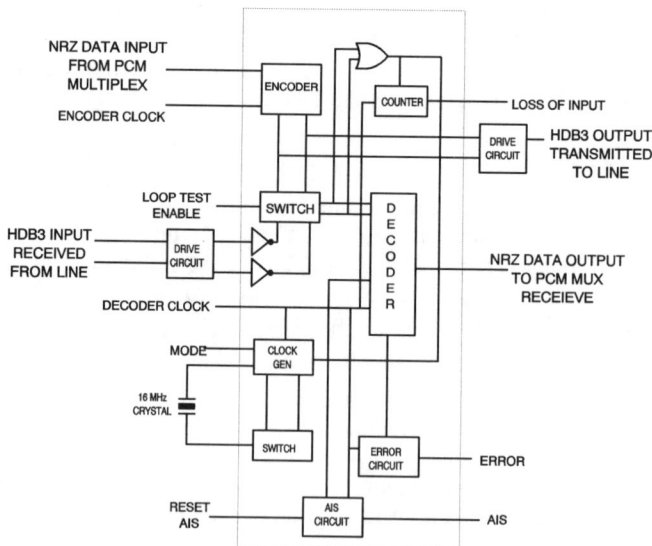

*Figure 5.20   The Plessey semiconductor MV1441 HDB3 encoder/decoder (Reproduced by permission of Plessey Semiconductors)*

❑ An HDB3 error monitor which will detect when the received signal does not conform to the encoding rule. This situation will generally only occur if a bit has been corrupted during transmission

❑ Loss of input detector. When the input signal is removed, for example as the result of transmission failure, an all zero condition appears The IC will produce an error signal which can be used to set the Alarm Indication Signal (AIS) in the associated transmit circuit.

❑ A loop test enable. The loop back control input enables a back to back test facility to be built into the equipment. This permits the testing of all the PCM multiplex functions without the need to physically connect a cable between output and input sockets.

## 5.11 Chapter Summary

Digital transmission is a key technology in today's telecommunications networks. This chapter has examined the primary rate time division multiplexing and pulse code modulation systems in use in Europe and North America.

In both systems, voice signals are converted to a digital form using PCM techniques, which have been specifically developed to obtain good quality speech reproduction while making efficient use of bandwidth. In Europe, A law PCM encoding produces a 64 Kbits/s signal for each voice channel. Thirty such channels are time division multiplexed with a framing channel and a signalling channel to produce an overall 2.048 Mbits/s system.

The American system, based on μ law PCM encoding, also produces a 64 Kbits/s signal for each voice channel; 24 such channels are then time division multiplexed with a framing channel producing a 1.544 Mbits/s system.

TDM systems require a framing signal, transmitted at regular intervals and detected by the receive equipment, so as to determine the significance of each subsequent bit in the payload area. Conventionally the framing signal is transmitted in the first timeslot of each alternate frame.

Switched networks also require the transmission of signalling information. The early networks were based on analogue channel associated signalling systems and therefore the digital systems which replaced them also incorporated this associated approach. In the 30-channel system TS16 was sub-multiplexed to provide 30 2 Kbits/s signalling channels, one for each traffic channel. In the 24-channel system, a bit stealing technique was used to provide a signalling channel at the expense of some degree of speech quality.

Both systems are now capable of handling common channel signalling, which enables modern processor controlled exchanges to transfer signalling information for all trunk circuits in the form of signalling messages bearing a label which indicates the traffic channel to which the message relates.

Both PCM/TDM systems described in this chapter form the basis for the high speed, high capacity networks described in Chapter 12.

PCM/TDM systems are also a fundamental element in every digital exchange. A digital exchange is in fact a TDM switch, and therefore this chapter provides a good introduction to the material on digital switching in Chapter 6.

# 6

# *Digital Switching*

## 6.1 The Concepts of Digital Switching

The digital telephone exchanges of System X, and almost all modern PABX incorporate digital switching systems. In order to describe these exchanges it is essential that an understanding of the concepts and techniques of this form of switching is gained first. The aim of this chapter is to introduce the concepts and technology of digital switching in order to provide the necessary basis for the remaining chapters of the book.

The term Digital Switching implies the switching of telephony signals while they are in digital form. We have seen that the most common method of digital transmission is in fact by time division multiplexing a number of PCM channels, typically 24 or 30. The architecture of a digital switching system will be based upon one of these two systems, and in fact the block diagram of any digital exchange will include a number of PCM multiplexers.

There are two forms of digital switching. One of these is analogous to a digital version of a crossbar switch, and is a form of digital space switching. The other type is known as time switching.

Both digital time switching and digital space switching are found in practical exchanges. The digital switch of a System X exchange, for example, comprises a space switching stage and two time switching stages. The basic concepts of both digital switching systems are relatively easy to understand, using a simple model such as that in Figure 6.1 which illustrates the concepts of time switching.

Figure 6.1 shows the two basic building blocks of a small digital exchange, a PCM multiplex, which in this case, is a 30-channel system, and a block labelled Time Switch (TSW). How the TSW operates is not important at this stage. The PCM multiplex is connected by two 2MBit highways to the TSW. One of these highways carries PCM signals into the TSW, while the other carries PCM signals from the TSW into the multiplex.

In this model there is only one multiplex to which up to 30 exchange subscribers may be connected. Each subscriber is allocated a specific channel and thus a specific time slot (TS), every 125 microseconds an 8-bit sample from each PCM channel is transmitted to the TSW, and an 8-bit sample from the TSW is received by each PCM channel.

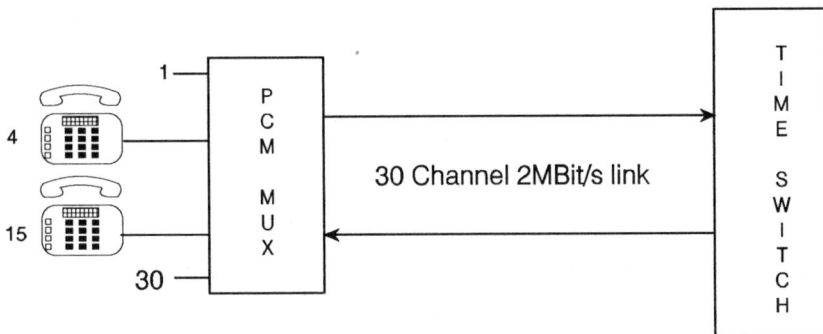

*Figure 6.1  The basic concept of time switching*

In order to understand the function of the TSW, consider that a call is progress between the subscribers allocated to channels 4 and 15. A digital connection must be made in the TSW so that speech signals transmitted by subscriber 4 are received by subscriber 15, and vice versa.

Speech from subscriber 4 is transmitted as a series of 8 bit PCM samples in TS4, of the MUX – TSW 2MBit highway. These samples are to be received by subscriber 15 and thus must be transmitted in TS15 of the TSW – MUX link. Within the TSW, the samples must be transferred from incoming TS4 to outgoing TS15, and similarly samples arriving during incoming TS15, must be transferred to outgoing TS4.

The transferring of samples from one TS to another is known as Time Slot Interchange and is the basis of digital time switching. The next section describes how time slot interchange is achieved quite simply using very basic digital hardware.

# 6.2 Time Switching Principles

Figure 6.2 shows two of the basic blocks of the time switch in Figure 6.1. The basic blocks are two small digital stores, labelled SPEECH STORE and CONNECTION CONTROL STORE. Both are of the same type of memory device used to provide volatile storage in digital computers, i.e. Random Access Memory (RAM)

## 6.2.1 The Speech Store

In order to keep this initial description simple, we will neglect the fact that the PCM multiplex is a 30-channel system and consider that all 32 time slots can be used for speech, and refer to them as TS1 to TS32.

The speech store is a digital read/write memory that contains a number of locations, each of which will be used to store speech samples from one particular channel of the incoming TDM system. In this model switch there are 32 TS in each multiplex frame, and so there must be 32 locations in the speech store. As each TS in the model system consists of eight bits, so each location in the speech store must also be eight bits wide.

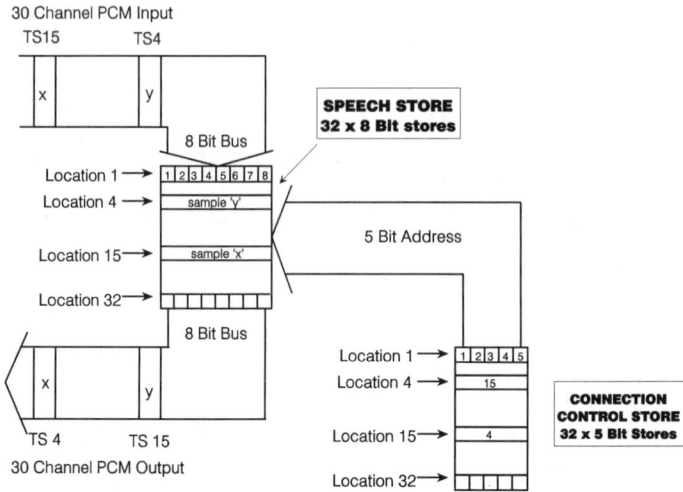

*Figure 6.2  A simple time switch*

## 6.2.2 The Control Store

The control store is also a digital read/write memory. This store will hold the information necessary for the TSW to connect one subscriber to another by time slot interchange. The model system has 32 time slots, and so the control store must also have 32 locations, one associated with each TS of the outgoing TDM system. The information which is held in each control store location is the binary address of one of the speech store locations.

Because there are 32 speech store locations to be addressed, a 5-bit address bus on the speech store is required, and so each location in the control store will also be five bits wide.

## 6.2.3 Time Slot Interchange

The incoming TDM system and the TSW are synchronised in some way (not shown in this diagram) such that as each incoming TS arrives from the multiplex, the eight bit sample in the time slot is stored in the next sequential location of the speech store. Thus the sample arriving in TS1 is stored in location 1 of the speech store, the sample arriving in TS2 is stored in location 2 and so on, until the sample arriving in TS32 is stored in location 32.

TS32 is of course the last in the multiplex frame, and therefore the next TS will be TS1 of the next frame. This will be stored in location 1, overwriting, and thus destroying, the 8-bit sample previously stored in that location. This is not a problem however, as the next part of the description shows that it is only necessary to store samples in the speech store for the duration of the frame time, i.e. 125 microseconds.

Returning to the example connection, the 8-bit samples representing the speech from subscriber 4 are stored in location 4, while speech samples from subscriber 15 are stored in location 15

The control store holds the information required by the TSW to identify which speech store location holds the 8-bit sample that is to be transmitted to the PCM multiplex in each TS of the outgoing TDM system. In the example, subscriber 4 is connected to subscriber 15. This information is actually stored twice in the control store, since it necessary to identify which speech sample is to be transmitted in both TS 4, and TS15.

In control store location 4 the number 15 is held, to identify that in TS 4, it is necessary to transmit the sample stored at location 15 of the speech store. Similarly, in control store location 15, the number 4 is held to identify that the sample held in location 4 is to be transmitted in TS15.

Consider the link between the TSW and the PCM Multiplex. During each TS on this link a two stage process is required to select from the speech store the correct speech sample to be transmitted. The identity of the required speech store location is held in the control store, thus it is necessary to synchronise events such that during outgoing TS 1, location 1 of the control store is accessed. The information stored at this location will be the address of the speech store location which must then be accessed so that the sample held in that location can be read out and transmitted in TS1.

In the example, during TS4, control store location 4 is read. This holds the address 15, so location 15 of the speech store is accessed, and the speech sample currently in this location is read out and transmitted in TS4. The same pattern of events happens during TS15, except that it is location 15 of the control store that is read first. As this control store location holds the address 4, the speech sample from location 4 in the speech store is read out and transmitted in TS15.

This technique of storing incoming speech samples into sequential locations of the speech store, and then transmitting samples from the speech store according to information held in sequential locations of the control store produces time slot interchange. The speech from the subscriber connected to channel 4 is transmitted as PCM samples by the multiplex in TS4, are switched to TS15 in the time slot interchanger, and then received by the multiplex channel 15. The PCM sample is reconverted to analogue and sent to the subscriber connected to channel 15, and vice versa for speech in the opposite direction i.e. from the subscriber on channel 15 to the subscriber on channel 4.

## 6.2.4 Practical Time Switching

In order to further develop this basic time slot interchange concept, the diagram of Figure 6.3 shows a time switch consisting of a speech store, a control store and some other items of hardware.

Since all switching in the TSW takes place with the speech samples in 8-bit parallel format, a serial to parallel (S/P) converter is required at the termination of the incoming TDM system. Likewise a parallel to serial (P/S) converter is required for the outgoing TDM system.

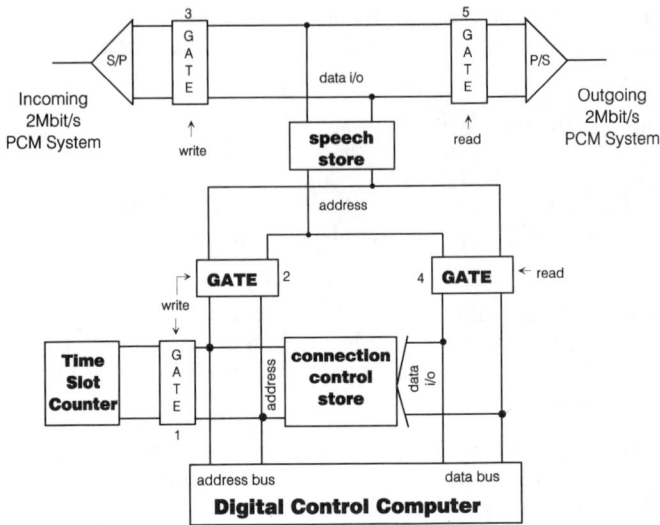

*Figure 6.3 A more complete time switch*

The other main additional item is a time slot counter (TSC). It is this TSC that provides the necessary synchronisation between the TDM systems and the time switch.

The other items are simply gates which are activated at different times in the switching process as will be described later.

It must be appreciated at this stage that the control of calls through this digital switch will be vested in a digital computer. The computer will interpret signalling information from the subscribers, and from this information will be able to write the necessary call set up data into the control store. This information is stored once only, at the beginning of a call, and will only be changed at the end of the call. Since new calls are generated at random, the control computer frequently accesses the control store to write in new call set up data.

The time switch must be synchronised with incoming and outgoing PCM systems. Bear in mind that the incoming and outgoing PCM systems actually terminate on the same multiplex and it is not hard to appreciate that during any TS period, say TS4, the multiplex is transmitting TS4 to the TSW, the TSW is transmitting TS4 to the multiplex, and the TSC is registering the number 4.

Two operations are thus required of the TSW during each TS. The first operation is a write phase during which the incoming sample in the TS is written into the speech store at the appropriate location. The second operation is a read phase during which the appropriate location in the control store is accessed and the address stored in this location used to select the speech store location holding the sample to be transmitted in this TS.

The speech store and control stores are in fact memory devices each with an address bus, and a data bus. When the control computer requires to set up a new call, it will inhibit the output of the TSC and access the address bus of the control store. The binary address of the required location will be placed on the bus by the computer. The computer will then write to that location the necessary data to effect one half of the call set up, by placing an appropriate binary number onto the data bus. In the example call the computer must write the number 15 into location 4 of the control store. To complete the call set up, it is necessary for the computer to address a second location in the control store and write the relevant data into that location. In the example call, this means that the computer must access location 15 of the control store and write the number 4 in that location.

The TSC is a binary counter which is incremented at each TS period. The TSC generates the necessary binary addresses required to access the speech store and the control store during the write and read phase of each TS.

During the write phase, gates 1, 2 and 3 are open. The output of the TSC is switched to the address bus of the speech store, thus the incoming sample will be written from the output of the S/P converter into the speech store location with that address.

During the read phase gates 1, 4 and 5 are open. Thus the output of the TSC is switched to the address bus of the control store, and the data bus of the control store is switched to the address bus of the speech store. In this way it is the data held in the addressed control store location which becomes the address of the required speech store location. The required sample is then passed to the P/S converter before transmission.

It is worth noting that apart from putting the necessary call set up data in the control store, the control computer plays no part in the switching process which is carried out purely in hardware.

The control store plays no part during the write, or input phase, it is only associated with the read, or output phase. Also since the speech store is accessed sequentially (i.e. location 1, then 2, then 3 and so on) during the write phases, and randomly during the read phases, this form of time switching is known as Output Associated Control, or Sequential Write, Random Read.

It is a simple matter to turn things around and produce an Input Associated Control system, in which case the switch would operate in a Random Write, Sequential Read manner. All that is required is that the control lines to gates 2 and 4 are reversed so that gate 2 is asserted during the read phase, and gate 4 is asserted during the write phase. This form of time switching is no more complex than output associated control, and there are some instances where this system would be preferred.

## 6.2.5 The Delay Effects of Time Switching

Time switching introduces one problem which must be considered when designing digital networks. The problem is that each time switching stage introduces a delay which is far greater than the delays introduced by the transmission stages.

To appreciate how this delay can occur, and why it can not be accurately determined consider what happens to a single sample arriving at the TSW. It will be converted from serial to parallel format and stored in the speech store. Let us assume that this sample arrives in time slot 12. If the sample is to be transmitted in TS13, it only remains in the speech store for the duration of one time slot, and hence suffers little delay. However it could be that this sample is to be transmitted in TS11. This will mean that the sample must remain in the speech store until the end of the current frame and be transmitted in TS11 of the next frame. The sample will thus remain in the speech store for almost the duration of a whole frame, i.e nearly 125 microseconds.

The average delay that can be expected is of course half the frame time, or 62.5 microseconds, but since the actual delay cannot be determined networks must be designed with worst case delays in mind.

Large networks with many exchanges present the greatest problem since any call may pass through a number of exchanges, and suffer an indeterminate switching delay in each exchange. These delays are cumulative and steps must be taken to ensure they do not have an adverse affect on overall speech transmission.

The main problem is not so much that there will be a delay between the speaker talking, and the listener hearing, although this can not be ignored, it is that some echo occurs at each end of the link because there is a hybrid terminating the 2-wire line and converting the transmission to the 4-wire system required in the multiplexer.

Unfortunately hybrids are not perfect and some of the received signal will pass across the hybrid into the transmit path and be returned to the distant end. The returned signal will be at a considerably lower level than the original but will encoded and transmitted back through the network with the wanted signal, arriving some considerable time after it was originally transmitted. To the talker this will sound as an unwanted echo, and if the delays are too great, make normal speech difficult . In its recommendation G164 the CCITT has recommended that the round trip delay through a national network should not be greater than 24 millisecs.

## 6.2.6 CCITT Recommendations for Time Switching Delays

The CCITT also recommended that the time switching delay through a single digital trunk exchange should not exceed 450 microseconds. This represents about three time switching stages if a worst case figure of 125 microseconds delay per time switch. In the design of the System X network, the preferred routeing of all calls should involve no more than four exchanges.

It should also be remembered that the telephone network is used for the transmission of computer data, a trend which will increase as more of the telephone network becomes digital. The effects of the time switching delay on the exchange of data between computers can become a problem in those computer systems which expect fast response times.

# 6.3 Typical LSI Time Switch IC

## 6.3.1 MA811 DSM

The Marconi Electronic Devices MA811 Digital Switch Module is a 28 pin DIL CMOS time switch capable of handling up to 256 incoming and outgoing channels organised as eight separate 32-channel PCM/TDM systems. The time switch is able to connect any incoming channel to any outgoing channel. A block diagram of the IC is shown in Figure 6.4 on the next page.

## 6.3.2 MA811 DSM Functional Description

The main functional elements of the MA811 are a 256-location Speech Store, a 256 location Connection (or control) Store, Input Buffer, Output Buffer and Control Logic. The Control Logic will include a Time Slot Counter and the circuits required to provide interfaces for three timing signals and a microprocessor control unit.

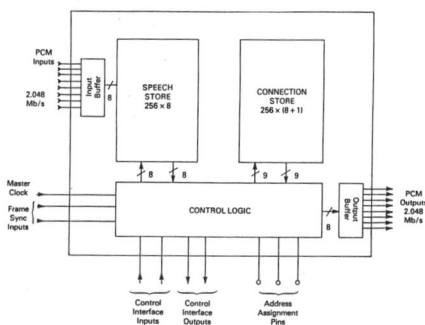

*Figure 6.4 The Marconi MA811 digital switch module (Reproduced by permission of Marconi Electronic Devices Ltd)*

To ensure the correct timing of all operations within the time switch the MA811 must be provided with a 4.096MHz master clock on pin 26 (Clk). In addition an input frame synchronisation pulse every 125 microseconds is required on pin 28 (FSP1). Optionally an output frame synchronisation pulse can be provided to pin 9 (FSP2) to ensure a timed relation between input and output frames.

The eight incoming PCM/TDM systems are connected as 2MBits/s serial inputs to pins 1 to 8 labelled DATA IN 0 (DI0) to DATA IN 7 (DI7). The eight outgoing systems are taken as 2MBits/s serial outputs from pins 10 to 19, labelled DATA OUT 7 (DO7) to DATA OUT 0 (DO0).

Since the time switch must be controlled by some form of microprocessor, an interface is provided via four pins labelled CONTROL INPUT 0 (CI0), CONTROL INPUT 1 (CI1), CONTROL OUTPUT 0 (CO0) and CONTROL OUTPUT 1 (CO1).

## 6.3.3 Speech and Connection Stores

Figure 6.5 shows the conceptual organisation of the speech and connection stores. The speech store and connection store are both read/write memories. The speech store contains 256 locations (numbered 0 through to 255). Each speech store location is eight bits wide and is allocated to one specific incoming PCM channel.

| Loc | Speech Store INCOMING | Connection Store OUTGOING | |
|-----|----------------|------------------|---|
| 0   | Sys 0  TS 0    | Sys 0  TS 0      | B |
| 1   | Sys 1  TS 0    | Sys 1  TS 0      | B |
| 2   | Sys 2  TS 0    | Sys 2  TS 0      | B |
| 3   | Sys 3  TS 0    | Sys 3  TS 0      | B |
| 4   | Sys 4  TS 0    | Sys 4  TS 0      | B |
| 5   | Sys 5  TS 0    | Sys 5  TS 0      | B |
| 6   | Sys 6  TS 0    | Sys 6  TS 0      | B |
| 7   | Sys 7  TS 0    | Sys 7  TS 0      | B |
| 8   | Sys 0  TS 1    | Sys 0  TS 1      | B |
| 9   | Sys 1  TS 1    | Sys 1  TS 1      | B |
| 10  | Sys 2  TS 1    | Sys 2  TS 1      | B |
|     | ..             | ..               |   |
|     | ..             |                  |   |
| 245 | Sys 5  TS 31   | Sys 5  TS 31     | B |
| 246 | Sys 6  TS 31   | Sys 6  TS 31     | B |
| 247 | Sys 7  TS 31   | Sys 7  TS 31     | B |
| 248 | Sys 0  TS 32   | Sys 0  TS 32     | B |
| 249 | Sys 1  TS 32   | Sys 1  TS 32     | B |
| 250 | Sys 2  TS 32   | Sys 2  TS 32     | B |
| 251 | Sys 3  TS 32   | Sys 3  TS 32     | B |
| 252 | Sys 4  TS 32   | Sys 4  TS 32     | B |
| 253 | Sys 5  TS 32   | Sys 5  TS 32     | B |
| 254 | Sys 6  TS 32   | Sys 6  TS 32     | B |
| 255 | Sys 7  TS 32   | Sys 7  TS 32     | B |

*Figure 6.5 Conceptual organisation of the speech and connection stores*

The connection store also contains 256 locations, one for each of the 256 outgoing PCM channels. Since each connection store location will hold the address of a speech store location it is necessary that each connection store is at least eight bits wide. In this IC however, the designers have included a ninth internal *busy* bit, B, which can be set or reset by the control system to indicate that an output channel is in use, or idle.

The conceptual organisation of the speech and control stores shown above allows sequential access to both during write and read operations respectively. The location addresses may be generated by an enhancement of the TS counter, in which an 8-bit binary counter is clocked at 2Mbits. The three least significant bits of the counter generate the PCM system number, while the five most significant bits generate the TS number. Taken together the eight bits provide the address of any single location in the speech or connection store

## 6.3.4 Control Interfaces

The Control Interface Inputs and Outputs provide the necessary interface to enable a microprocessor to set up and release calls through the DSM. Using these interfaces it is also possible for the microprocessor to inject and extract certain codes into any channel.

## 6.3.5 Limitations on the Capacity of Time Switches

Using this particular time switch as an example, this section will show that the number of channels that can be handled by a time switch is dependent upon the access times of the memory technology in use, rather than the number of memory locations that can be fabricated on a single LSI circuit.

In the MA811, during each TS there are at least 24 memory accesses required:

❏ Eight to write the eight incoming PCM words to the relevant locations in the speech store

❏ For the eight outgoing channels, eight accesses to the relevant locations in the control store to ascertain which speech store locations are to be accessed

❏ To read out from the eight relevant locations in the speech store the required PCM words for the eight outgoing PCM channels.

(This neglects any control interface instructions and response which will probably inhibit normal switching accesses.)

Bearing in mind that the duration of a TS is approximately 3.9 microseconds, the access times for the speech store and connection store memories must be less than 162.5 nanoseconds.

The reason for highlighting the requirement for fast access times is to illustrate that the switching capacity of a digital time switch is not dependent upon how much memory can be provided, but on the speed with which each memory location can be accessed. If the capacity of this switch was to be increased to 16 input and output systems (i.e 512 channels) the size of the speech store would need to be doubled. The size of the connection store would need to be more than doubled to 512 locations, each of 10 bits (9 speech store address bits, and an internal busy bit). However, during each 3.9 microsecond TS, 48 memory accesses would be required, increasing the demand for fast access times.

A 256-channel digital switch would provide adequate performance in a wide range of small to medium PABX applications, however in large PABX and public exchanges greater switching capacity is required. This may be achieved by using the DSM in a large switching matrix with other DSM, or by using digital space switching as described in the next section.

# 6.4 Digital Space Switching

## 6.4.1 The Requirement for Digital Space Switching

The previous section of this chapter showed how several TDM/PCM systems could be interconnected on the same time switch, so that any TS on any incoming system can be switched to any TS on any outgoing system. It was shown that the maximum number of connections possible on such a time switch is dependent upon the memory access times. This fact means that a time switch terminating a large number of PCM systems would need to be constructed with very fast memory devices to operate correctly.

High capacity memory devices with fast access times are now relatively cheap compared with the devices that were available a decade or so ago. However in the early days of digital switching, high memory costs meant that large time switches would have been prohibitively expensive. For this reason many early digital exchanges included a digital space switching stage, to allow smaller sized time switch units to be used than would otherwise be the case.

## 6.4.2 The Concept of Digital Space Switching

Digital space switching can be likened to the operation of an extremely fast matrix of crosspoints. A digital space switch will provide time divided cross connections between any two of a number of PCM/TDM systems. Figure 6.6 shows the concept of a digital space switch. In this diagram four TDM systems each with 32 TS, incoming and outgoing, terminate on the space switch.

The model space switch of Figure 6.6 can be considered as a matrix of 16 cross-points, with four incoming lines and four outgoing lines. The important point to realise is that there is a 32-channel TDM system on each incoming line. The switch is able to connect any incoming system to any outgoing system for the duration of one TS. Since there are four input and output systems, up to four separate connections are possible during each TS.

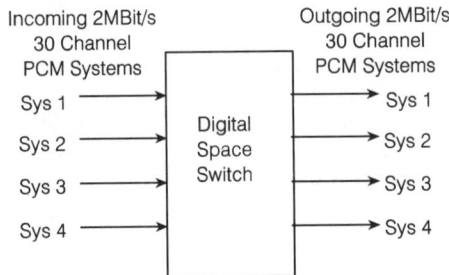

*Figure 6.6  The concept of digital space switching*

The duration of a multiplex frame is 125 microseconds, thus the duration of each TS is 125/32 microseconds, or 3.9 microseconds. During this 3.9 microsecond period the eight bits associated with the current TS on each system pass across the switch and are transmitted on the appropriate outgoing system.

Note that apart from a few nanoseconds propagation delay across the switch, there is no delay associated with digital space switching.

At the end of each TS, the switch reconfigures the connections for the duration of the next TS.

Figure 6.7 shows how the space switch may be configured during TS1. In this case, the following connections are made, and held for the duration of TS1 only.

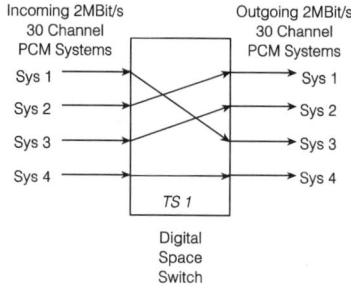

*Figure 6.7 Possible configuration of digital space switch during TS1.*
*Notes: a. Connections are not necessarily symmetrical; b. Connections between same incoming and outgoing system are possible.*

Figure 6.8 shows how at the end of TS1 the digital space switch is re-configured to provide a completely different set of connections. This new combination of connections is again only held for the duration of one TS, in this case TS2. Since there are 32 TS in each multiplex frame, the switch must be able to cater for 32 different combinations of connections, and will reconfigure itself at the end of each TS. At the end of each multiplex frame the switch repeats the sequence of connections. This process will continue until the exchange control system releases a call in progress or sets up a new call. In either case the connections made in one TS will be changed to reflect the new call status.

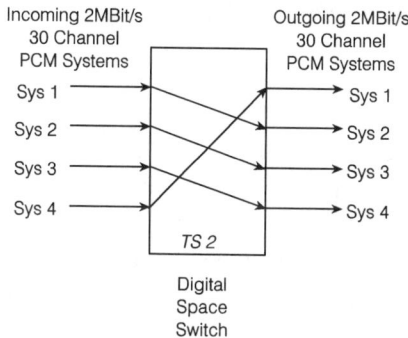

*Figure 6.8  Possible configuration of digital space switch during TS2*

## 6.4.3 Realisation of Space Switching Stage

The two major hardware items required for a simple digital space switch are shown in Figure 6.9. These are a multi-line to single line data selector and a control store. The model space switch shown in this diagram has four TDM/PCM incoming systems and only one outgoing system.

The control store has 32 locations, one for each TS of the outgoing TDM/PCM system. Each control store location will hold a number which represents the incoming system which must be connected to the output during the time slot associated with that location.

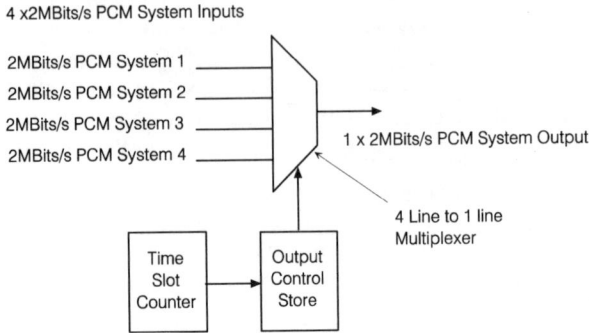

*Figure 6.9 A simple digital space switch*

As an example, consider that during TS1, the sample arriving in TS1 of system number 3 is to be switched to the outgoing system TS1. Location 1 of the control store will hold the number 3. Similarly if during TS2, the sample arriving in TS2 of system 1 is to be switched to TS2 of the outgoing system, control store location 2 will hold the number 1.

This simple space switch would not be practical in a real system because it only has one outgoing system. What is required of course is a switch with as many outgoing systems as incoming systems. In the case of the model space switch of Figure 6.9, there are four incoming and four outgoing systems. Such a switch could be realised by using the simple space switch of Figure 6.9 as a building block and connecting four such switches together as shown in Figure 6.10.

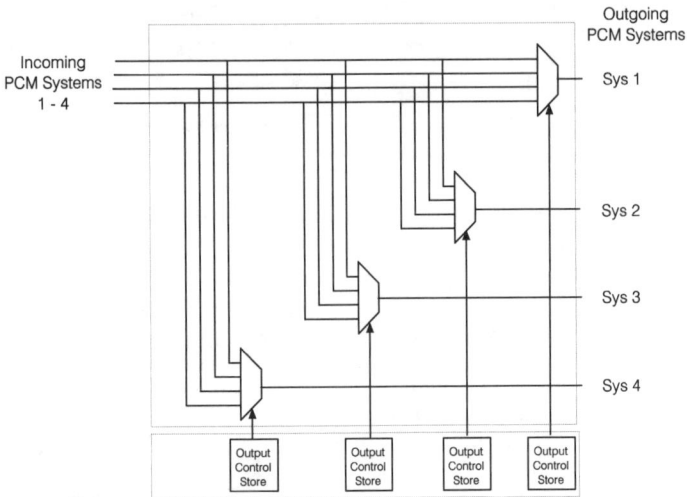

*Figure 6.10  A more practical digital space switch*

The practical realisation of a digital space switch includes a number of identical switches, with their inputs effectively connected in parallel. Each switch has its own dedicated control store, the contents of which will be updated as required by the control system of the exchange as new calls are set up and released. Intelligence within the software of the exchange will ensure that at no time is an attempt made to connect an incoming system to more than one outgoing system simultaneously.

While many large digital exchanges incorporate space switching stages, these are always used with time switching stages as it would be impractical to make an exchange from space switching stages only. The next section describes how time and space switching stages may be combined to produce large switching systems such as those used in System X.

# 6.5 Time and Space Switching Combinations

## 6.5.1 Time-Space-Time Switching

Many large exchanges are configured as time-space-time (TST) switches. This section describes such a configuration as an introduction to the Digital Switching Subsystem of System X.

Figure 6.11 shows how a digital exchange terminating four PCM/TDM systems (i.e. 128 channels) may be constructed from four PCM/TDM multiplexers, and digital switch consisting of eight time switch modules and one space switch module.

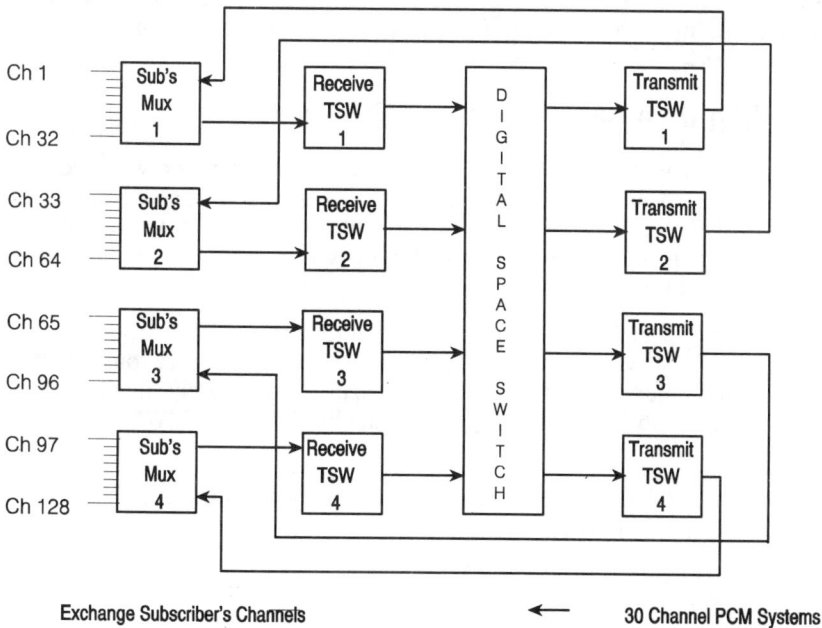

*Figure 6.11 Simple digital exchange employing a time-space-time Switch*

As with any exchange, this digital exchange must be capable of connecting any subscriber to any other. In other words the digital switch must be capable of providing a connection between any two of the 128 PCM channels, and the maximum number of simultaneous connections. In reality the Marconi MA811 DSM described in Section 6.3 could be used on its own in this application, but this simple example is designed to show the concepts which are applied to the construction of large exchanges.

## 6.5.2 Description of the Time-Space-Time Switch

Subscribers' telephones are connected to the analogue inputs of four PCM multiplexers based on the standard 30-Channel PCM multiplex system.

For reasons of simplicity this example ignores the fact that two time slots would be used for framing and signalling, and assumes that all 32 time slots are associated with subscribers' telephones. Also in the interests of simplicity, the diagram does not show the control system of the exchange.

The digital output of each subscriber's multiplex is connected to the input of a receive time switch, while the digital input to each multiplex is taken from the output of a transmit time switch.

The eight time switches are identical modules based on the simple time slot interchanger of Figure 6.1. They are labelled as receive time switch or transmit time switch depending upon their role within the exchange. The receive time switches are so called due to the fact that they receive the incoming digital signal from the subscribers' multiplex. Conversely the transmit time switches transmit the switched PCM signals back to the subscribers' multiplex.

## 6.5.3 Setting Up a Call Through the TST Switch

To understand how this configuration works, and is able to provide the required connection between any two of the 128 subscribers, consider some of the actions which the exchange control system must perform to set up a call through the 3-stage digital switch. Bear in mind that there will already be several calls in progress at the time the call is set up.

The example call will be between subscriber no 6 and subscriber no 92. This call requires that PCM signals from TS6 of subscribers' multiplex 1 are routed to TS28 of subscribers' multiplex 3, and vice versa. This will obviously involve time slot interchange from TS6 to TS28, and space switching from Rx TSW 1 to Tx TSW 3, and vice versa.

To set up the call the control system must store the necessary call set up data in the control stores of the following five switches:

❏ The Space Switch

❏ Rx TSW 1

❏ Tx TSW 3

❑ Tx TSW 1

❑ Tx TSW 3

As no other switches are involved, the control system disregards them during the set up of this call.

Because there are already calls in progress it may not be possible for the space switch to connect Rx TSW 1 to Tx TSW 3 during either TS6 or TS28. The control system must therefore select a time slot during which the space switch is free to connect Rx TSW 1 to Tx TSW 3, and likewise a time slot when the space switch is free to connect Rx TSW 3 to Tx TSW 1. There is no reason to suggest that the same time slot must be chosen for both these connections.

For this example, it is assumed that the following free time slots have been selected by the control system.

❑ TS10 to connect Rx TSW 1 to Tx TSW 3

❑ TS15 to connect Rx TSW 3 to Tx TSW 1

The control system must now update the necessary control stores to effect the connection.

Firstly the control store of the space switch must be updated so that during TS10 the output of Rx TSW 1 is connected to the input of Tx TSW 3, and that during TS15 the output of Rx TSW 3 is connected to the input of Tx TSW 1. The next step is to update the control stores of the four time switches involved.

As the space switch will be connecting Rx TSW 1 to Tx TSW 3 during TS10, the only role of Rx TSW 1 in this call is to switch the incoming TS6 to outgoing TS10, and so one entry to this effect is required in the control store of Rx TSW 1. To complete this direction of the call the control store of Tx TSW 3 must be updated so that Tx TSW 3 switches its incoming TS10 to outgoing TS 28. The complete connection for this direction of the call is shown in Figure 6.12.

The reverse direction of the call involves Rx TSW 3 switching incoming TS28 to outgoing TS15, while Tx TSW 1 must switch incoming TS15 to outgoing TS6. Thus the control stores of these two switches must be updated to effect these connections. The connection for the reverse direction of the call is also shown in Figure 6.12.

## 6.5.4 Comparison of TST and STS Arrangements

This section has introduced the concept of TST arrangements for large digital switches. STS designs are feasible, in fact several of the British Post Office's early digital exchanges used such a design.

This explanation of the TST switch has demonstrated that a complete two-way connection requires two separate switched routes to set up through a digital switch. In both switched routes the switched channel passes through two time switching stages and a space switching stage. The time switching delay effects of two time switching stages are obviously greater than had a Space-Time-Space (STS) configuration been used.

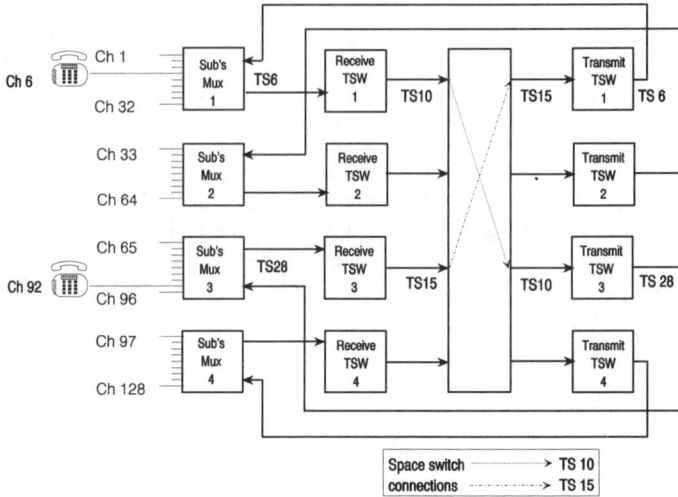

*Figure 6.12  Connection of the example call through the TST switch*

There are, however, several reasons for using the TST design. These are based on the ability of this design to make use of further high speed multiplexing of the PCM/TDM systems, to reduce the physical number of crosspoints required in the space switch. Further explanation is beyond the scope of this book, however the chapter on the System X Digital Switching Subsystem will illustrate this point.

Now it is generally accepted that large multistage exchanges should be of the TST variety. But as memory technology improves we are likely to see more large exchanges built using only time switching modules employing architecture similar to that of the Marconi MA811 IC.

# 6.6 Chapter Summary

The stated aim of this chapter was to provide an introduction to digital switching concepts and technologies. The chapter commenced by introducing, in relatively simple terms, the concept of switching PCM signals from one time slot to another by time slot interchange. This idea was developed further to show that PCM signals could also be switched from one PCM system to another, and create the basis of a digitally switched exchange.

A relatively simple 256-channel, i.e. 8 PCM system, digital switch module from Marconi was used to introduce the time switching technology that is available. Digital switches in large systems however do not rely on time switching stages alone, and thus it was necessary to introduce the digital space switch.

Many large digital switches are actually fabricated from more than one switching stage. For instance several time-space-time switches are in existence. The final part of the chapter briefly described such a switching system. This description will be amplified in Chapter 9, when the digital switching subsystem of British Telecom and GPT's System X exchanges are dealt with.

# 7

# *Control Systems in Digital Exchanges*

## 7.1 Functions of Control

The two basic elements of any exchange are a switching unit and a control unit. The function of the switch unit is to connect incoming and outgoing speech channels under control of the control unit. As well as controlling the switch unit, the control unit has many other functions. In Table 7.1 some of the main functions of the control unit are listed.

*Table 7.1   The functions of an exchange control system*

Recognition of new calls from subscribers' and other exchanges
Reception of dialled digits from subscribers' telephones and other exchanges
Translation of received digits into network routing digits
Determination of required connections in the switch unit
Transmission of address information through the network
Recording call details such as destination and duration for charging purposes
Fault detection and clearance
Maintaining traffic statistics
Provision of enhanced telephony facilities for users

This list is by no means exhaustive, in fact it would probably be possible to write an entire book solely on the subject of exchange control systems. Just as the switch unit has evolved over the last two decades into a fast digital unit, considerable developments have been made in the area of control. For example in the early Strowger exchanges in which each selector had its own switch and small control unit, the functions of control were limited to the basic actions of call set, charging and call release. There was little scope to offer any extra facilities in an environment based on electromagnetic relays and wired logic.

## 7.2 Control Systems

### 7.2.1 Progress Towards Computer Control

The introduction of switching technologies such as reed relay and crossbar provided an opportunity to reduce the cost of the control unit, by using a single unit to control several hundred switches. The single control unit was used in a time sharing role to operate and release the switches in the switch unit. Also the provision of a single control unit meant that extra features could be included in the wired logic at a cost that was no longer prohibitive.

Although initially the control unit was based on electromagnetic relays, semiconductor and computer technologies were rapidly introduced into the field of telecommunications by telephony engineers who had an ideal background to seize the opportunities these technologies offered.

## 7.2.2 Common Control

Common Control was the term that was used to describe systems employing a single centralised control unit. Strowger switching, in which the *intelligence* of the exchange was contained in each small control unit associated with every selector in the exchange, became identified as a distributed control system.

## 7.2.3 Reliability Considerations

The distributed control system, exemplified by Strowger switches was inherently extremely reliable. Although individual selectors could, and often did fail, the risk of a complete exchange failure, due to one or more faulty units, was negligible. Simple facilities existed to permit a technician to isolate the faulty unit, so that its failure did not interfere with the continued operation of the exchange. In fact it is unlikely that the failure would have been noticed by more than a handful of users who may have experienced a loss of service.

In contrast the adoption of common control and its relatively complex control units, was not inherently as reliable as Strowger. Although the reliability of individual components may have been greater, the reliability of the exchange as a whole was based on a large number of components in a single control unit.

Since all functions of the exchange depended upon the correct operation of the control unit, even small faults could have a catastrophic effect, perhaps resulting in complete failure of the exchange.

To secure operation in such an environment requires building a measure of redundancy into the system. This would normally involve duplication of equipment and there is therefore a cost penalty in respect of the extra equipment which is not strictly required for the operation of the exchange.

The degree of redundancy in any particular system will depend upon whether the extra costs involved are considered to be worthwhile when compared to the inconvenience of losing the use of the exchange when some critical unit fails.

## 7.2.4 Stored Program Control

Stored Program Control (SPC) is a natural progression from common control. The common control unit becomes in fact a digital computer (or processor) which operates according to a set of instructions, the program, stored in some form of computer memory. This stored program must make use of information regarding customer numbers, network routing, charging rates etc. This information, or data, is also held in the computer memory.

The SPC digital computer is operating in a considerably different environment from its counterpart in the world of data processing. The SPC computer is, for instance, operating in real time. It has to react to varying levels of demand for service with response times that are both fast and constant. The exchange must to be able to set up and release each call just as quickly when it is handling 10 calls an hour as when it is handling 1,000 calls an hour.

During the time the call is in the speech phase, there is little work for the computer to do, so the number of concurrent calls is limited by the size of the switch. However the number of calls that can be handled by the exchange in a given time depends more upon the ability of the computer to process the call set and release phases speedily. A figure known as The Busy Hour Call Attempt (BHCA) is a measure of this aspect of an exchange's traffic handling capacity

### 7.2.5 Reliability Issues in SPC Systems

In general the SPC computer is performing simpler tasks than the data processor. It is, however, repeating the same operations many times an hour.

Unlike the data processor where faults may cause acceptable delays of hours (or even days), the SPC computer in the exchange should be in operation 24 hours a day, 365 days a year, with no unscheduled downtime. The issue of reliability is thus especially important when SPC is employed.

Computers consist of many thousands of components, and although each component may be designed to give high reliability, the sheer number of components involved reduces the reliability of the computer to level at which it would not be acceptable to use a single computer in the exchange control task. Just as in the case of common control, redundancy techniques involving the duplication of critical units, are necessary to achieve the required levels of availability.

Within the field of SPC, reliability issues are not constrained to preventing faults in the computer hardware causing complete failure. The program and the data, on which the program operates, are both potential causes of disaster.

The next sections of this chapter describes some basic reliability theory and considers some techniques which are used in various exchange designs to improve reliability of both hardware and software.

## 7.3 Reliability Techniques in SPC Systems

### 7.3.1 Reliability Theory

Reliability theory is an area of operational research which has many applications in the world of electronics. A very brief introduction to the subject will provide sufficient background for the remainder of this section.

Reliability can be defined as the probability that a component, or a system of components, will not fail in a given time. The reliability (R) of a component is normally given as number between 0 and 1. A reliability of 1 means there is no possibility that the unit will fail, needless to say this ideal is never reached in practice. One aspect of calculating the reliability of a component, or a system of components, is a knowledge of the failure rate of the item concerned.

LSI circuits can be life tested experimentally to measure the failure rate, which is defined as the fraction of all components of the same age which fail at that age. For example, if 1,000 units are life tested, and 10 fail after 2,000 hours, the failure rate at 2,000 hours is given by:

Failure rate = <u>No of failed units</u> = <u>10</u> = 0.01
(2000 hrs)     No of units tested    1000

The failure rate can be measured for similar components of all ages and a graph such as that in Figure 7.1 produced.

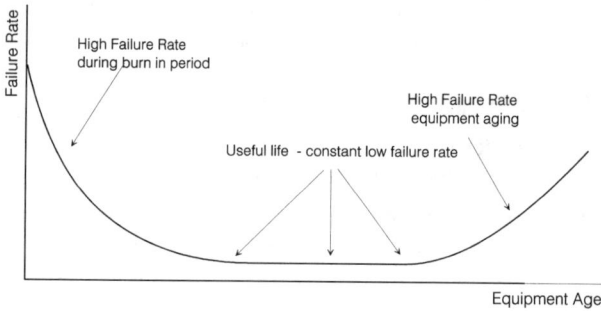

*Figure 7.1 Typical failure rate against age graph for LSI circuits*

The shape of this graph is typical of many components in all areas of engineering, and for obvious reasons is known as the bath tub curve. The curve has three important areas:

❏ The burn in area; where due to teething trouble or faults in manufacture, the failure rate is relatively high

❏ The useful life area, which is almost a straight line indicating that the failure rate is constant with increasing age

❏ The wear out area, where the failure rate increases with age.

Our concern is with the second area of the curve. Manufacturers of components and systems test their products before releasing them to customers in an attempt to catch all failures in the burn in area. This can not guarantee to prevent a failure in the early life of the product, but should ensure that of all products delivered the percentage which fail in early life is low.

The third area of the curve indicating high failure rate with old age can be predicted to a certain extent, and marks the end of the designed life of the system. During this period, it will become increasingly expensive to repair faulty items, and there will come a time when it would be prudent to replace the system in its entirety. Since the design life can be predicted, steps should be taken to procure a new system before it becomes uneconomic to repair the old.

Since the processor controlling the telephone exchange, and in fact the majority of the exchange hardware, is built from components which will all exhibit failure characteristics similar to that of the bath tub curve, it is important to consider what will happen during the second part of this curve. The failure rate curve shows that during the useful life of a system, there is a small probability that a failure will occur in one or other components. As previously stated in Section 7.2 steps must be taken to ensure that if a failure occurs it does not result in failure of the whole exchange. i.e. the system must be designed to be *fault tolerant*.

The reliability of any unit, such as a processor, depends upon the reliability of all its subsystems or components. Generally if any single subsystem fails, the unit itself will fail. This can be represented by the reliability diagram in Figure 7.2, in which the reliability of the unit is shown as three subsystems in series. The life line of the system running from left to right will be broken if any of the three units fails.

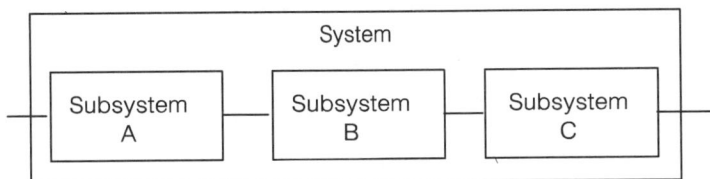

Correct operation of the whole system dependes upon operation of all
3subsystems

*Figure 7.2 Reliability diagram for series systems*

If the reliability of each subsystem is known, the reliability of the whole unit can be calculated simply by multiplying the individual reliability figures together. Note that as the reliability figure for any item can never be greater than unity, so the reliability of the whole is less than the reliability of any of the individual units.

Here's an example. In the diagram above, the subsystems have the following reliability figures:

| Subsystem 1: | $R1 = 0.95$ |
| Subsystem 2: | $R2 = 0.9$ |
| Subsystem 3: | $R3 = 0.85$ |

The reliability of the whole unit (RT) is the product of R1, R2 and R3.

$$RT = R1 \times R2 \times R3$$
$$RT = 0.95 \times 0.9 \times 0.85$$
$$RT = 0.72675$$

It can be seen that even if the reliability of each component was very high, say, in the order of 0.9999, the reliability of a unit containing 1,000 such components in a series reliability combination is reduced to about 0.9.

The reliability of a system can be improved by connecting a number of similar units in a parallel combination, such that operation is ensured even if only one of the units is working. A good example of this is the case of a three-engined aircraft which can still fly with only one engine.

In the diagram in Figure 7.3, each subsystem may itself contain a number of series connected elements. Now the reliability of the whole system is greater than the reliability of the individual subsystems.

The process of calculating the increased reliability is somewhat more complex than that for a series case, but example figures will help to appreciate the improvement. If each unit in the diagram is identical, it is a safe assumption that they all have the same failure mechanisms, and equal reliability figures, say R = 0.95. It can be shown that the reliability of the whole RT = $(1 - R) = (1 - 0.95)^3$ or 0.999875. Thus the reliability of the whole approaches, but can never equal 1. Note that if four units were used, the reliability increases slightly to 0.9999938.

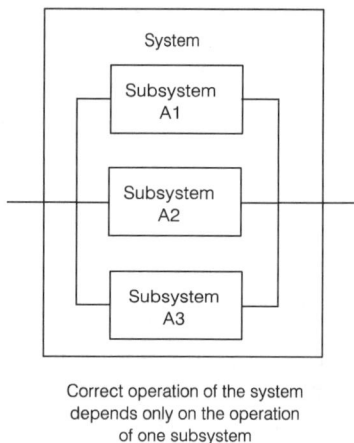

```
┌─────────────────────────────────────┐
│              System                  │
│        ┌─────────────────┐           │
│        │   Subsystem     │           │
│   ┌────┤      A1         ├────┐      │
│   │    └─────────────────┘    │      │
│   │    ┌─────────────────┐    │      │
│   │    │   Subsystem     │    │      │
───┼────┤      A2         ├────┼───────
│   │    └─────────────────┘    │      │
│   │    ┌─────────────────┐    │      │
│   │    │   Subsystem     │    │      │
│   └────┤      A3         ├────┘      │
│        └─────────────────┘           │
└─────────────────────────────────────┘
```

Correct operation of the system
depends only on the operation
of one subsystem

*Figure 7.3 Parallel combination*

## 7.3.2 Redundancy

The term redundancy when applied to engineering really means providing more equipment than is strictly necessary to carry out some function or operation, typically by duplication of critical components. Usually the aim is to take account of the fact that a failure may occur as no component is completely reliable. While the simple provision of duplicate equipment improves the reliability of the whole system, there are methods by which this improvement can be more economically achieved. Some of the methods used will be illustrated after the discussion of basic duplication techniques.

In the remainder of this section we will use the term processor rather than computer, if only because the word computer tends to mean the total hardware of a processing system.

## 7.3.3 Active Parallel Redundancy (Synchronous Operation)

Figure 7.4 shows a typical parallel redundancy situation based on the reliability diagram in Figure 7.3. Two processors, each capable of controlling the whole exchange on their own, are coupled together in such a way that they are operating in synchronism, each controlled by the same clock unit, which is itself duplicated for security. Both processors are carrying out the same operations, on the same data, at the same time, and are said to be operating in active parallel redundancy.

The operation of both processors is validated by comparing their outputs, and at all times they should be the same. If the validation fails, both processors must carry out some form of self diagnosis to determine which of the two is in error. The faulty processor is then automatically isolated, leaving one processor to continue on its own, while a maintenance alarm is raised to ensure that the faulty unit is repaired and brought back into service.

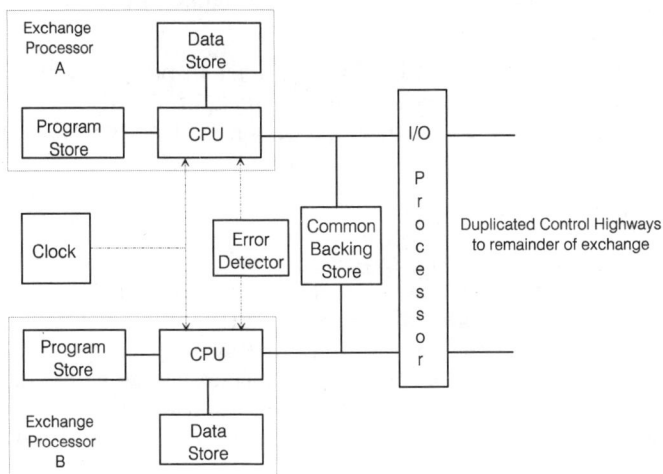

*Figure 7.4 Active parallel redundancy or synchronous operation*

Duplicated control systems in themselves are not the sole key to reliable operation, as such systems pose several problems which need to be overcome. The list below gives some idea of the extra problems which need to addressed if a duplicated system is to be successful.

❑ Immediate detection of a fault in one of the processors

❑ Determination of which processor is faulty, and ensuring it is isolated from the rest of the exchange

❑ Transfer of full control to the working processor

❑ Diagnosis of the actual fault on the failed processor

❑ The repaired processor must be reconnected and returned to service without disrupting the exchange operation.

In the system shown in Figure 7.4, fault detection is immediate because the outputs of both processors are constantly being compared. The problem is to decide which processor is faulty. This requires each to run the self diagnostic software, which must first be loaded from the backing store into the processors. This process will take time and calls which are in progress, or being set up may be lost.

It should also be appreciated that some part of the whole system is not duplicated. In this case, the failure of the error detect unit may cause an erroneous fault indication.

## 7.3.4 Triplicated Systems

Figure 7.5 on the next page shows a triplicated control system. This is not simply running a third processor in parallel with the other two. The equipment which detects a fault is now also able to detect which processor is at fault by using majority vote logic. Although the cost of three processors is higher than two, the benefit of this system over duplicate synchronous control is that the faulty processor can be isolated immediately a fault is detected.

Neither of these two approaches is, however, tolerant to program error, as in both cases the processors are carrying out the same instructions simultaneously. Thus if a program fault occurs, the outputs from the processors will be the same, and not be detected by the comparator or majority vote logic.

## 7.3.5 Worker-Standby Systems

Worker-Standby systems, as the name suggests, involve utilising one processor to run the exchange, while keeping a second in reserve, to be brought into use when the first fails. The engineering term for this type of arrangement is passive parallel redundancy. The degree of readiness at which the standby processor is kept will depend upon the operating environment of the exchange. The reliability diagram of a worker standby arrangement, shown in Figure 7.6 is similar to that of active parallel redundancy, except lines from the standby units are shown dotted to indicate that they are not used until the worker fails.

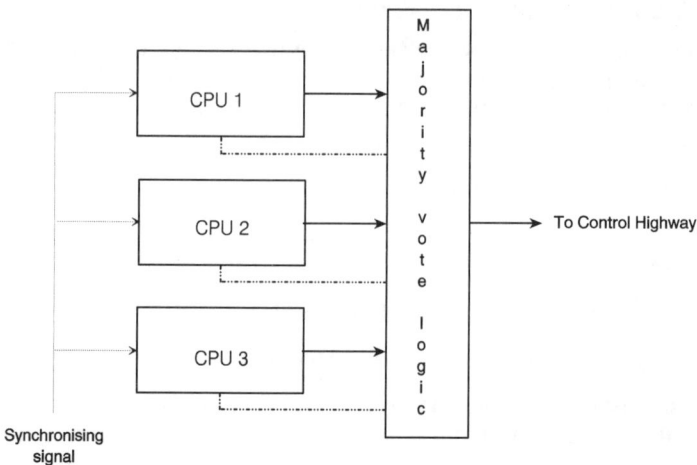

*Figure 7.5 Triplicated control with majority vote logic*

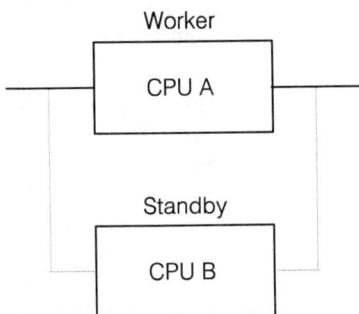

*Figure 7.6 Reliability diagram for a worker-standby arrangement*

## 7.3.6 Cold Standby

Figure 7.7 illustrates the concept of cold standby control, in which the standby processor is idle until required. Additional complexity is required to detect the presence of a fault on the single working processor, as the comparison techniques used in duplicate synchronous control can not be applied. A practical method of fault detection involves the working processor periodically carrying out a set of self checking procedures which are timed by a watch-dog circuit. If the routines are not completed within a predetermined time frame, a fault is deemed to have occurred in the working processor. The standby is switched on automatically, and commences operation by loading its operating software and the exchange database from the shared backing store.

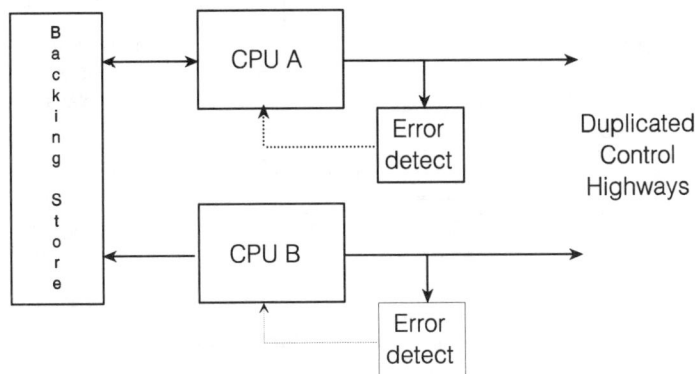

*Figure 7.7 Cold standby control system*

Because details of current calls are held in the working processor's own store the standby processor will have had no knowledge of the calls in progress at the time the failure occurred. Calls in progress may well be lost, and calls in the process of being set up will certainly be lost.

However if this inconvenience can be tolerated, this is a practical system which is adopted in some small exchanges as cost savings in power consumption and storage requirements can be made.

There are several points regarding cold standby operation which should be borne in mind. There is a risk that a fault could develop in the standby processor, and remain undetected until the standby processor is required to be operational. The risk of such latent faults can be minimised by operating a regular change-over policy. For example at some predetermined time each day, the role of the two processors is reversed, and the working processor becomes the standby. Regular change over between worker and standby systems has the added benefit that both systems are operational for approximately equal amounts of time, and thus age at the same rate.

A second point is that both processors share a common backing store, probably a magnetic disk which holds the operating programs and exchange database. Since this unit is not duplicated its failure can cause failure of the exchange.

## 7.3.7 Hot Standby Systems

In a hot standby system, such as illustrated in Figure 7.8, the standby processor is kept switched on, with its operating programs and exchange database loaded into its memory. The working processor keeps the standby informed of the current status of all calls on the exchange, by sending call status messages over the inter-processor communications link. When the working processor fails, no calls will be lost as the standby is completely ready to take over the task of running the exchange.

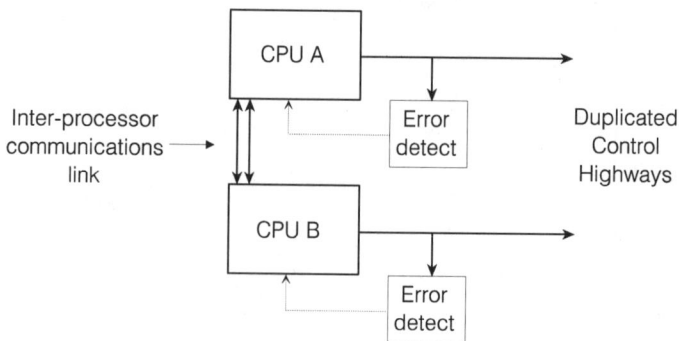

*Figure 7.8 Hot standby control system*

The main points to note here are that the cost of preserving calls during the change over period is the additional hardware and software complexity required to operate the communications link between the two processors. Handling the inter-processor communication places additional tasks on both processors. Therefore they must have sufficient processing power, so that during periods of peak demand they are able to handle not only the tasks of setting up calls, but also informing the standby of the new calls.

The vulnerable part of this type of system is of course the single inter- processor link, although steps which are expensive can be taken to reduce the risk of exchange failure due to problems in this part of the system.

Exchange manufacturers tend employ standby systems which are somewhere between hot and cold, so called *warm standby*, in which the standby system can take over the task of running the exchange with some limited knowledge of call status. This information may be slightly out of date and so a balance is struck between cost and degree of *warmth*, or readiness. A typical system will attempt to preserve calls which are in progress, i.e. have reached the speech phase, but accept that calls in the process of being set up will be lost, and users will have to redial.

## 7.3.8 Load Sharing Control Systems

In all the duplicated systems considered so far, two identical processors have been required to operate the exchange with sufficient reliability. In each case, one of the two processors is able to run the exchange on its own. In terms of processing power, as each processor must be capable of handling all call attempts during the periods of peak demand, the total system has twice as much processing power than is actually required.

In financial terms, there is a general relationship between the power of a processor and its cost. Thus these duplicated systems could be considered to be more costly than necessary, and any steps which would reduce the overall cost while still maintaining reliability would be welcomed. Dual load sharing provides an opportunity to reduce the hardware costs of the processing system at the expense of more complex software. It has already been stated that the functions of the control system are many and varied. There are, for example, several tasks involved with each stage of a call set up, and each task will be catered for by a module of software within the main set of programs.

In the systems considered so far, in which only one processor is handling all the tasks, or both machines are carrying out the same tasks in synchronism, an important piece of software called the scheduler ensures that the processor carries out all the tasks required of it in the correct sequence.

The essence of dual load sharing is that two processors work together, carrying out different tasks, in order to run the exchange. Since the processing load on each is approximately half the total load, the power of each processor, and hence the cost, can be reduced.

Dual load sharing relies heavily on fast, efficient two-way inter-processor communication as the processors keep each other informed about current call status. The scheduler now allocates tasks to both processors.

If a either processor fails to perform a task, the scheduler can arrange to test the suspect processor using self diagnostic routines, and re-run the task on the other. In any case, periodically the scheduler will run the diagnostic routines on one processor, while the other is still operating the exchange. Should the diagnostics fail on either processor, the failed processor is isolated, and all tasks are then scheduled on to the working unit.

The clear benefits of such an arrangement are that less costly processors are required, and no calls are lost when a fault is detected. The problems are that a much more complex scheduler is required, and a not inconsiderable amount of processing power is required to maintain the two way inter- processor communications, so although some cost savings have been made on hardware, the cost of software will be increased. However since software can be duplicated at virtually no cost, the cost of this more complex software can be shared by all systems using it.

The load sharing system can cope with most failures – certainly failure of a single processor during low traffic periods would probably not be noticed by users, as the processing capacity of the working unit is well able to handle the demand. If however a processor fails during peak periods, although no calls will lost, calls will be handled less efficiently and will take longer to set up. Users would probably notice an increased delay in receiving dial tone, and a longer than usual wait for ring tone after having dialled.

### 7.3.9 Multiprocessing Control Systems

When SPC systems were first introduced, they were based on single large machines, similar to main frame computers, normally arranged in pairs in some way for reliability. The processor was at the centre of the exchange control system, and carried out all call processing tasks. The technology available at the time favoured this centralised approach as processing power was expensive, and only came in large units. The introduction of modern microprocessor technology has permitted two important developments in SPC systems: Multiprocessor Control and Dispersed Control.

Multiprocessor control involves using a processing system which comprises a number of relatively small processors, operating in a load sharing manner, to provide the total processing power requirements of the exchange. The distinct advantage of this approach over dual load sharing is that the processing power of the control system can be tailored to the requirements of the exchange. Thus a particular exchange design can be used to build exchanges of different capacities using a suite of common modules. This is in fact one of the main philosophies in the design of System X.

It is important to note that all processors in a multiprocessing system are capable of running any of the exchange tasks, so reliability is also preserved, as in the event of the failure of one processor, the total processing load can be taken on by those remaining.

### 7.3.10 Storage Requirements in Multiprocessing Systems

Each CPU in a multiprocessing system will have its own high speed store which will contain its operating system software, and other programs which are used often. All CPUs will have access to a common main store containing the exchange database and a copy of all programs in general minute to minute use. There will also be a third level of memory storage, a backing store, typically a high capacity Winchester magnetic disk. This will contain all programs and data, including those which are used infrequently.

### 7.3.11 Multiprocessing Control in System X

The control system of System X is based on multiprocessing and dispersed control principles. The central processing subsystem of a System X exchange consists of a number of processor *clusters*. Figure 7.9 on the next page shows a basic System X cluster as a multiprocessor of up to four individual central processor units (CPU) connected to a common store by a duplicated data bus. Input/Output processors handle communications between the cluster and the peripheral units such as console VDUs and backing stores.

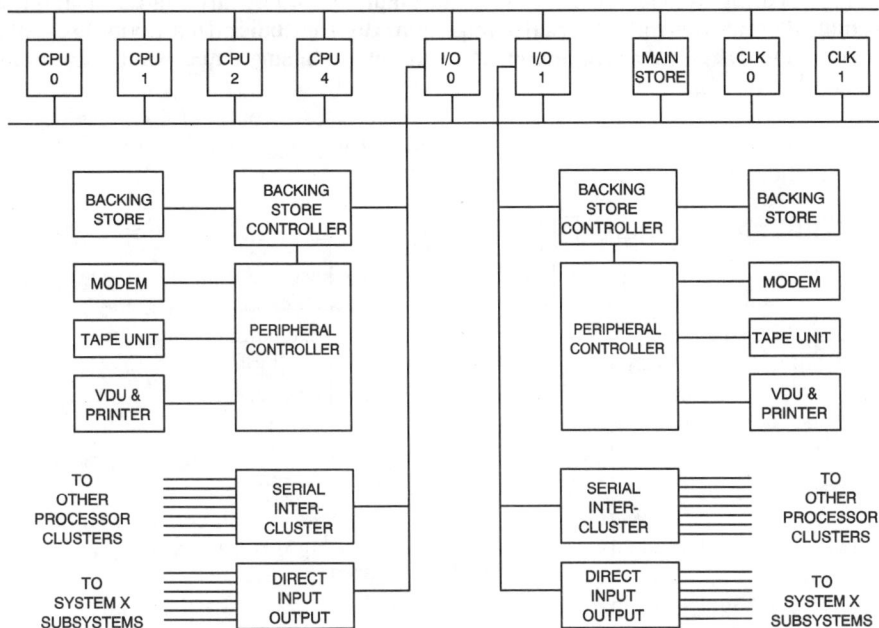

*Figure 7.9 A multiprocessing cluster in System X. (Reproduced by permission of GEC Plessey Telecommunications Ltd)*

As each CPU in the cluster can handle one million instructions per second (mips), the total processing power of the cluster is four mips. This gives a cluster the ability to handle 250,000 call attempts in the busy hour (250,000 BHCA). If this capacity is not sufficient, the architecture of the system includes a Serial Intercluster which allows up to seven extra clusters to be added. The major advantage of this approach is that each exchange in the system can be built from standard modules and be equipped with only as much processing power as is required. It is not thought necessary at present to use more than four clusters on even the biggest and busiest exchanges.

## 7.3.12 Dispersed Control

The centralised approach to control, which all the previously mentioned systems represent, requires that the central processing facility be involved with the actual processing of every activity in the exchange. Thus as well as controlling what happens, the processing facility does the work. The dispersed approach sets out to remove much of the work load from the central processing system and give it to small microprocessors in individual units in the exchange. The central processing system is still in overall control, ensuring that each task is carried out in sequence.

However the actual tasks are carried out by smaller processors more closely related to the actual task involved.

The Subscribers' line unit of a modern PABX is a simple example of basic dispersed control in practice. The subscribers line unit is a single PCB which can normally deal with 8 or 16 subscribers' lines as shown in Figure 7.10. For all lines terminating on the unit, the functions of new call recognition, digit decoding (for Loop disconnect dialling), and call release recognition are dealt with by a single microprocessor on the line unit PCB.

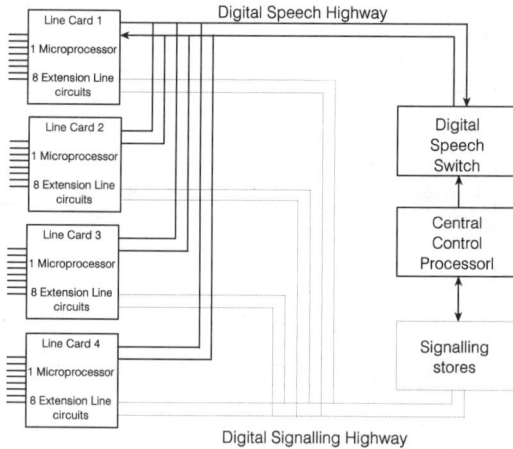

*Figure 7.10 Dispersed control example 1 – PABX subscribers' line unit*

The microprocessor carries out all the work involved with these tasks, and informs the central processing system accordingly by sending messages over the digital signalling link between the line unit and the central processing system.

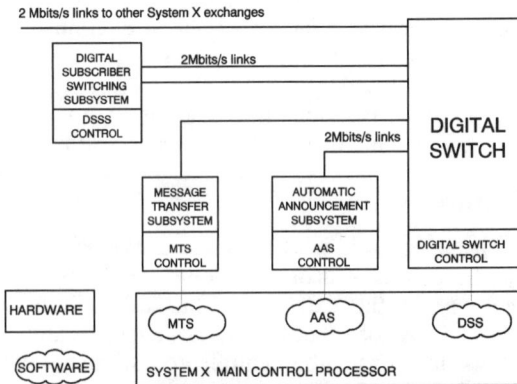

*Figure 7.11 Dispersed control example 2 – System X exchange*

## 7.3.13 Dispersed Control in System X

System X is another practical example of dispersed control. The block diagram of a System X trunk exchange in Figure 7.11 shows that the exchange consists of a number of modules, all linked to the central processing system. Each of these hardware modules has its own processor and software handler to enable it to carry out its functions when instructed to do so by the multiprocessing central control unit.

System X, like many systems, does not use a single approach to control architecture. The approach taken is to use multiprocessing in the central processing system to allow for reliability and growth, while taking advantage of microprocessor technology to distribute processing power through out the exchange to control individual hardware units. These dispersed controllers do not act autonomously, but are under command of the central processing system.

Most exchange manufacturers adopt a control system based on one or more of the concepts described. However they do not use standard terminology to describe their systems. In a typical case the term *distributed* is used, where perhaps *dispersed* would have been more appropriate. It is also the case that manufacturers adopt control systems which are hybrids of those covered in this section.

# 7.4 Control Software

## 7.4.1 The Requirement of Software

Whatever hardware configuration is chosen for the control system of an exchange, there is a requirement for control software, that is programs and data, to complete the control system. As in any computer based system the software will include two categories of program. These are known as :

❑ The Applications Programs

❑ The Operating System Software

## 7.4.2 The Applications Programs

In a traditional computer system, the applications programs are those pieces of software that carry out real tasks, such as database enquiries, payroll runs, inventory stock reports and such like. In an SPC exchange, the applications programs carry out tasks such as call processing, call logging, call charge accounting, diagnostic testing and compiling traffic statistics.

## 7.4.3 Operating System Software

The Operating System is the suite of programs which provide an interface between the hardware of the processing system and the Applications Programs. Figure 7.12 shows the hardware, which includes the processor, storage devices, VDUs and printers being surrounded by a shell that is the operating system. The application programs surrounding the operating system interact with each other, the hardware, and users via this *shell*.

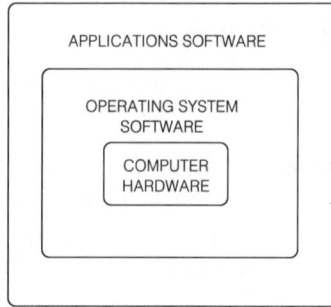

*Figure 7.12*

The main components of an operating system are:

❏ Process Scheduler

❏ Input/Output Handlers

❏ Memory Manager

❏ Man Machine Interface

## 7.4.4 The Process Scheduler

The operating system and applications programs contain many functional processes which must be run on a processor in order to carry out the necessary tasks involved. Once a process has been invoked, it will have to execute code on a processor. It may also invoke other processes and will probably have to pass information to other processes. There will also be times when a process is itself waiting for a response from another process or data from a peripheral device.

The role of the process scheduler is to maintain lists, or queues, of all current processes waiting to be run, and to allocate processor time to them. Although simply stated, the task of the processor is very complex. For example different types of process will have different priorities and so the lists of processes to be run are not simple first in, first out queues. Other complexities will exist as the scheduler may be able to allocate processes to more than one processor, as is the case in the cluster arrangement of the System X Processor Utility.

Probably the most important function of the scheduler is to make as efficient use of the processing power available as possible, to enable the maximum number of call set ups to be made. This will mean suspending tasks which are waiting for external responses, and running other tasks during this waiting time. Once the required response has been received, the scheduler can resume the suspended task. To enable the scheduler to suspend and resume tasks midway through their execution, it must be possible to save details of the task, such as its data and status, in memory so that the task can be restored and continue to run as if it had not been stopped.

In June 1989, at a special demonstration in London, a System X Local Exchange successfully handled over 1.5 million calls in one hour, an average of over 400 new calls per second. Although significant processing power was available in hardware terms, these figures could not have been achieved without very efficient software, particularly a powerful scheduler to match the call processing load to the processing power available in real time.

## 7.4.5 Input/Output Handlers

This part of the Operating System handles all communication with peripheral devices such as VDUs, printers and backing storage devices. There are several reasons for providing device handlers in the operating system. Of these the most important is that the task of writing applications programs which will use these devices is made easier, because the programmer does not need to know the details of the interface between the computer and the device. Following on from this, only one copy of the device handler software needs to exist on the system. If all the applications programs had their own device handler, they would be more complex, and since they would be written by different programmers, would probably not treat the devices in a uniform way.

Another very important benefit of having I/O drivers in the operating system is that it is possible to upgrade a peripheral device, and change the device driver in the operating system to work with the new device, without the need to rewrite the applications programs that will use that device.

There is another approach to handling I/O, and that is to use a hardware I/O processor. I/O processing software in the operating system uses processor time, and can provide a significant proportion of the total load on the processing system. In small PABX systems this may not be a problem, as the microprocessor system controlling the exchange may have sufficient capacity to cope with the demand.

Rather than provide software in the operating system to handle I/O, in large systems it is not uncommon to devolve all I/O communications tasks to a completely separate processing system. In the computing world, the Front End Processors (FEP) attached to main frame computers, are in fact minicomputers whose sole tasks are providing a communications interface between the computing facility, i.e. the mainframe, and the users i.e. the terminals and printers. The architecture of the System X exchange adopts a similar approach using duplicated I/O processors (IOP) with their own software to interface the CPUs in the cluster to the peripheral devices as shown in Figure 7.9.

Another major function of the System X IOP is Direct Memory Access (DMA). DMA refers to the direct movement of data between the main stores and the backing store. Although DMA tasks are initiated by the scheduler in the main operating system, the actual transfers of data are handled by the IOP. Without DMA such transfers would have to be processed by the cluster CPUs, significantly increasing the workload of the processors.

## 7.4.6 Memory Management

Memory management is required to control the use of the main store and the backing store by other processes such as the scheduler.

The total memory requirement of all the programs and data needed to run the exchange will usually far exceed the amount of main memory that can be accessed by a CPU. A technique known as virtual memory permits the size of the actual programs and data to far exceed the size of the available main store. Programs and data are held in the backing store and copies of them are brought into main memory when required by the scheduler, thus allowing many processes to apparently be active concurrently.

The memory management system must ensure that time critical processes and their respective data are always held in main memory so that they are constantly available. Other processes, which are not time critical must be held in main memory while they are executing code on a processor, however when these processes are suspended, they can be "swapped" out of main memory onto backing store. This de-allocates areas of main store which can then be used for other processes. An important point to note here is that processes are continually being swapped in and out of main store, with little chance that they will occupy the same area of main store each time.

The memory management system must provide two specific mechanisms for dealing with swapping, the first to translate the memory addresses in the program code in to the actual address of the required item in the main store, the second to protect each process from interference by other processes. This is achieved by preventing access to each process's current area of memory by other processes. Such *illegal* read accesses could possibly produce erroneous data for the interfering process, or worse, in the case of illegal write accesses, obliterate the program or data of the process that is being interfered with.

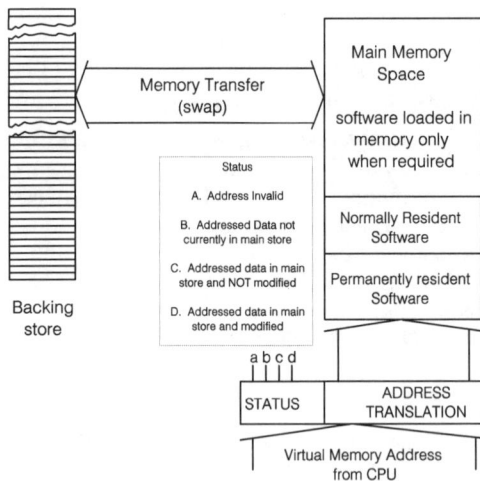

*Figure 7.13 Memory management and the use of memory*

In Figure 7.13 each memory address generated by the CPU is tested to determine if the required data is currently in main store. If this is the case a translation is produced to give the actual address of the required item in the store. If the address is invalid for some reason a status flag is set to prevent illegal access to any area of the store.

If the required item is not in the main store, a flag in the status register is set to prompt the memory management system to bring the block of data containing the required item from backing store into main store. If there is not enough free space in main store, one of the blocks currently there must be overwritten, and there are several algorithms to select which block will be so deleted. If the selected block contains data which has been changed during its time in the store, it is of course essential that this data is stored on the backing store (i.e. swapped out) before it is overwritten in main store. The status register will indicate if any changes have been made, i.e. the block has been written to. If the block has not been altered, there is no need for it to be swapped out to the backing store, and the required block can simply be allowed to overwrite the old.

## 7.4.7 Man Machine Interface (MMI)

Very few computer based systems operate with no manual intervention. In fact nowadays providing the ability for humans to communicate with the system is one of the most important areas of software development. In an exchange environment, software must be provided to allow engineers to bring the exchange into operation, alter details of subscriber records, query the system to diagnose problems, and close the exchange down in very extreme cases. The term MMI is used by many manufacturers to describe the software interface to the engineer with its screen format and command language.

The MMI software provides a set of standard messages, or commands, with which the exchange staff can communicate with the system software. These commands are keyed in at a terminal device, normally a VDU, and are interpreted by the MMI, which can execute suitable program modules to carry out the required task, or provide the required response to a query. In the case of the MMI within the operating system, the range of commands will probably be limited to a few basic instructions associated with running up the exchange control computer from scratch, various tests, and closing it down. There will also be an MMI within the applications programs for dealing with exchange functions, such as making changes to the subscribers' database.

## 7.4.8 Developing Exchange Software

One of the major attractions of SPC systems is that the nature of software control provides an opportunity for increased flexibility in the exchange design process. In theory it should be possible to develop software for an exchange without a knowledge of the hardware on which the programs will be run. In practise this ideal is not achieved as it often is necessary for the programmers to be aware of how the processor will deal with certain pieces of code, for example processing time critical activities. This knowledge allows the programmers to produce the most efficient code for particular tasks.

Development of any software product is a costly process, in terms of time and money. The software for a telephone exchange is no different, it is an extremely complex mix of program modules which must work together to run the exchange. Just as hardware and software design should be separable it should be possible to separate the design of individual software modules. This means each module can be designed, programmed and tested individually before the modules are brought together as a whole.

The design of modular software involves carefully specifying the task that the module is to perform, and its interfaces to other modules. So long as the integrity of the interfaces is preserved, each module can be designed and programmed with no regard as to how the associated modules have been written. Later in the system life time, it should then be possible to redesign a module, perhaps to make it more efficient, or add new user features. So long as the interface characteristics have been maintained, the redesigned module can be put into the existing system without it being necessary to re-work any of the other modules.

Modular software is just one of a number of software engineering principles which are being adopted to improve the production of major software products.

## 7.4.9 Languages for Applications Software

As already stated the applications software of the exchange will contain the programs and data necessary for the computer system to correctly control the exchange i.e. carry out the tasks associated with call processing, call accounting, maintenance, collation of traffic statistics, and so on.

Generally these programs will be written in a computer language that permits the programmer to write programs without taking into account the processing hardware on which the software will eventually run. Such languages are known as high-level languages (HLL), and are characterised by the fact that they look more like natural English than low-level languages (e.g assembly language) which reflect the architecture of the actual processor.

A further characteristic of HLLs is that they are able to reflect the applications for which they were designed. For example, the language COBOL Common Business Oriented Language is designed to allow programmers to write programs for business applications. Languages such as CORAL (Computer Oriented Real-time Applications Language) were developed for time critical systems such as those found in the defence environment for missile guidance systems etc. This language proved to be suitable for other real time applications such as SPC exchange control. The software of System X has been developed in CORAL, as has the software for a major military switching system.

The main benefits of using high-level languages are:

❑ Program Portability, that is the ability to run the same program on several machines, so long as compilers are available for the target processor.

❑ They can be developed for specific types of application, i.e. they are problem oriented, rather than machine oriented.

❑ They tend to use symbols and phrases which are like natural language. This leads to an improvement in programmer productivity, and reduced training costs when compared to programming in low-level languages such as assembler.

❑ Due to the fact that program code written in a HLL is easier to understand, it is also easier to maintain, thus making software changes easier to implement.

In an effort to bring about some standardisation in the area of SPC software, the CCITT has developed its own high-level language, CHILL, for use in computer based switching systems. It has been used for several systems, for example the Siemens

EWSD range of switches in which it was used to produce the applications programs and some of the Operating System components.

In recent years there has been a growing interest in Ada, a high-level language developed for the US Department of Defence for use in military real time applications. Ada bears some resemblance to CHILL, and other high-level languages such as Pascal. However Ada was developed specifically to meet the needs of modern real time systems which are complex to program and test. Some work has been done to evaluate whether there are benefits to be gained from writing SPC software in Ada, though the results seem to be inconclusive at present. (*Experience in the use of Ada for a Digital Switching System*; Ada User **Vol. 7 No 4 1986.**)

Some exchange vendors have chosen to write their SPC software in languages such as C, a high-level language designed for writing system software. For example the Unix operating system is written in C. In many respects it is suitable for SPC software which tends to involve little actual processing but considerable amounts of data transfer or (I/O).

Generally then, high-level languages are used to produce SPC software. There is likely to be an increasing tendency to use languages such as Ada in the future, as such languages include features which make software re-usability a feasible proposition in the fight against increasing software costs. However there may be occasions, when using HLLs, the SPC software design team can not produce a program module that is sufficiently efficient, either in terms of execution time or memory utilisation. In these cases it will be necessary to write the relevant program modules in the assembly language of the processor involved.

## 7.4.10 Applications Software

The Applications Software can be considered to consist of several systems which must all interact with each other, usually by passing information, or tasks via the operating system to produce the total exchange software.

As an example some of the more important software systems to be found in a System X exchange will be briefly discussed. To use the terminology of System X, these applications are known as subsystems. Some are listed in Figure 7.14:

Call Processing Subsystem (CPS)

Call Accounting Subsystem (CAS)

Signalling Interworking Subsystem (SIS)

Automatic Announcements Subsystem (AAS)

Digital Switching Subsystem (DSS)

Management Statistics Subsystem (MSS)

Maintenance Control Subsystem (MCS)

*Figure 7.14 The application programs of a System X local exchange*

Some of these subsystems consist of hardware and software, such as SIS, AAS and DSS. Others consist of software only, e.g. CPS and CAS. In this section it is appropriate to cover both types for completeness, and to provide the basis for the case study of a System X exchange in a later chapter.

## 7.4.11 Call Processing Subsystem

The Call Processing Subsystem is a software only system found in both the System X trunk exchange and local exchange. In a trunk exchange, CPS must be capable of receiving signalling information from other exchanges in the network so that it can correctly connect a call through the exchange. In its turn CPS must pass signalling information to other exchanges so that they can continue to process the call.

The trunk exchange CPS software must carry out the following tasks:

❑ Decode digits passed from other exchanges, especially existing analogue exchanges which are still connected to the network.

❑ From all the available network routes, select a suitable route for the call.

❑ Having selected a suitable route, a free circuit on that route must be selected.

❑ The incoming circuit must be switched to the selected free circuit in the digital switch. This requires CPS to tell the DSS which circuits to switch together. The actual switching is carried out by DSS software and hardware.

❑ Maintaining a record of the call set up data, for the duration of the call.

❑ Transmission of the call set up data to an external device such as a call logger, as each call is completed.

❑ Recording information regarding faulty routes and circuits.

The CPS within the local exchange must carry out all the above tasks, but it has additional tasks which relate to the direct handling of subscribers.

The local exchange CPS must set up both local and trunk calls, and the additional user features, i.e. the star services such as abbreviated dialling, call diversion, call waiting and so on.

This entails the extra tasks of:

❑ Recognising whether subscribers are free or busy.

❑ Storing details of subscriber's Class of Service e.g. outgoing calls barred, call diversion in operation.

❑ Setting up star services features.

❑ Communicating with other software systems, not least the Call Accounting System, which provides the revenue for the system.

The description of the System X Call Processing Subsystem illustrates that call processing software will contain many modules, each having a dedicated task to perform. As the call progresses through its various stages each module must pass information to other modules within the system.

## 7.4.12 Call Accounting Subsystem

Another software only system, the Call Accounting Subsystem (CAS), is invoked by CPS. When a call is set up CPS passes to CAS the necessary information to enable the correct charge for the call to be calculated and recorded. The list below gives examples of the type of information which CAS requires from CPS:

❏ Identity of the calling subscriber, since it is this subscriber's call record that will be updated.

❏ Identity of the called subscriber – necessary for itemised billing and so on.

❏ Time of called subscriber answer, since this is when charging commences and indicates the time band for charging.

❏ Time the call is cleared; since this is when charging ceases. (Note: some networks calculate charges using intervals of less than a minute.)

❏ Charge Band: usually a function of distance, although some call types are charged at local rates to the calling subscriber other charges being levied against the called subscriber.

## 7.4.13 Signalling Interworking Subsystem

As System X exchanges replace analogue units in the British Telecom network, it is essential that the ability for subscribers on all exchanges to call each other, remains. In the old analogue network a variety of different signalling systems exist. It is possible to think of a signalling system, rather as an interface between one exchange and another, both exchanges having to use the same interface (signalling system) on the trunks or junctions which join them together. Thus if the digital System X exchanges are to be able to work with existing exchanges, some form of signalling conversion must be provided. This is the role of the Signalling Interworking Subsystem (SIS).

SIS is a hardware and software system that provides the ability to connect junctions and trunks from a System X exchange to analogue exchanges which may use any of the following signalling systems:

❏ Loop Disconnect

❏ DC Systems e.g. DC2

❏ AC Systems e.g. AC8, AC9, AC11

❏ MF Systems e.g. MF2, MF4

❏ Digital System i.e. 30-channel PCM with Channel Associated Signalling

Both the System X local and trunk exchanges have an SIS module. Eventually the whole network will be digital, and then the SIS will no longer be required.

## 7.4.14 Automatic Announcements Subsystem (AAS)

This is a hardware and software system found only in the System X local exchange. AAS provides *voice guidance* with digitally recorded messages such as: *Sorry, the number you have dialled is unavailable.* AAS is invoked by CPS, particularly in response to subscribers' requests for star services.

## 7.4.15 Digital Switching Subsystem (DSS)

The Digital Switching Subsystem (DSS) carries out all switching in a System X exchange. The Digital Switch is a Time-Space-Time switch capable of switching over 95000 PCM channels, although some of these channels will be used for signalling. The DSS has its own internal microprocessor control unit, triplicated for security, which operates on instructions from the DSS handler software in the main processor.

## 7.4.16 Management Statistics Subsystem (MSS)

This software only system collects statistical data such as traffic volumes, and records of failed calls, to enable the network management to make decisions regarding provision of equipment in the network.

## 7.4.17 Maintenance Control Subsystem (MCS)

This hardware and software system enables maintenance activities to be carried out automatically, or remotely by the engineering staff. For example using a Visual Display Terminal (VDT), an engineer can test facilities at a remote exchange site, and if necessary take a faulty equipment out of service.

MCS interacts with hardware only subsystems and non System X Alarms (e.g. fire alarms) via an Access Utility Subsystem which provides an interface between the hardwired alarms and the software of MCS. Other software subsystems such as CPS, the DSS handler and the SIS handler are able to interact directly with MCS by sending fault reports in the form of inter-process messages.

# 7.5 Control Data

## 7.5.1 Categories of Data Required by the Control System

The software of an exchange control system will operate on a wide range of data items. How each item of data is stored and processed depends to a large extent on how permanent the data is, i.e. how likely it is to be changed during the lifetime of the exchange. Although different philosophies are adopted by exchange manufacturers, generally data can be categorised as one of three main types:

❏ Permanent

❏ Semi Permanent

❏ Transitory

## 7.5.2 Permanent Data

The term Permanent Data refers to those items which do not change during the lifetime of the exchange. For example, information regarding connection of exchange hardware such as dial tone port, ring tone port, MF receivers, conference bridges, the attendant's console and MMI terminals.

Normally this data will be installed by the manufacturer when building the exchange and will probably form part of the resident system software. The same permanent data will be installed on all exchanges of the same model, and will only be changed by the manufacturer if a major upgrade to the exchange is carried out. Such an upgrade will usually require the ROM chips holding the software to be replaced.

## 7.5.3 Semi-permanent Data

The term Semi-permanent Data refers to those items which are likely to be changed occasionally during the life of the exchange. Facilities will exist to permit exchange staff to make changes to this data easily via an MMI terminal. It is this data which reflects the customers' requirements of each individual exchange, and will include the subscribers' database and information about trunks and junctions connected to the exchange.

After consultation with the customer this data will be installed by the exchange supplier, normally on some form of non volatile erasable memory, e.g. floppy disk or EPROM. When the exchange is initially switched on, software within the exchange transfers the database from non volatile storage to RAM. The exchange control software then uses the database stored in RAM when processing calls. Any subsequent changes made by the exchange staff will alter the database held in RAM only, so it is imperative that an archive copy of the database is made each time any data is changed. Thus in the event of a power failure the new data is not lost, and the exchange can power up with an up-to-date version of the database.

## 7.5.4 The Subscribers' Database

The subscribers' database which will include the following information about each extension:

❑ Extension Number

❑ Equipment Number (or Port)

❑ Class of service

❑ Call Diversion in effect

❑ Type of Telephone

The range of extension numbers available will depend upon the manufacturer of the exchange. Any number in the available range can be allocated to any particular telephone, so long as no two extensions are given the same number (the software will prevent attempts to do this). It is also often possible to allocate two numbers to the same telephone.

Each telephone will be connected to a port on a subscribers' line card. The physical location of the line card, and the actual port to which the telephone is connected will determine the Equipment Number (EN) relating to that extension. (Note: some manufacturers refer to Port Numbers rather than ENs.)

The control software works with Equipment Numbers only, and has no specific knowledge of the extension numbers allocated to each port except by reference to the subscribers' database. Thus whenever calls are set up, the software must perform the necessary translation between the dialled digits received by the exchange and the EN of required extension number using the subscribers' database.

Note, that ENs relate not only to telephone line ports, but also to other physical pieces of equipment on the exchange, e.g. the trunk ports, the dial tone port and the MF receiver port.

The Class of Service of an extension will determine the facilities given to the user of that extension. For example, selected types of outgoing trunk calls may be barred, or the user may not be able to invoke call diversion from his own telephone. From time to time, users will require changes in the facilities available, and so the facility exists for such changes to be made by the exchange staff.

Call Diversions may be invoked by the user, or by the exchange staff, and will be subject to frequent change. The record of call diversions in effect for a particular extension must be checked by the call processing software every time a call is made to that extension number.

The subscribers' line card is often designed to terminate both types of telephone currently in use, i.e. Loop disconnect or Dual Tone Multi-frequency instruments (this excludes the digital telephones for ISDN and ISPBX exchanges). Since the exchange software deals with calls from these types of telephone differently, it is important to record in the subscribers' database which type of telephone is connected to each port. Note that as the subscribers' line card is designed to cater for both types of instrument, if the instrument is changed it is only necessary to alter the record in the subscribers database. On some exchanges, the software can deal with both types of telephone by detecting which type of signalling is being used.

## 7.5.5 Typical Subscribers' Database

Figure 7.15 shows an extract from the subscribers' database of a typical PABX.

```
                    EXTENSION ATTRIBUTES

                    EXTN 250

     LINE 1    SECOND      PILOT DN      MEMBER OF
               EXTN        OF GROUP      PICKUP GROUP

                           300           3

     LINE 2    DAC  DNR  DOB  DOU  NDC
                    251  251

     LINE 3    I/C CALLS BARRED   DATA    INTRUDE   CAMPON   DIVERT   LEVEL 9 BLK
               OE PRI PUB         LINE    INHIBIT   OPTOUT   BARRED   OVERIDE


     LINE 4    ROUTE RESTRICTION CATEGORIES ACTIVE
               1 2 3 4 5 6 7 8
                       1

     LINE 5    MASTER      MF4   10PPS      LN    EN
               EXTN
               1           1     1          37    100


          ROUTE RESTRICTION TABLE  4

               010     bars access to international numbers
               191     bars access to Directory Enquiries
               192     bars access to Directory Enquiries
               0898    bars access to "chatline numbers"
               0836    bars access to "chatline numbers"
```

*Figure 7.15 Typical subscribers' database entry*

The subscriber record is printed on five lines. Each line gives information about specific attributes of the subscriber which must be checked by the call processing software during call set ups involving this extension number. Many of these attributes affect the special user features of the exchange which are explained in Chapter 9, so only a very brief description of each attribute is given below. This particular exchange uses the terms Directory Number (DN) to denote the telephone extension number, and Equipment Number (EN) to indicate the physical line circuit assigned to that extension number.

## Line 1

**SECOND EXTN:** Any extension may be given two Directory Numbers (DN). If an extension has two directory numbers, the second DN will appear here. In this case Extn 250 is also Extn 297.

**PILOT DN OF GROUP:** Any extension may be a member of a group of extensions. If a DN is assigned to the group as a whole, rather than a particular extension, it becomes known as the PILOT DN. Any calls to that DN will be routed by the call processing software to one of the extensions in the group. Extension 250 is in a group whose pilot DN is 300.

**MEMBER OF PICK-UP GROUP:** If the extension is a member of a pick-up group, the group number will appear in this position. Extension 250 is a member of pick up group 3.

**ACTIVE:** Equipped extensions do not become live until the exchange staff has issued the MMI command Activate Extension. Once activated, the extension may be de-activated for maintenance purposes. The current status of the extension is shown at the end of the first line; in this case Extn 250 is active.

## Line 2

This line shows any call diversions which may be current for this extension. The DN to which calls are to be diverted is given under the diversion heading. If a particular type of diversion is not in operation a space appears under the heading. The five types of call diversion possible are:

❏ DAC Divert All Calls to this extension

❏ DNR Divert On No Reply (after some user variable time delay)

❏ DOB Divert on Busy

❏ DOU Divert on unobtainable (i.e. de-activated)

❏ NDC Non Dialled Call (not strictly a diversion, this attribute permits the extension to use telephones without dials or keypads. As soon as the handset is lifted, the call processing software routes the call to the number shown.

## Line 3

This line shows various attributes associated with class of service. If a particular attribute is set a 1 will appear under the headings, in this case none are set.

**I/C BARRED – Incoming Calls Barred:** Three types of incoming call may be barred to this extension.

**OE:** Calls from Own Exchange.

**PRI:** Calls from the PRIvate network to which the exchange is connected.

**PUB:** Calls From the PUBlic network to which the exchange is connected.

**DATA LINE:** When set this attribute is used to inform the call processing software that the extension is used to handle computer data, and has a modem connected to it. This may also be used for FAX terminals.

**INTRUDE INHIBIT:** When set this attribute informs the call processing software that operator assisted calls may not be made to this extension when it is busy.

**CAMPON OPTOUT:** When set this attribute prevents other users making use of the WAIT ON BUSY facility when making calls to this extension.

**DIVERT BARRED:** This attribute can be set to prevent the extension user setting up call diversions from his own telephone.

**9 BLOCK OVERIDE:** When set the user of this extension may make level 9 calls (i.e. public network calls) even if a level 9 block is in operation.

### Line 4

**ROUTE RESTRICTION CATEGORIES:** There are eight categories (or tables) of numbers which may be applied to an extension. Each table has a fixed list of prohibited numbers which may not be dialled by the extension if the Route Restriction is in force. By combining various categories, the extension can be barred from making several types of trunk call.

### Line 5

**MASTER EXTN:** This attribute will be set on one or two extensions only to allow MMI commands to be made from the telephone. This is particularly useful on small installations which do not have an attendant's console.

**MF4:** When set the extension may be equipped with a DTMF telephone.

**10PPS:** When set the extension may be equipped with a rotary dial telephone. (Note that both types of instrument may be connected as both these attributes are set.)

**LN – Logical Number:** This number relates to the logical position of this record in the subscriber database and has very little other significance.

**EN – Equipment NUMBER:** This number is very important as it provides the translation between the abstract DNs and actual physical circuits on the exchange. In this case extension 250 is a telephone connected to Equipment Number (or port number) 108 on the exchange.

## 7.5.6 Transitory Data

This term refers to the data which is stored only for the duration of each call. For each call set up by the exchange, the call processing software will maintain a call record. This record will typically contain the extension number of the calling party, the dialled digits, time of answer, time of clear and the call status.

The call status refers to the current phase of the call, i.e. dialling, ringing, call answered, calling party clear or called party clear. In the case of a PABX, where there is no requirement to maintain charging information, the call record is erased from memory at the end of the call, and unless a call logging terminal is connected all record of the call will be lost.

## 7.5.7 Typical Call Record

Figure 7.16 shows some of the detail of a call record from the PABX. This part of the record is printed on the MMI terminal in response to the command *List Port Status*, or *LPS* for a particular EN.

**PORT STATUS**

| | | | | |
|---|---|---|---|---|
| LINE 1 | **PORT**<br>**108** | **PORT–TYPE**<br>**EXTN 2W** | **STATUS**<br>**0F** | **MANUAL BUSY**<br>**00** |
| LINE 2 | **ENCG**<br>**108** | **ENCD**<br>**109** | **CALL–TYPE**<br>**EXT–EXT** | **DIALLED DIGITS**<br>**251** |

*Figure 7.16 Call record*

The printout of the call record is over two lines. The first line will always be printed for an LPS enquiry. The second line is only printed if a call is in progress. This printout is decoded as follows:

**EN:** This field holds the EN of the port about which the query was made.

**PORT-TYPE:** This field denotes whether the port is an extension, a trunk, or the attendant's console. In this case the extension is a standard 2-wire telephone.

**STATE:** This field indicates the current state of the call. The hexadecimal digits are used to indicate whether the call is in the off-hook, dialling, ringing phase or speech phases. In this case HEX 0F denotes the call is in the speech phase. 00 in this position would have indicated that the port was idle and no second line would have appeared.

**MANBUSY:** This field is altered when the port is taken in and out of use. the Hex digits 00 indicate that the port has been activated.

**ENCG:** This attribute gives the EN of the CallinG extension. In this case the call originated from port 108, the port about which the query was made.

**ENCD:** This field gives the EN of the CalleD extension. In this case, the call was made to the extension connected to port number 109.

**CALLTYPE:** This field indicates the type of call in progress. In this case it is an extension to extension call. Other call types include extension to trunk, extension to console and vice versa.

**DIALLED DIGITS:** Self explanatory. This field shows the digits which were dialled by the calling extension.

# 7.6 Load Control

Before we leave the subject of control, it is important to consider an aspect known as load control, or perhaps more correctly overload control. The hardware and software of an exchange are designed to cope with a particular maximum level of traffic, i.e. the traffic expected during the busiest periods of operation. From time to time abnormally high traffic loads will occur, for example during phone in TV programs, or in the event of a major disaster the load on some exchanges and parts of the network will be considerably higher than normal peak loads. It is thus very important to ensure that the exchange can cope with the overload, even if it can not process all the calls being offered at any one time.

The concept of load control is to regulate the load actually placed on the exchange system, so as to ensure that increased loading above the designed maximum does not cause the exchange control system to crash, and that call processing is maximised without exceeding the processing capacity available. To consider what may happen if load control was not imposed, bear in mind that there is only so much processing power available in any system, and that the scheduler allocates processing power to tasks according to a prioritised system. During periods of abnormally high demand, the scheduler may not be able to complete call set ups, if it has to continually interrupt tasks in progress, to add new calls, and thus new tasks, to the task queues. Thus as the number of calls offered to the exchange increases above the design maximum, the actual number of calls which are completed reduces.

In a multiprocessing, or load sharing system, one of the processors may fail thus preventing the system as a whole from operating at maximum capacity. This will have the effect of reducing the maximum load which can be placed on the system before overload occurs.

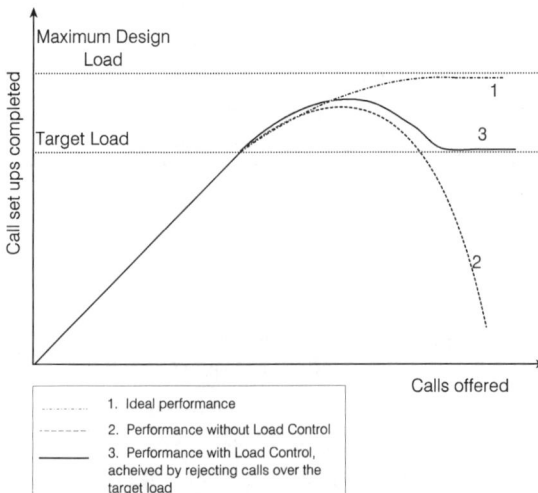

*Fig 7.17 The effects of load control*

Figure 7.17, Line 1 shows the ideal performance of the exchange, where once the design load has been reached, the throughput remains constant irrespective of how many more calls are offered.

Line 2 shows how the number of calls processed will start to fall once the number of offered calls exceeds the design limit if load control is not imposed. This clearly shows that the exchange is not coping with the overload situation.

The third line on the graph shows that with load control, if the number of calls offered continues to increase, the number of calls processed remains constant at a level which is just below the maximum.

In this case the exchange is coping with overload, by effectively rejecting an increasing proportion of the new calls offered.

Load control involves continually monitoring the traffic load offered to the exchange, the lengths of the task queues, and number of tasks within the processors, and using a suitable algorithm calculating how much processing power is available.

This permits new processing limits for applications such as the CPS to be set. The processing limit is effectively the current maximum number of tasks allowed to be queued for a particular application, During extremely busy periods, maximum queue lengths will be reduced to take account of the reduced processing power available.

## 7.7 Chapter Summary

This chapter has covered many aspects of Stored Program Control. After the progress towards centralised computer control in the early days of SPC systems, it became evident that these systems were only as reliable as the computer system at the heart of the exchange. Techniques were required to increase the overall reliability of the exchange as a whole to within tolerable limits. Various redundancy techniques have evolved with most manufacturers adopting an approach which is a hybrid of several differing techniques.

The software systems have also evolved to provide more features. The cost of producing such complex software has lead manufacturers to adopt techniques which as well as reducing the programming effort required, also improve the reliability of the software by making it less prone to errors, and simpler to maintain. Generally this involves the use of sophisticated software engineering principles coupled with the use of a high-level programming language suitable for real time applications.

The software systems of a System X exchange that were briefly described in this chapter are typical of those that will be found on any SPC exchange. Different types of data will exist within the exchange. Some of this data will be installed by the manufacturer when the exchange is built. This permanent data will reflect the architecture of the exchange by providing information about the configuration of items such as tone ports, conference bridges, MMI terminals and consoles.

The customers' requirements will be reflected in the subscriber database, and will be subject to change as the users of the exchange change offices, appointments and so on, and require different exchange facilities. The description of the subscribers' database within a typical modern PABX provided several illustrations of this type of semi-permanent data.

As well as providing a means to make these changes easily, the manufacturer must provide a system to archive changes to the database. Failure to archive after database changes will result in users not having the service they expected when the exchange is returned to service after having been shut down.

The exchange must keep track of calls in progress, and be aware of the stage that each call has reached. Call records are held for all calls in progress. At the end of the call this transitory data is erased as new calls are set up, and new call records established. Unless call logging equipment is provided, no information regarding previously made calls will be available.

For the exchange to be able to cope in periods of high demand for service, it is important that some overload control strategy exists. Failure to provide adequate load control will result in the throughput of the exchange being reduced considerably below maximum capacity when the number of call attempts exceeds the designed number of Busy Hour Call Attempts.

# Common Channel Signalling

## 8.1 Introduction to Common Channel Signalling

A brief introduction to the subject of Common Channel Signalling (CCS) was given in Chapter 5. This chapter provides a more detailed coverage of the topic. dealing in some depth with the CCITT Signalling System No 7, as this is the basis for several current systems. At this stage it is worth repeating the definition of Common Channel Signalling.

Common Channel Signalling is a signalling technique in which signalling information relating to a multiplicity of circuits, and other information such as that used for network management, is conveyed over a single channel by addressed messages.

### 8.1.1 The Problems of Channel Associated Signalling (CAS)

To better understand the reasons for the interest in CCS systems, it is necessary to appreciate the problems associated with using CAS systems especially in a digital telephony network. Channel Associated Signalling was inherited from the analogue era, and used in the early days of digital multiplexing. There are, however, four main problem areas associated with CAS systems, which made it necessary for telephony engineers to consider other alternatives. These problems are listed in Figure 8.1.

❏ Slow signalling speed

❏ Limited signalling repertoire

❏ Difficult to modify to meet changing requirements

❏ Uneconomic

*Figure 8.1 The problems of Channel Associated Signalling*

### 8.1.2 Slow Speed of CAS Systems

CAS is relatively slow. Compare the possible speed of the CPUs in the control unit of a digital exchange, operating at anything up to four mips, with the 10 pulses per second of Loop Disconnect and E&M signalling. Although the 64KBit/s signalling channel in the PCM/TDM multiplex scheme can not match CPU speeds, it is much faster than that which can be achieved by transmitting VF tones or other forms of CAS. The results of using slow CAS include long call set up times and long post-dial delay, these times can be substantially reduced using fast CCS systems, giving users the impression of almost instantaneous call set up, even on long trunk calls.

### 8.1.3 Limited Repertoire of CAS Systems

In the digital CAS system described in Chapter 5, four bits are used to convey signalling information for each channel. Ignoring the combination 0000, which is used for the Multi-frame Alignment Signal, there are only 15 possible combinations of four bits. As these combinations have to be used to convey information relating to signalling conditions, the number of possible conditions, and hence the repertoire of the signalling system is limited. In today's world, users have come to expect many features from their digitally controlled exchanges, these features can only be included if a signalling system with a capability of large repertoire of signalling messages is used.

### 8.1.4 Difficulty of Modifying CAS Systems

Digital CAS systems must provide conversion between various analogue systems and the 4-bit CAS signalling words. These analogue systems include relatively complex hardware interfaces designed to meet specific telephony requirements. It is not easy to change these interfaces to produce a system capable of meeting new requirements. CCS systems do not have any requirement to convert from one signalling system to another, and as they use a formatted message system to pass signalling information between the processors of the exchanges, it is a relatively easy task to incorporate new messages into the system to meet new user requirements.

### 8.1.5 Economic Considerations

By its very nature CAS must provide a unique signalling circuit and interface for every single voice channel. When you consider that these signalling circuits are idle for a vast majority of the time, this is an expensive over provision if some form of equipment sharing system would carry out the task more efficiently. CCS systems provide one signalling circuit and interface to cope with the signalling traffic for a large number of voice channels. Typically one 64 KBit/s CCS circuit is sufficient to handle all the signalling for up to 960 trunk circuits.

## 8.2 The Advantages of Common Channel Signalling

There are several benefits to be gained from using a CCS system. Some are of course the direct opposite of the disadvantages of CAS, and were introduced in the previous section. However there are other advantages from using CCS which are not perhaps quite so obvious at first. A brief list of advantages is given in Figure 8.2.

### 8.2.1 Some Problems with Common Channel Signalling Systems

CCS does however have some problems which need to be addressed if a reliable system is to be realised. The four main problems are based on the need to provide a fast, reliable and secure signalling system, in an environment in which speech and signalling may travel by totally different routes between the same two exchanges.

❏ Increase in speed, thus a reduction in call set up times and post dial delay

❏ A large repertoire of signals possible providing the ability to offer many user features

❏ The signalling system is easily modified to cope with new requirements by introducing new messages

❏ As the system uses a completely separate digital channel it may be enhanced without being constrained by the need to meet switching and voice channel requirements

❏ One signalling channel handles signalling for many traffic circuits, thus the signalling system is economical

❏ On low capacity routes where a dedicated CCS link can not be cost justified, a CCS circuit may be provided via a third-party exchange

❏ The signalling system can be used for other purposes such as network management and maintenance

*Figure 8.2 The advantages of Common Channel Signalling*

We have already seen in an earlier chapter that digital transmission is subject to bit corruption in the presence of noise. If the error rate is very low, these corruptions will have negligible effect on the speech channels of a link. But if, in the signalling messages being passed between two SPC exchanges, such corruptions are allowed to occur unchecked, the result will be wrong connections or call failures. To overcome this problem the CCS system must include a form of error detection and correction. As the probability of errors is normally very low, error detection can be provided by the use of a CRC checksum added to each signalling message. In the event that an error is detected in a received signalling message, a message retransmission request is made to the exchange which originated the message.

Another problem that must be addressed is that of providing a reliable signalling link. The problems of a single signalling channel are not dissimilar to the problem of a single centralised control unit in an SPC exchange: it is an *all the eggs in one basket* situation. If the signalling channel fails, traffic on all circuits which depend upon that channel is affected. The obvious step is to duplicate the signalling channel, and this is the minimum that is required. We shall see various methods of providing a reliable signalling channel later in this chapter.

Given that a reliable signalling channel is available, and that this channel handles all signalling for a large number of channels, there is still a finite limit to the information carrying capacity of this channel. The density of signalling traffic will depend upon the number of new calls to be set up in any given period. (The reader should notice more similarities here with the problems of sizing the processors to control exchange functions.)

In periods of heavy demand the signalling channel may become congested, and it may not be possible for the processor to transmit all signalling messages as they are produced. Thus some messages may have to wait in a message queue to be transmitted when the channel is free. This will obviously impact on call set up times and post dial delays. An alternative strategy to queuing is to permit the signalling messages to be routed via other exchanges in the network, but this has the disadvantage that it may cause congestion on these routes.

The final problem to be discussed in this section relates to the fact that as the signalling system is carried over circuits which are completely separate from the speech channels, and indeed may be over entirely separate routes, it is imperative that some check of the serviceability of speech channels is made before calls are connected over them. In the analogue world of CAS, this problem did not occur, especially when using VF tone signalling, as the voice channel was used to carry the signalling tones. Thus if the circuit could be signalled over, it was almost certainly suitable for voice traffic.

The CCS system must include protocols for testing speech circuits in both directions prior to calls being connected. These protocols can be used to inform the call processing software of unusable speech channels and to raise alarms to the maintenance systems.

The four main hurdles which need to be overcome in a CCS system are summarised in Figure 8.3.

---

❑ Transmission errors will cause signalling messages to be corrupted. Some form of error detection is required.

❑ Failure of the signalling channel will result in loss of all traffic on the route covered by that channel. A strategy is required involving a back-up channel.

❑ High demand levels may cause the signalling channel to become congested, resulting in call set up delays if messages are queued. A strategy is required involving the use of alternate routes for signalling traffic.

❑ CCS links are completely separate from the voice channels they control. Voice circuits may fail. A protocol involving the testing of these circuits before use for calls is required.

*Figure 8.3 Problems to be overcome in CCS systems*

---

# 8.3 Common Channel Signalling Systems

## 8.3.1 Introduction to the Various Systems in use

There is a variety of CCS systems in use in different parts of the telephone network. It is true to say however, that there is a degree of standardisation between the systems which permits good inter-operability. In Figure 8.4 the uses of some of the more common systems are shown. The diagram is based on the current situation in UK and includes the following:

❑ CCITT Signalling System 7 in the public network between System X local and trunk exchanges.

❑ DPNSS (Digital Private Network Signalling System) in private networks linking digital PBXs (see Chapter 12).

❏ DASS (Digital Access Signalling System) linking PBXs in private networks to System X exchanges in the public network (see Chapter 13).

❏ CCITT Q931 linking digital subscribers' Network Terminating Equipment to System X local exchanges (see Chapter 13).

❏ Other channel associated signalling systems for completeness, for example Loop Disconnect and MF signalling on analogue subscriber and extension lines. Manufacturers' proprietary systems for providing digital extension lines to their respective PBXs.

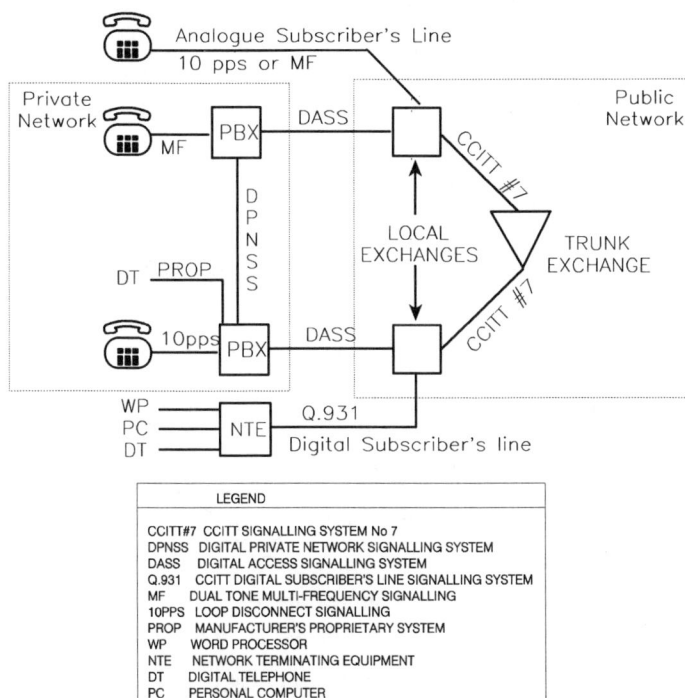

*Figure 8.4  Signalling systems used in UK digital networks*

## 8.3.2 CCITT Signalling System No 6 (CCITT #6)

As long ago as 1968 the first specification for a common channel signalling system to be used on international links was produced by the CCITT. This system, known as CCITT Signalling System No 6, was designed to interconnect SPC exchanges and used digital modems and an analogue circuit as the signalling channel. At this time the switch block of the SPC exchange was still analogue, digital switching and transmission were still in their infancy. The specification was improved over the years to include a version for use on PCM/TDM links in which the signalling channel was digital, thus the modems were no longer required and signalling speed was increased.

Eventually in 1978 the first CCITT #6 was brought into service between USA, Australia and Japan. In 1980 CCITT #6 came into service on links between UK and USA. CCITT #6 was the forerunner of all CCS systems in use today, but does not feature in the diagram in Figure 8.4.    (See *The CCITT No 6 Common Channel Signalling System;* **POEEJ Vol 73**, Jan 81)

### 8.3.3 CCITT Signalling System No 7 (CCITT #7)

CCITT #7 was conceived during the period 1972-1976 when the impact of digital switching and transmission on common channel signalling was being studied by the CCITT. The first specification of CCITT #7 was produced in 1980, at which time BT, and exchange manufacturers GEC and Plessey in the UK, were actively involved in the design of the new integrated digital network, System X. CCITT #7 is a message based system in which the processors controlling digital exchanges communicate using TS16 of a 2Mbit/s Digital transmission path to send call set up, and other information.

CCITT #7 is the signalling system used within the System X network, and eventually became the basis for several other systems. We cover more detail later in the chapter.

### 8.3.4 Digital Private Network Signalling System (DPNSS)

As its name suggests, DPNSS has been designed to allow digital PBXs to be linked together over 2MBit digital transmission paths to form a private telephony network.

DPNSS is a common channel signalling system and bears some similarity to CCITT #7. DNPSS provides the call set up and other basic telephony signalling messages for use on 30-channel inter-PABX links.

DPNSS however goes further than basic call set up and release. Telephony features such as *call backs,* and *call transfers* can be provided over a network of digital PBXs using this system. Because DPNSS is designed to be used on a large range of PBX from different manufacturers it is specified in such a way that in a network of different exchanges on which different features are supported, the system can cope with requests for facilities which may not be supported on the target exchange.

### 8.3.5 Digital Access Signalling System (DASS)

DASS is the signalling system designed for use with British Telecom's ISDN service which is based on its network of System X exchanges. It is important to appreciate that CCITT #7 is used on the links between System X exchanges, and that DASS is used between the System X Local Exchange and those subscribers who are connected by digital rather than analogue lines.

## 8.4 CCS Terminology

### 8.4.1 New Terminology

The introduction of CCS systems has introduced several new terms into the telecommunications vocabulary. These terms are applied to CCS systems in general, and not to any particular system. It is therefore important to understand the meaning of these terms, before a detailed study of any CCS system is made. The aim of this section is to define some of these terms and where necessary provide examples from CCITT #7.

## 8.4.2 Signalling Network

A modern digital telecommunications network consists of digital exchanges, with its network management and maintenance centres, all of which are linked by digital transmission systems. As well as telephony call information, the CCS system will carry other data, e.g. for network management and maintenance. Thus the signalling network must extend to all these locations. The signalling network can thought of as being very similar to a special purpose packet switched network which overlays the main circuit switched telecommunications network.

## 8.4.3 Signalling Points and Point Codes

The nodes of the signalling network will be:

❑ Telephone Exchanges

❑ Network Management Centres

❑ Operations and Maintenance Centres

These nodes are defined as Signalling Points (SP) and will usually have a user function as well as a signalling function. For example an exchange has a switching function and a signalling function. It is however possible to for a signalling point to have only a signalling network function, e.g. an SP may be established with the sole function of routeing, or distributing signalling messages. In this special case the SP is known as a Signalling Transfer Point (STP).

Figure 8.5 shows three symbols used to represent signalling points in a signalling network diagram. The circle represents a Signalling Point with at least a user function, this will generally be an exchange, though it could be a network management or maintenance centre. The square represents a Signalling Transfer Point. A circle within a square represents a signalling point with a user function and an STP function.

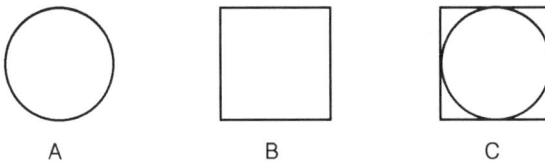

A      B      C

*Figure 8.5 Symbols used to represent signalling points on signalling network diagrams. a) Signalling point with user function. b) Signalling transfer point. c) Signalling point with user function and STP function*

Each SP in a signalling network must be capable of being uniquely identified. This is achieved in CCITT #7 by giving each SP a unique Point Code. A signalling point that generates a signalling message is known as the Originating Point. Similarly the signalling point to which the message is addressed is known as the Destination Point.

Point codes are used to label signalling messages with the originating and destination signalling points.

## 8.4.4 Signalling Relation

A signalling relation exists between any two signalling points if there is possibility of traffic between their user functions, as this produces a need for signalling between the two SPs. For example, there is a signalling relation between two exchanges. On a signalling network diagram this signalling relation is represented by a dotted line joining the two SPs, as shown in Figure 8.6.

*Figure 8.6   Representation of a signalling relation between two signalling points*

## 8.4.5 Signalling Link

A single digital channel over which CCS messages are passed between two signalling points is called a Signalling Link. In the case of most CCITT #7 systems this signalling link is of course TS16 of a 2Mbit/s system.

In many cases a single signalling link between two SPs will not be sufficient. At least two links will be required for security reasons, though more signalling links may be required to cope with the traffic on certain routes. All the signalling links between two SPs form the Signalling Link-Set, (which could be a single signalling link). Irrespective of the number of signalling links involved, the signalling link-set is represented by a single line joining the two SPs. This is shown in Figure 8.7 by adding the signalling link-set in parallel with the signalling relation.

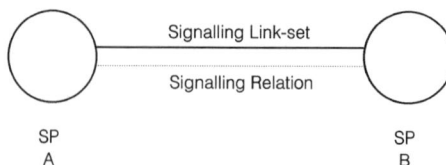

*Figure 8.7   Representation of a signalling link-set between two signalling points having a signalling relation*

## 8.4.6 Signalling Modes

In most cases a signalling link-set will directly connect two signalling points between which there is a signalling relation. This is the case represented by the diagram in

Figure 8.7, and is known as the Associated Mode of signalling. In Figure 8.8 the associated mode is used between SPs A and C, and between SPs C and B.

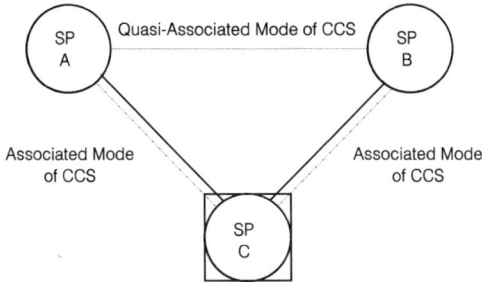

*Figure 8.8  Associated and quasi-associated modes of signalling*

If there is not a signalling link-set directly connecting two signalling points which have a signalling relation, signalling messages must be routed via a third signalling point which is connected to the two signalling points in question. This situation is also represented in Figure 8.8 as SP A and SP B have a signalling relation, but no direct signalling link-set. This case is known as the Non-Associated signalling mode. In fact the term non-associated would also apply if it were necessary to route signalling messages by two or even more intermediary signalling points.

All signalling points, other than the originating point and the destination point, through which a signalling message passes, must carry out a signalling transfer function. that is they must receive the message on one signalling link-set, and retransmit it on another. This is shown in Figure 8.8 as one of the roles of SP C, but as SP C is also an exchange, it is shown as having a user function and an STP function.

A further signalling mode, Quasi-Associated, refers to a special case of the Non Associated mode in which the route of the signalling messages via other signalling points is fixed. In a true non-associated signalling mode, signalling routes between SPs are not fixed, and consecutive signalling messages between two related SPs could be sent over different routes in a meshed signalling network. One disadvantage of such a system is that the time taken for messages to travel from originating point to destination point is indeterminate, and that signalling messages may not arrive at their destination in sequence.

There are however situations when this is satisfactory. A current military mobile switching system uses an approach similar to the non-associated mode, as it is relatively simple to implement in such networks, which are subject to constant planned changes in connectivity, and in which traffic and signalling links tend to be rather tenuous.

Note that although CCITT #7 does not permit the use of the true Non- associated mode, the quasi-associated mode is allowed. Within a CCITT #7 system all signalling routes must be defined, and thus all relevant routeing details are known to the SPs and STPs involved. Since the diagram in Figure 8.8 shows only one possible route between SP A, and SP B, it is more correct to describe the signalling between A and B as Quasi-Associated CCS.

## 8.4.7 Signal Units

Signalling information is passed over signalling link-sets in the form of short digital messages known as Signal Units (SU). Within the CCITT #7 system there are three types of SU:

❑ Message Signal Units (MSU)

❑ Link Status Signal Units (LSSU)

❑ Fill In Signal Units (FISU)

The basic format of a CCITT #7 Signal Unit is shown in Figure 8.9. It contains three main parts, a header, a signalling message and an error check.

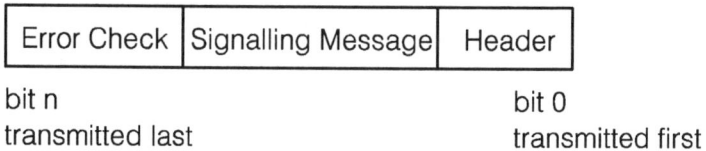

| Error Check | Signalling Message | Header |
|---|---|---|

bit n                                                          bit 0

transmitted last                                    transmitted first

*Figure 8.9 The basic format of a CCITT #7 signalling unit*

## 8.4.8 Fill In Signal Units

In the case of Fill In Signal Units the signalling message part will not be present as this type of SU is sent when there are no actual signalling messages to be transmitted. The transmission of FISUs maintains synchronism between the signalling terminals, and provides a ready check that even though no signalling messages are being received the signalling link is still serviceable. In this text the term Signalling Terminal means the hardware and software at a Signalling Point associated with the transmission and reception of Signalling Units.

## 8.4.9 Link Status Signal Units

LSSUs do not include a message part, but will instead contain a one or two byte Status Field. When MSUs can not be transmitted over the signalling link, this field is used by the signalling terminals at either end to indicate to each other the condition, or status, of the link. Several conditions are possible for which a status indication can be transmitted, these are:

### Status Indication :Out of Alignment (SIO)

Before the signalling link is initially brought into service, or after a failure, it must be proved that the link will operate above certain performance criteria concerning bit error rates. A proving period during which $2^{16}$ bytes are checked is normally used. The initial LSSU contains SIO when an alignment procedure is commenced.

### Status Indication: Normal (SIN)

LSSUs containing SIN are sent after the alignment procedure has commenced, to indicate, not that the link is in alignment, but that normal alignment is taking place between the terminals on the link. Normal alignment involves proving for $2^{16}$ bytes, or approximately 16 minutes.

### Status Indication: Emergency (SIE)

In some circumstances it will be necessary to prove the link for a shorter than normal period, for example if the signalling traffic rate is high. In such cases an emergency proving period of $2^{12}$ bytes (approximately one minute) is used. During this time LSSUs with SIE are sent to indicate that the link is in the emergency alignment state.

### Status Indication: Out of Service (SIOS)

An LSSU containing SIOS is transmitted to indicate that the terminal can not transmit or receive any Message Signal Units, for example due to a local fault. It would, of course, be normal for the link to be aligned when the fault has been cleared.

### Status Indication: Processor Outage (SIPO)

LSSUs containing SIPO are sent when the transmitting terminal can not receive or transmit any MSU due to a processor outage, rather than any other condition such a localised fault in the terminal.

### Status Indication: Busy (SIB)

Should a signalling link become congested, SIB is transmitted to indicate the fact to the distant terminal. During periods of congestion positive and negative acknowledgements are withheld, and SIB transmitted so that the terminal can distinguish between failure of the link, and congestion.

This mechanism provides a degree of message flow control over the link. Note that the link is still in alignment. However upon receipt of an SIB, a time out is started. If the link has not restored to normal operation before the time out expires, a failure indication is given.

## 8.4.10 Flags and Bit Stuffing

The signalling terminals at each end of the two-way signalling link are continuously transmitting and receiving signalling units. As each type of SU is of different length it is imperative that the receiving signalling terminal can recognise the end of one SU and the beginning of the next. Figure 8.10 shows how an 8-bit Begin Flag containing the unique bit pattern 01111110 is used to mark the end of one SU, and the start of the next.

Since it is possible for the flag pattern (01111110) to occur naturally in the SU, some provision must be made to ensure that the receive signalling terminal only responds to real flags and ignores imitations occurring within the signalling traffic. Before the flag is added to the SU, a technique known as bit stuffing is employed in the transmitting signalling terminal. Bit stuffing involves breaking up all strings of five or more consecutive 1s in the SU by inserting an extra redundant zero after each group of five 1s. This is done irrespective of whether the next bit is actually a 1 or a 0. The preceding flag is then added, and the SU transmitted on the signalling link-set.

At the receiving signalling terminal, the flag is detected and removed first. Then the SU is bit stripped i.e. any zero following a string of five 1s is removed, to reproduce the original SU.

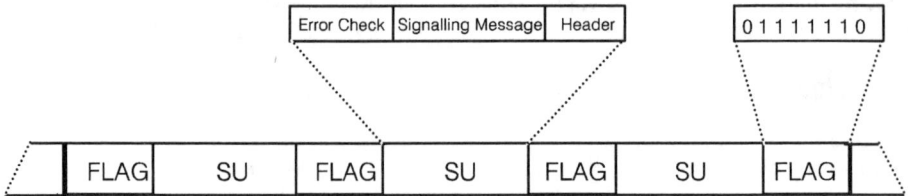

*Figure 8.10 The use of flags to separate consecutive signalling units*

## 8.4.11 Signal Unit Header and Error Check Field

Figure 8.11 shows the formats of the three bytes which form the SU header, and the two bytes which form the SU Error Check. The three header bytes contain information which is used for *housekeeping* purposes. They are described in the next sections:

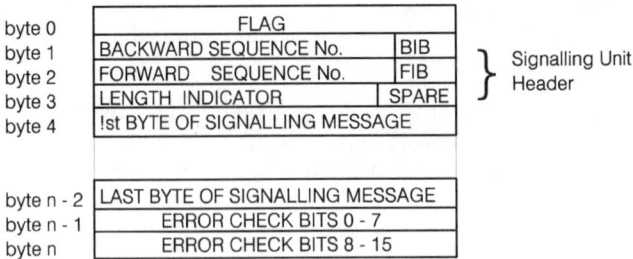

*Figure 8.11 Basic format of a signal unit showing the header fields and error check fields*

## 8.4.12 Signal Unit Backward Sequence Number

If, when each SU is received, the error check indicates that no errors occurred during transmission, and the FSN field contains the next expected sequence number, it is assumed that the SU has been correctly received. A positive acknowledgement is then returned by sending the FSN of the received SU in the Backward Sequence Number (BSN) field of the next SU to be sent in the opposite direction. To complete the acknowledgement the Backward Indicator Bit (BIB) of this SU is set so that it is the same as the BIB of the previous SU transmitted in this direction.

Signalling Terminals include a store in which they will keep a copy of all transmitted SUs until a positive acknowledgement for each is received. Once the SU has been positively acknowledged the copy of the SU will then be erased.

### 8.4.13 Signal Unit Forward Sequence Number

Each Signal Unit is transmitted with a Forward Sequence Number (FSN) in the range 0 to 127. The actual value of the FSN transmitted in any SU will normally be one greater than the value of the FSN transmitted in the previous SU. Three exceptions to this are:

❑ The FSN following FSN = 127, will be FSN = 0.

❑ The FSNs will not be in sequence when an SU retransmission occurs

❑ The FSN of an FISU will be the same as that of the previous SU

### 8.4.14 Length Indicator Field

Fill in Signal Units are of fixed length, that is, zero bytes neglecting the flag and header. Link Status Signal Units may be one or two bytes long. The length of a Message Signal Unit will depend upon the nature of information contained in the MSU, and may be between 3 and 63 bytes. A Length Indicator (LI) field included in each SU provides a means of identifying the type of SU to a receiving signalling terminal, and of informing the receiving terminal of the number of bytes in the signalling message part.

### 8.4.15 Signal Unit Error Check

A 16-bit check field is included at the end of every SU. The check field is generated by using a Cyclic Redundancy Code check on all bits in the SU, in a manner similar to that described in Section 5.6.12. At the receive signalling terminal the SU and the Check field are tested to detect whether or not any errors have occurred during the transmission of the SU.

# 8.5 Correction of Errors Occurring in Signalling Units

### 8.5.1 The Principles of Error Detection and Correction in CCITT #7

If the error check indicates that a received SU has been corrupted during transmission it is discarded by the receiving terminal, which also ceases to receive any new signalling messages. The receiving terminal indicates that an error has occurred by changing the Backward Indicator Bit (BIB) of the next outgoing Signal Unit (a process referred to as *toggling*, i.e. if the BIB was set to 1, it is set to 0, and vice versa), and loads its BSN field with the FSN of the last correctly received SU.

When the signalling terminal which originated the corrupted SU recognises that the BIB from the distant terminal has been toggled, it takes this as a request for a SU retransmission and will then cease transmitting new SUs and retransmit the SU with a FSN one greater than that received in the BSN field, and all other SUs in sequence after it. To indicate that the originating terminal has received the SU retransmit request, the Forward Indicator Bit (FIB) of the retransmitted SU is toggled.

When the retransmitted SU, with the FIB toggled is received correctly at the distant terminal, it is acknowledged as before by sending the FSN in the BSN field and leaving the BIB as it was. Signalling Units can then continue to be exchanged between the two signalling terminals until a subsequent error occurs.

## 8.5.2 An Error Detection and Correction Example

The diagrams in Figure 8.12 show the exchange of a number of SUs between two signalling terminals, A and B.

The situation at the start is that SU number 76 has just been correctly received by Signalling Terminal A, which sends a positive acknowledgement of SU 76 in its next outgoing SU, number 34. The process continues until SU 36 is corrupted on its way from A to B. B detects the corruption and requests a retransmission of SU 36 by toggling the BIB of its outgoing SU number 79 to indicate an error has been detected. The BSN of SU 79 is loaded with the FSN of the last correctly received SU (i.e. No. 35).

The toggled BIB from B is recognised by terminal A, which ceases transmitting new SUs, and retransmits from store the SU with an FSN 1 greater than that indicated in the BSN field (i.e. SU 36). To indicate that this SU contains a retransmission of a previous SU, the Forward Indicator Bit (FIB) is toggled.

Once SU 36 is correctly received by terminal B, the process can continue as before, until another transmission error is detected.

## 8.5.3 Error Detection and Correction in Practice

This explanation serves to illustrate the principles of signal unit error detection and correction but neglects an important point. There will be a finite time between the transmission of a particular signalling unit, and its reception at the distant signalling terminal. Thus it is probable that several SUs are in the pipeline as it were at the same instant in time. If an error is detected, there will also be a finite time for the SU retransmission request to reach the originating terminal.

The diagram in Figure 8.12 shows only the corrupted SU having to be retransmitted, but in practice several SUs must be transmitted again. The actual number in any circumstance will depend the length of those SUs in the pipeline when the error occurred, and the transit delay between the two signalling terminals.

There are in fact two types of error detection defined. The basic type described above, in which the one-way transit delay on a signalling link is less than 15mSecs, is used in the BT national trunk network.

The second method, Preventive Cyclic Retransmission, is used for signalling links in which the one-way propagation delay exceeds 15mSecs, as will be the case in links provided by communications satellites.

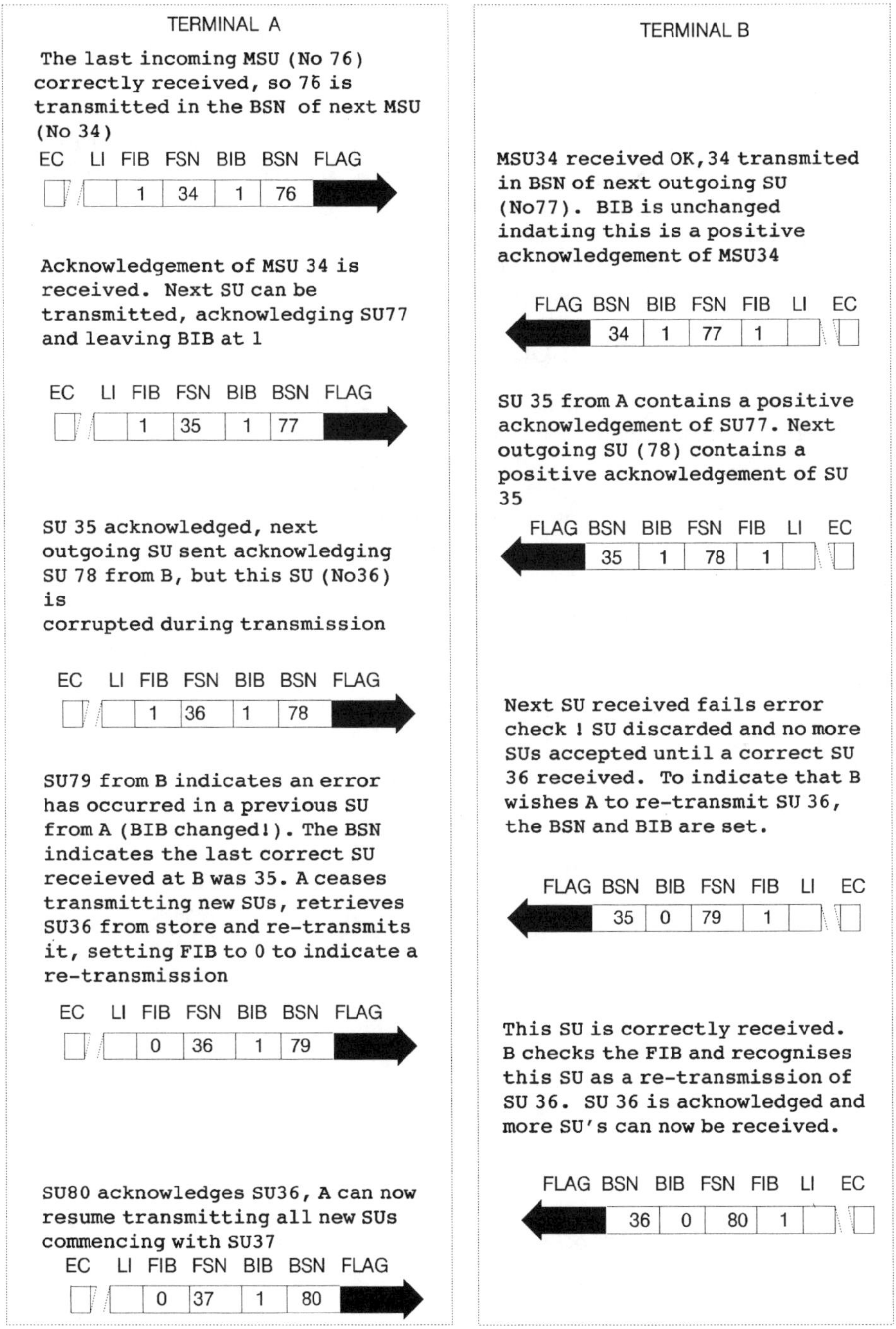

**TERMINAL A**

The last incoming MSU (No 76) correctly received, so 76 is transmitted in the BSN of next MSU (No 34)

| EC | LI | FIB | FSN | BIB | BSN | FLAG |
|----|----|-----|-----|-----|-----|------|
|    |    | 1   | 34  | 1   | 76  |      |

Acknowledgement of MSU 34 is received. Next SU can be transmitted, acknowledging SU77 and leaving BIB at 1

| EC | LI | FIB | FSN | BIB | BSN | FLAG |
|----|----|-----|-----|-----|-----|------|
|    |    | 1   | 35  | 1   | 77  |      |

SU 35 acknowledged, next outgoing SU sent acknowledging SU 78 from B, but this SU (No36) is corrupted during transmission

| EC | LI | FIB | FSN | BIB | BSN | FLAG |
|----|----|-----|-----|-----|-----|------|
|    |    | 1   | 36  | 1   | 78  |      |

SU79 from B indicates an error has occurred in a previous SU from A (BIB changed!). The BSN indicates the last correct SU receieved at B was 35. A ceases transmitting new SUs, retrieves SU36 from store and re-transmits it, setting FIB to 0 to indicate a re-transmission

| EC | LI | FIB | FSN | BIB | BSN | FLAG |
|----|----|-----|-----|-----|-----|------|
|    |    | 0   | 36  | 1   | 79  |      |

SU80 acknowledges SU36, A can now resume transmitting all new SUs commencing with SU37

| EC | LI | FIB | FSN | BIB | BSN | FLAG |
|----|----|-----|-----|-----|-----|------|
|    |    | 0   | 37  | 1   | 80  |      |

**TERMINAL B**

MSU34 received OK, 34 transmited in BSN of next outgoing SU (No77). BIB is unchanged indating this is a positive acknowledgement of MSU34

| FLAG | BSN | BIB | FSN | FIB | LI | EC |
|------|-----|-----|-----|-----|----|----|
|      | 34  | 1   | 77  | 1   |    |    |

SU 35 from A contains a positive acknowledgement of SU77. Next outgoing SU (78) contains a positive acknowledgement of SU 35

| FLAG | BSN | BIB | FSN | FIB | LI | EC |
|------|-----|-----|-----|-----|----|----|
|      | 35  | 1   | 78  | 1   |    |    |

Next SU received fails error check ! SU discarded and no more SUs accepted until a correct SU 36 received. To indicate that B wishes A to re-transmit SU 36, the BSN and BIB are set.

| FLAG | BSN | BIB | FSN | FIB | LI | EC |
|------|-----|-----|-----|-----|----|----|
|      | 35  | 0   | 79  | 1   |    |    |

This SU is correctly received. B checks the FIB and recognises this SU as a re-transmission of SU 36. SU 36 is acknowledged and more SU's can now be received.

| FLAG | BSN | BIB | FSN | FIB | LI | EC |
|------|-----|-----|-----|-----|----|----|
|      | 36  | 0   | 80  | 1   |    |    |

*Figure 8.12 Error detection example*

# 8.6 The Structure of CCITT Signalling System 7

## 8.6.1 Introduction

Sections 8.1 to 8.5 have provided a broad introduction to the principles of CCS, and have already cited some examples from CCITT #7. In this short section the basic structure of CCITT #7 is introduced. It will be seen that CCITT #7 consists of several functional levels which have some resemblance to the ISO seven layer model for Open Systems Interconnection, although there is no direct equivalence between the two.

## 8.6.2 Functional Levels of CCITT #7

Figure 8.13 shows the four functional levels of CCITT #7. The first three levels are specifically concerned with the transmission of signalling messages between signalling points in the network and form the Message Transfer Part (MTP). The MTP is used by various user functions at a signalling point to transfer messages to the equivalent user function at other signalling points. For example the call processing subsystem at one exchange uses the MTP to transfer call related messages to the call processing subsystem at other exchanges.

Those subsystems using the MTP are thus known as User Applications and are the concern of the fourth level of CCITT #7, and typically include call processing, operations, maintenance and so on. For this reason Level 4 is known as the User Part, and as there are several different user applications, several different Level 4 User Parts are defined.

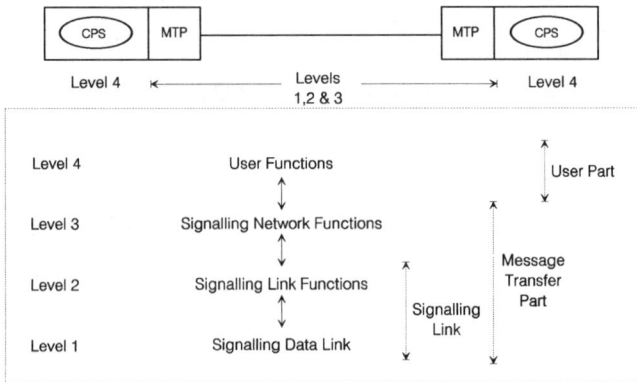

*Figure 8.13 The four levels of CCITT #7*

## 8.6.3 Level 1: The Signalling Data Link

The function of Level 1 is to provide a bi-directional digital link between two signalling points. This digital link will typically be TS16 of a 2Mbits 30 Channel

PCM system, and will thus operate at 64Kbit/s. Although some multiplex equipment includes facilities for the insertion of signalling messages into TS16, and their extraction from TS16 at the distant end of the link, there are cases in which the complete end to end signalling link includes semi-permanent switched paths across a digital switch at each end of the link, as shown in Figure 8.14.

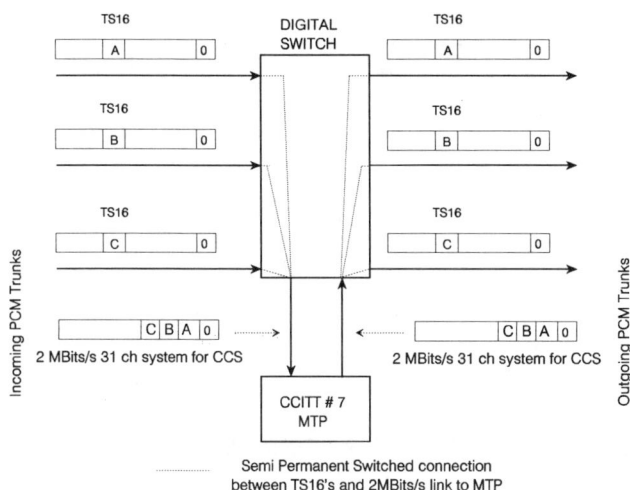

*Figure 8.14 A signalling link which includes a path through a digital switch*

Although usual, a 64 Kbit/s capacity is not mandatory and data rates as low as 4.8 Kbit/s can be used. In some cases low data rate links can be provided over analogue systems so long as suitable modems are available. Whichever physical implementation is used, it is inevitable that some bit errors will occur due to the nature of the link. Level 1 does not provide error detection and correction.

The main function of level 1 is to provide a suitable physical signalling link between two signalling points over which CCS messages can be passed.

## 8.6.4 Level 2: Signalling Link Functions

The functions of Level 2 are to operate the signalling link provided by level 1, and to ensure that messages are passed correctly between the two Signalling Points at each end of the link. Level 2 functions ensure that messages are delivered in sequence, without errors or duplication.

Thus Level 2 must provide the sequencing mechanisms discussed in Section 8.4, and the Error Detection and Correction protocols discussed in Section 8.5. Figure 8.15 shows that Level 2 receives basic signalling messages in sequence from Level 3 and formats them into transferable signalling units which can be transmitted over a physical signalling link (i.e. Level 1). At the distant end of the signalling link the Level 2 functions remove the fields that were added to get the message across the link, and pass to the Level 3 just the basic message as transmitted by Level 3 at the originating signalling point. Level 2 functions includes the addition of sequence numbers, error check field and the flag to the basic signalling message. The previous sections of this chapter have mainly been concerned with Level 1 and 2 functions.

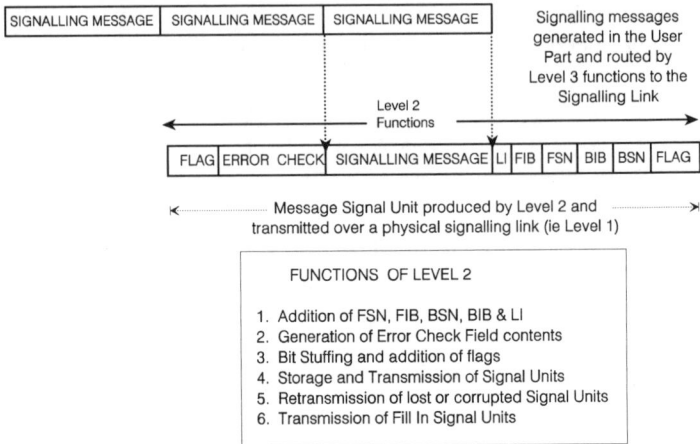

*Figure 8.15 The functions of Levels 1 and 2*

## 8.6.5 Level 3: Signalling Network Functions

Level 3 contains the Signalling Network Functions, and is concerned with two aspects, namely, Signalling Message Handling (SMH) and Signalling Network Management (SNM).

## 8.6.6 Level 3: Signalling Message Handling

SMH functions ensure that signalling messages from a Level 4 User Part are routed over the correct signalling link, and likewise signalling messages received from a signalling link are routed either to the appropriate Level 4 User Part, or outwards on another signalling link. Signalling messages contain sufficient information to identify both the destination SP, and the required User Part at that SP. The three basic SMH functions provided in Level 3 are:

❑ Message Routeing

❑ Message Discrimination

❑ Message Distribution

The format of a signalling message produced by the Level 4 User Part is shown in Figure 8.16. Note that some of the fields used by Level 4 are also used by Level 3. The use of these fields will be explained in the following sections.

Figure 8.17 shows the SMH functions of Level 3 and how they interact with the functions of Levels 2 and 4. In this diagram, two Level 4 User Parts within the exchange are identified, one for telephony calls, the other for data calls. The exchange is directly connected to five other exchanges, thus signalling messages originating in either Level 4 User Part could be addressed to any of these exchanges, or indeed any other exchange in the network. For each connection to another exchange only one link is shown. In practice there will be several links in the link-set, and each will have its own level two functions.

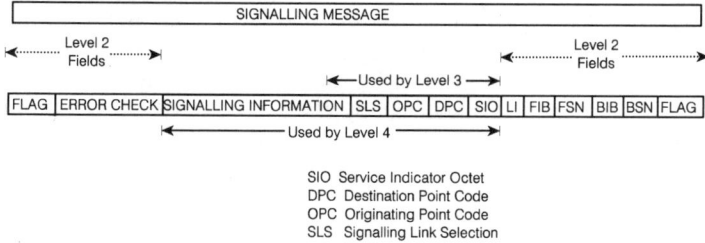

*Figure 8.16 Format of a signalling message showing fields used at Level 3*

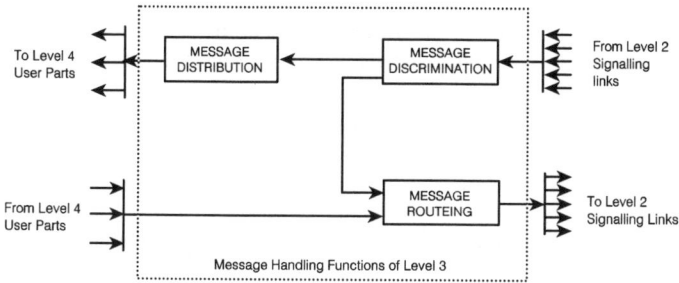

*Figure 8.17 The signalling message handling functions of CCITT #7 Level 3*

## 8.6.7 Message Discrimination

This function deals with all incoming messages from other signalling points. After Level 2 information (flags, error check etc) have been removed from the Message Signal Unit, the resulting signalling message as shown in Figure 8.15 is delivered to the Level 3 Message Discrimination function.

The DPC of the message is examined to determine whether the message is addressed to the host SP, or it is for onward transmission to another SP.

In the first case the message is delivered to the message distribution function. In the latter case the SP is acting purely as an STP, and the message is passed to the message routeing function.

## 8.6.8 Message Distribution

The message distribution function examines the Service Information Octet (SIO) to determine the Level 4 User Part to which the message is addressed. The SIO contains two 4-bit fields as shown in Figure 8.18 overleaf.

The Service Indicator identifies the User Part to which the message has to be finally delivered (e.g. the Telephony User Part, the Data User Part). The Sub-Service field includes two bits which form a Network Identifier (NI).

Most exchanges are part of only one network, and are thus part of only one signalling network. Several exchanges will however be gateways to other networks, for example an international exchange will be part of the national network, and the international network. Thus signalling messages can be routed to either the national user part, or the international user part as indicated by the Network Identifier within the Sub Service Field.

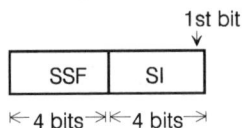

SI   Service Indicator, determines User Part
     to which message is to be delivered.

SSF  Sub-service Information Field, includes
     a 2bit Network Identifier

*Figure 8.18 The service information octet of a CCITT #7 signalling message*

## 8.6.9 Message Routeing

This function deals with messages originating at the SP Level 4, and with incoming messages that Message Discrimination has identified as not being addressed to the host SP.

The exchange can not be directly connected to all other exchanges in the network so routeing tables are held from which Message Routeing can identify the appropriate outgoing route for any particular destination point code.

## 8.6.10 Signalling Network Management (SNM)

Although the main functions of Level 3 are the SMH functions mentioned in the previous sections, SMH can only properly be carried out over a reliable signalling network. It has already been stated that the signalling links of the network, like any other communications system, will be subject to outages due to equipment failures, etc. Other problems which may be encountered in the signalling network include:

❑ Complete failure of a signalling point

❑ Congestion of signalling routes

❑ Congestion at signalling points

Signalling Network Management is required to ensure that the SMH functions can continue to operate effectively even when problems such as these occur. There are three basic areas covered by SNM:

❏ Signalling Traffic Management

❏ Signalling Route Management

❏ Signalling Link Management

Figure 8.19 shows, in a simplified form, how these functions interact closely with each other and the equivalent functions at other SPs. To enable each function to communicate with other SPs, special signalling messages known as signalling network management messages are transmitted using the SMH functions of Level 3.

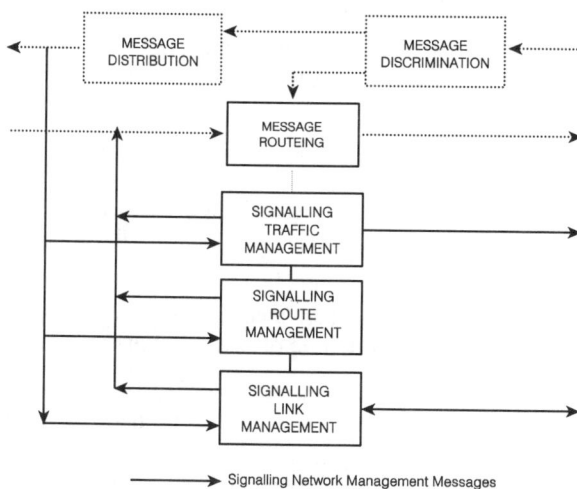

*Figure 8.19 Interaction of the signalling network management functions*

## 8.6.11 Signalling Traffic Management (STM)

The term *signalling traffic* refers to all the signalling messages generated at signalling points for transmission over the signalling network. STM incorporates procedures to ensure that signalling traffic is diverted from failed signalling links in a particular route, and that in the event of failure of a complete link-set, signalling traffic is diverted to another route.

STM also includes procedures to restart traffic on signalling links which were previously unavailable when the links are returned to service, and to restart traffic on routes that were unavailable as individual links on the route become serviceable.

In the event of congestion on a signalling route, or at a particular signalling point STM invokes signalling traffic flow control procedures to regulate the transmission of signalling messages to congested parts of the network.

All STM procedures should ideally take place without loss or mis-sequencing of signalling messages. However when emergency procedures have to be adopted some signalling messages may be lost, and some calls will fail to be established.

Briefly, STM involves the following five procedures:

❑ **Link change-over:** to divert traffic from failed links

❑ **Link change-back:** to restore traffic to repaired links

❑ **Forced re-routeing:** to divert traffic to alternate routes when all links on a certain route have failed.

❑ **Controlled re-routeing:** the equivalent of route change-back

❑ **Signalling traffic flow control:** to regulate flow of messages through congested parts of the network.

## 8.6.12 Signalling Link Management (SLM)

SLM is involved with the management of individual links and provides procedures for the activation and deactivation of signalling links, and the restoration of failed signalling links.

Links which have become unavailable due to high error rates etc are restored to service only after Level 2 alignment has been successful and it has been proved that the link can carry traffic. Similar procedures are adopted to bring into service, links which are idle for some other reason, for example the activation of newly commissioned signalling links.

Operational signalling links can be deactivated so long as no signalling traffic is actually being carried on the link at the time of deactivation.

Circumstances will occur when it is necessary for the signalling network management to effectively deactivate a link so that it can not be used for signalling traffic, but leave it operational so that test traffic may be passed. A procedure known as Management Inhibiting is provided for this purpose. When links have been so inhibited, they will be restored to service automatically by SLM in the event of failure of all other links in the same link-set.

The main SLM procedures are:

❑ **Signalling link activation:** to bring inactive links into use

❑ **Signalling link restoration:** to bring failed links back into use

❑ **Signalling link deactivation:** to take an operational link out of use as long as it is not carrying signalling traffic

❑ **Link-set activation:** to bring new signalling routes into use for the first time, and to restart traffic on failed routes which have been repaired

❑ **Management inhibiting:** to prevent operational links being used for signalling traffic, while permitting their use for test purposes.

A further procedure defined within the CCITT #7 recommendations permits the automatic allocation of data links and signalling terminals to allow more flexibility in the provision of signalling links.

## 8.6.13 Signalling Route Management (SRM)

SRM contains procedures for the transmission of information concerned with the connectivity and status of the signalling network, so that the signalling network management may react to problems and reconfigure the network to achieve the most efficient signalling message routeing. There are four main procedures within the SRM functional block. One of these provides for the broadcasting to all signalling points, of messages detailing routes over which the transfer of signalling messages is prohibited. Similarly broadcast messages are sent to all SPs giving details of routes which are reopened for traffic. When either type of message is received at an SP, the internal routeing tables of the SP are updated to show the current status of the network.

When a signalling message for a congested destination is received at an STP, the STP sends to the originating SP a Transfer Controlled Message. When this message is received by the originating SP, the level 4 user parts are informed of the congestion so that signalling traffic flow towards the affected destination may be controlled.

Test procedures are included to provide the signalling point with information regarding the status of the network. An SP will transmit a route set test message to other STPs in the network in order that changes in network status can be identified, and the routeing tables of the SP updated accordingly.

Summarising, the four main procedures of Signalling Route Management are:

❑ **Transfer Prohibited Procedure:** to inform all SPs of unavailable routes

❑ **Transfer Allowed Procedure:** to inform all SPs that previously unavailable routes have become operational

❑ **Transfer Controlled Procedure:** to inform individual SPs as necessary of congestion in the network

❑ **Route Set test Procedure:** to permit SPs to gain information regarding changes in the status of the signalling network.

## 8.6.14 Level 4 Applications User Parts

In discussing the MTP, i.e Levels 1 to 3 of CCITT #7, we have seen that the signalling network consists of signalling links connecting signalling points. Signalling messages are routed over these links without error or loss, and the signalling network is managed to maintain it at its most effective during times of network problems. It is now time to discuss how this signalling network is used by exchange applications such as call processing, i.e. the Level 4 applications or User Parts.

CCITT #7 includes three User Parts:

❑ Telephony User Part (TUP)

❑ ISDN User Part (ISUP)

❑ Signalling Connection and Control Part (SCCP)

The TUP is provided for the control of telephony calls, while the ISUP provides basic telephony call set up and release, along with other facilities required by ISDN users, such as the set up and release of data calls. The ISUP also provides the facilities required for the transfer of user to user signalling. Such facilities are required for several purposes including the transfer of calling line and called line information between the two parties to a data call. More about these topics in the chapter devoted to ISDN.

The SCCP provides facilities which enable non call related information to be transmitted over signalling links. The SCCP allows the signalling network to be used as a data communications medium for use by the various nodes in the network, for example the interrogation of centralised databases by call processing systems, or the transfer of data between network management centres. The data communication service provides by SCCP is similar in several respects to that provided by packet switched networks.

BT has developed its own Level 4 User Part. Known as the National User Part (NUP), it is heavily based on the CCITT TUP but includes facilities required by BT's ISDN service. A brief look at how telephony calls are set up by the BT NUP will illustrate how a message based common channel signalling system operates.

## 8.6.15 An Example Call Set Up

Figure 8.20 on the next page represents the set up of an example call between two subscribers connected to different local exchanges. This example call is routed through two trunk exchanges, A and B. In this diagram forward signalling messages are represented by solid arrow headed lines pointing left to right. Arrows pointing right to left represent backward signalling messages, while the dotted lines represent the actual multiplex channel (or TS) selected by each exchange for the call.

**A:** The call request signal generated by calling party going off hook is detected at the originating local exchange, which prepares to receive the dialled digits.

**B:** Local exchange sends dial tone to the caller.

**C:** The caller then proceeds to dial.

**D:** When sufficient digits have been received by the local exchange, it is able to determine the trunk exchange to which the call is to be routed,

**E:** The local exchange then selects a free channel on an outgoing route to the trunk exchange and sends an Initial Address Message (IAM) to trunk exchange A over the signalling network. The IAM will contain information which includes the identity of the channel selected for the call and the dialled digits.

**F:** This information allows the trunk exchange A to determine that the call is to be routed to another trunk exchange, B. A similar IAM is sent to trunk exchange B, once trunk exchange A has selected a free channel on the route connecting the two trunk exchanges.

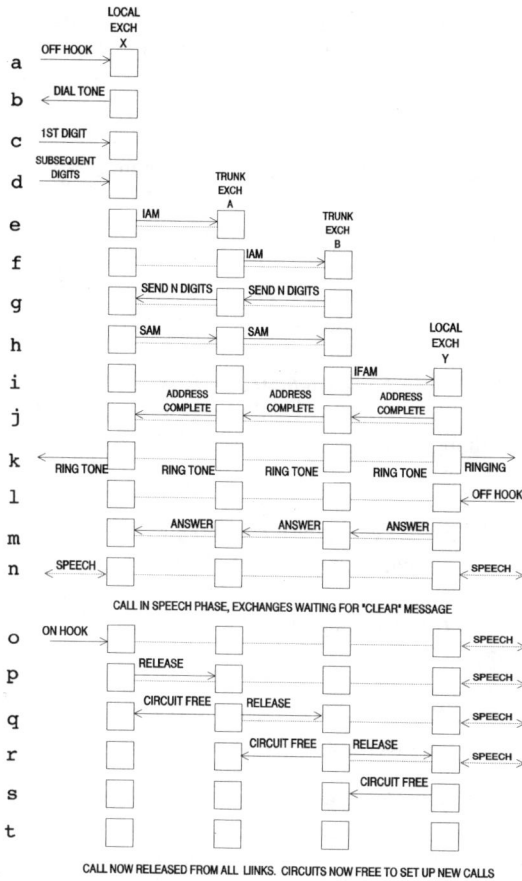

Figure 8.20 Example call set up within CCITT#7 environment

**G:** Trunk exchange B determines that the dialled digits received so far are insufficient to complete the call, and so requests the transmission of further digits by transmitting a "send N digits" message back through to the originating exchange. (N is the number of digits required to complete the called subscriber number).

**H:** The originating local exchange responds by sending a Subsequent Address Message (SAM) to trunk exchange B via Trunk exchange A. The SAM contains the required digits to allow trunk exchange B to determine to which local exchange the call is to be finally routed.

**I:** Trunk exchange B selects a free channel to the destination local exchange, and sends an Initial and Final Address Message (IFAM) detailing the channel selected and all the dialled digits.

**J:** When the IFAM is received by the called party's local exchange, if the called line is free, an Address Complete Message (ACM) is sent back to the originating local exchange which can then complete the transmission path between the calling party's line and the selected channel.

**K:** As well as sending the ACM the destination exchange will connect ringing current to the called line, and ring tone back toward the calling party, since the transmission path is complete at the originating exchange the ring tone will be heard by the caller.

**L:** When the called party answers, the off hook condition on his line is detected by his local exchange which will cease transmitting ringing current and ring tone.

**M:** An Answer message is sent back to the originating exchange which can start charging for the call. Simultaneously the transmission path is completed at the destination local exchange by the connection of the incoming channel to the called party's line

**N:** The call is now in the speech phase and no further signalling relating this call is required until either of the parties to the call clear down.

**O:** When the calling party clears down, the on hook condition on his line is detected by his local exchange.

**P:** The local exchange frees the channel used for the call and sends a release message to trunk exchange A. This message will include the reason for the release.

**Q:** Trunk exchange A also frees the channel between the local exchange and itself, and then frees the circuit selected between itself and trunk exchange B. Trunk exchange A now sends B a Release message, and simultaneously acknowledges the initial Release message from the local exchange by sending back a Circuit Free message.

**R, S and T:** Similar actions are carried out by trunk exchange B and the destination local exchange as they in turn receive Release messages, until eventually the call is completely cleared and all channels used to route the call are free.

A few points which are not covered in this diagram are worth noting.

At step J the destination local exchange may not have been able to complete the call because the called line was busy, engaged or unobtainable. In this case a suitable message would be sent back to the originating local exchange.

At any time during the set up of the call the calling party may have abandoned the attempt and cleared down. This would have resulted in Release messages being sent to those exchanges which had become involved in the call set up thus far.

In some cases the called party may clear first. This does not result a backwards cleardown of the established call path. Instead a Called Party Clear message is sent back through the network to the originating local exchange. When this message is received a timer is started and should this timer mature before the calling party clears, the release sequence shown in steps P to T is followed, except that the release message will identify the reason for the call being cleared as being due to called party clear.

## 8.6.16 Format of Level 4 Signalling Message

Figure 8.16 showed the format of a signalling message in order that those parts used by level 3 could be identified. Figure 8.21 shows the format in a little more detail so that we can outline the main items of interest at Level 4.

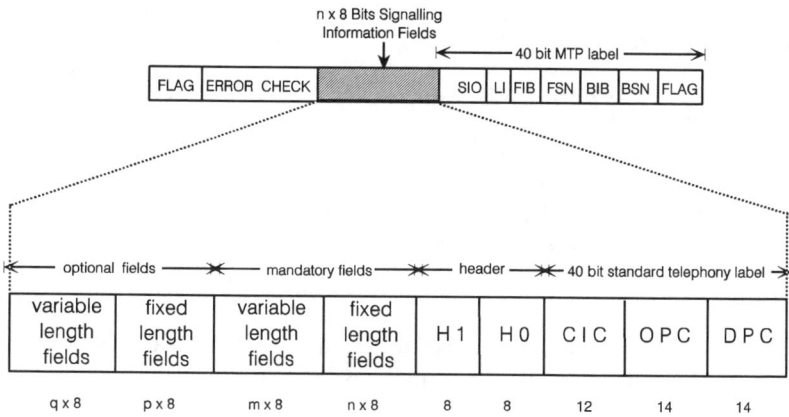

*Figure 8.21 Standard signalling message format showing Level 4 fields*

The functions of the OPC and DPC have already been discussed. The Circuit Identity Code (CIC)is a 12-bit field which identifies the actual channel to which the signalling information contained in the message refers. It is worth remembering there will usually be several 2Mbit systems, each with 30 channels available, connecting two exchanges.

The OPC, DPC and CIC form a 40-bit standard telephony label which is present in all signalling messages. The next two octets, H0 and H1 form a message header. H0 identifies the group of messages to which the message belongs, there are eight possible groups:

**Group 0:** Forward Address Messages e.g. IAM, IFAM, SAM, FAM

**Group 1:** Additional set up information messages

**Group 2:** Backward Set Up Request Messages e.g Send Additional Digits

**Group 3:** Backward Set Up Information Messages e.g ACM, Sub engaged, Sub Out of order and so on

**Group 4:** Call supervisory messages, .e.g. Answer, Clear, Release etc

**Group 5:** Circuit supervisory messages, e.g. Circuit free, blocking etc

**Group 6:** Reserved for inter-exchange use only

**Group 7:** ISDN messages

The actual message type within the group is identified by the H1 Octet. Although many of the possible combinations of H0 and H1 are not used at present, their use in the future is not precluded.

The present maximum length of a signalling message in the BT NUP is 62 octets, although this may increase to 272 octets in line with CCITT specifications in the future to support services using the SCCP.

## 8.6.17 Example Signalling Messages

Not all messages require the maximum 62 octets and thus messages are of variable length. One of the longest and most complex messages is the IAM as this contains a considerable amount of information that will be used by the exchanges processing the call. The format of an IAM is shown in Figure 8.22.

The purpose of each of the 13 fields (a – n) is given briefly below.

**a.** Calling Party Category indicates whether the calling line is a business or residential customer, and whether or not the subscriber has ISDN facilities.

**b.** Originating Node UP Version No. is used in ISDN calls to identify the version of CCITT #7 in use at the originating exchange

**c.** Message Indicator indicates whether the call has originated in another network, and whether analogue interworking is involved.

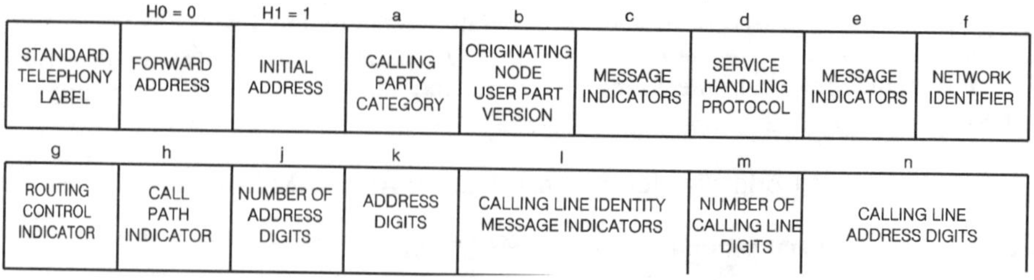

| | H0 = 0 | H1 = 1 | a | b | c | d | e | f |
|---|---|---|---|---|---|---|---|---|
| STANDARD TELEPHONY LABEL | FORWARD ADDRESS | INITIAL ADDRESS | CALLING PARTY CATEGORY | ORIGINATING NODE USER PART VERSION | MESSAGE INDICATORS | SERVICE HANDLING PROTOCOL | MESSAGE INDICATORS | NETWORK IDENTIFIER |

| g | h | j | k | l | m | n | |
|---|---|---|---|---|---|---|---|
| ROUTING CONTROL INDICATOR | CALL PATH INDICATOR | NUMBER OF ADDRESS DIGITS | ADDRESS DIGITS | CALLING LINE IDENTITY MESSAGE INDICATORS | NUMBER OF CALLING LINE DIGITS | CALLING LINE ADDRESS DIGITS | |

*Figure 8.22 Initial address message format*

**d.** Service Handling protocol used mainly in ISDN calls, and calls involving the operator.

**e.** Message Indicators indicate release procedures to be used

**f.** Network Identifier indicates the network from which this call is being routed and processed.

**g.** Routeing Control Indicator indicates whether or not alternate call routeing is allowed.

**h.** Call Path Indicator indicates for certain types of call, whether or not the call path can include ADPCM or other speech processing systems.

**j.** Number of Address Signals indicates how many digits are being sent in this message.

**k.** Address Signals, the dialled digits are transmitted in 4-bit BCD format.

**l.** Calling Line Identity Message Indicator indicates whether or not the identity of the calling subscriber can be given to the called subscriber.

**m.** Number of Calling Line Address Digits, the number of digits in the calling line number.

**n.** Calling Line Address Digits, the calling line number in 4-bit BCD format.

Contrast the IFAM message with the Answer message, which contains only one information field, as shown in Figure 8.23. The Type of Answer field indicates whether or not this call is to be charged and the remaining field is spare.

| | HO = 4 | H1 = 0 | a | b |
|---|---|---|---|---|
| STANDARD TELEPHONY LABEL | CALL SUPERVISION | ANSWER | TYPE OF ANSWER | SPARE |

*Figure 8.23 Answer message format*

Figure 8.24 shows that the Clear message contains only the HO and H1 octets as no further information is required. The Clear message may be sent in the forward or backward direction when a clear down indication has been received from one of the parties to a call. Note that the call is only released when sufficient indications have been received by the exchange responsible for the call.

|                                    | HO = 4               | H1 = 1 |
| ---------------------------------- | -------------------- | ------ |
| STANDARD TELEPHONY LABEL           | CALL SUPERVISION     | CLEAR  |

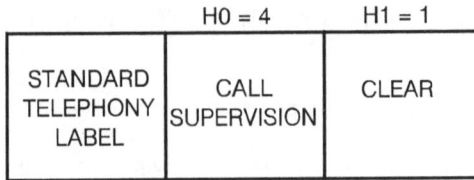

*Figure 8.24 Clear message format*

# 8.7 Chapter Summary

The problems of Channel Associated Signalling such as slow speed, limited repertoire and cost led to an interest in alternative techniques. Common Channel Signalling was introduced to overcome these problems and produce flexible systems capable of evolving to meet new user requirements.

Several CCS systems are found in practice, though in general these are based on the CCITT System 7 specification, adapted where necessary to meet the requirements of the particular network administration.

Within the UK, the public telephone networks of BT, Mercury and Kingston use CCITT #7, while more private networks are beginning to use DPNSS. With the advent of ISDN more organisations will be using DASS links between their ISPBXs and the public telephone network. Both DPNSS and DASS are similar to CCITT #7.

The introduction of CCS systems has brought many new concepts and terms into use. CCITT #7 is structured into four levels: Levels 1 to 3 form the Message Transfer Part and are responsible for the orderly transmission of signalling units through the network. Level 4 encompasses the applications which rely on the MTP for the delivery of signalling units. BT has developed its own Level 4 User Part, the NUP. The NUP is based on the TUP of CCITT #7, but includes ISDN facilities.

The CCITT ISUP and the NUP are able to deal not only with call set up and release related signals, but are able to transfer signalling information between two network terminals connected during a call.

This chapter has provided an introduction to the complex subject of CCS. It has attempted to show that CCS systems are not simply a replacement for link by link CAS systems, but provide a comprehensive signalling network, complete with signalling network management functions.

# 8.8 Further Reading

For those readers wishing to read further into this subject, the following are recommended:

a. POEEJ, Vol 72 **July 79** *The principles of System X*

b. POEEJ, Vol 73 **Oct 80** *System X Subsystems Part 4 Common Channel Signalling and the Message Transmission Subsystem*

Although some detail contained in these articles may be out of date, they still make an interesting introduction to the subject.

c. British Telecom Engineering Vol 7 Part 1 **April 1988**. The whole issue is devoted to CCITT Signalling System No. 7.

d. CCITT Signalling System No.7 Telephony User Part. CCITT Red Book VI Fascicle VI.8

e. British Telecom National Requirement No 167 (BTNR 167). This contains full details of the formats and codes used in the NUP.

f. Signalling in Telecommunications Networks by Prof. Sammy Welch. Published by Peter Peregrinus; ISBN 0 90 604 846 X. This book is now out of print, but it is well worth checking to see if your library has a copy.

# 9

# System X

## 9.1 Introduction

The preceding chapters have dealt with the principles of digital transmission and switching, stored program control and common channel signalling. This and the following chapter build on these principles as we describe public and private digital exchanges. This chapter deals with a typical public digital exchange by briefly describing the System X Network and, in more depth, one of the exchanges in the System X family. Chapter 10 deals with a typical digital private digital exchange.

### 9.1.1 What is System X?

System X is the name given to a family of exchanges and associated products developed by British Telecom and GPT to replace the majority of analogue exchanges in the PSTN. The first System X exchange entered service in London in 1980. Since then many exchanges have been commissioned, but with a total of 6,000 to be replaced it will be a few years before the every subscriber in the country is connected to a modern digital exchange.

British Telecom is, however, not the only customer for System X products, GPT has sold (or is about to sell) System X equipment to China, Columbia, Kenya, Gibraltar and the Falkland Islands. Mercury Communications and Kingston Communications have also purchased System X exchanges for their networks in the UK. An interesting point to note is that although System X was developed for use in public networks, it is also finding use in private systems such as that owned by the International Stock Exchange which plans the first System X private network to link the regional stock exchanges in Glasgow, Liverpool, Manchester, Birmingham and Dublin.

### 9.1.2 The System X Network

The System X Network is a hierarchical Integrated Digital Network (IDN) consisting of digital trunk and local exchanges connected by 2Mbit digital transmission links.

Operator assistance, such as directory enquiries, is provided by an Operator Service System (OSS) which can be parented on to a number of exchanges within the network. Management of the network is undertaken at Operations and Maintenance Units (OMUs) which are connected to the exchanges for which they are responsible by packet data links. Figure 9.1 shows a number of exchanges in an area with an OSS and an OMU.

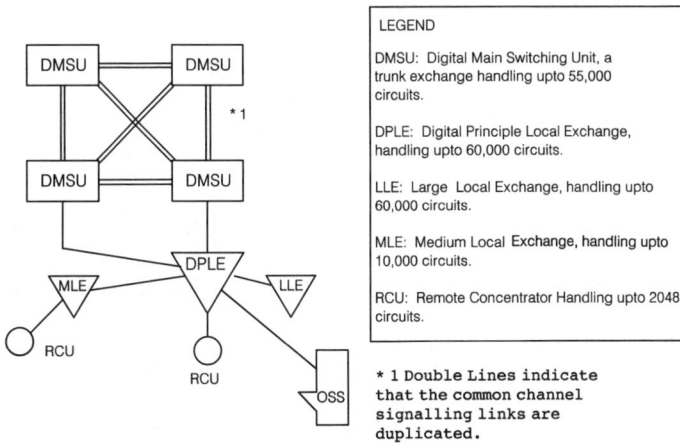

```
LEGEND

DMSU: Digital Main Switching Unit, a
trunk exchange handling upto 55,000
circuits.

DPLE: Digital Principle Local Exchange,
handling upto 60,000 circuits.

LLE: Large Local Exchange, handling upto
60,000 circuits.

MLE: Medium Local Exchange, handling upto
10,000 circuits.

RCU: Remote Concentrator Handling upto 2048
circuits.
```

* 1 Double Lines indicate
that the common channel
signalling links are
duplicated.

*Figure 9.1   A System X network*

The two basic types of exchange shown in Figure 9.1 are The Trunk Exchange or Digital Main Switching Unit (DMSU) and the Local Exchange. In many respects trunk exchanges and local exchanges are similar and are built from a common set of exchange modules. The few differences that do exist result mainly from the following points:

❏ A trunk exchange has no subscriber connections

❏ The network of trunk exchanges forms the top level of the network hierarchy.

There are three versions of local exchange, the largest being the Digital Principle Local Exchange (DPLE). There is also a Large Local Exchange (LLE) and a Medium Local Exchange (MLE), but plans for a Small Local Exchange were dropped some time ago. All three are built from the same modular design and there are no significant differences between them other than size.

## 9.1.3 The System X Trunk Exchange Network

By the end of July 1991 the System X Network consisted of around 60 DMSUs covering the whole of the United Kingdom. The remainder of the network consists of various local exchanges each of which is said to be parented on to a specific DMSU. Unlike the trunk network of the old PSTN which was based on a three level hierarchy, the System X trunk network has only one level.

In the diagram four DMSUs are shown, fully interconnected by a number of 2Mbit PCM systems. The actual number of PCM systems used to connect any two particular DMSUs will depend upon the amount of traffic between the two exchanges. The reason for showing these connections as double lines is to indicate that on every route the common signalling channel is at least duplicated for security reasons.

### 9.1.4 The System X Local Exchange Network.

Within the catchment area of a DMSU there will be a number of DPLEs which are directly connected to the area DMSU. In this way the DMSU is said to be the parent of these DPLEs, and that the DMSU is primary route for all trunk calls from any of them. For security reasons each DPLE also has a secondary connection to the DMSU in an adjacent area.

The smaller local exchanges are securely connected into the network by having links to a DPLE and a DMSU. In areas where there are insufficient customers to justify a whole digital exchange, service is provided by a Remote Concentrator Unit (RCU which can handle approximately 2,000 subscribers and can be connected to a local exchange by up to 16 30-channel PCM links. We will describe later how the RCU actually forms a remote part of the local exchange to which it is connected.

Two or more RCUs may be located in the same premises to provide more lines than would be available from a single RCU. Such a configuration is termed a Remote Concentrator Centre (RCC). Note that a concentrator can only be connected to one local exchange.

### 9.1.5 Operations and Maintenance

Day to day control and management of System X exchanges is carried out from a number of OMUs. Each DMSU will have its own OMU and its associated equipment, while a single OMU will probably serve all the local exchanges in its area. Not all OMUs are manned 24 hours a day, facilities are provided to enable all the indications and alarms to an OMU that is to be unmanned during silent hours to be diverted to a working OMU.

### 9.1.6 Operator Services System

This system will handle all calls to the operator for assistance and enquiries for a particular area. All calls for such services will be initially directed by the network to the appropriate OSS. Usually the operator will be able to give sufficient information for calls to be dialled by the subscribers, but where necessary the operator will be able to connect subscribers together using OSS.

## 9.2 The System X Local Exchange

### 9.2.1 Description of the Local Exchange Block Diagram

In this section we examine in some detail the architecture of the local exchange. Figure 9.2 shows the block diagram of the local exchange. We have already seen part of this diagram before in the chapter on Digital Control. Now we will be putting all the principles together to see how the System X exchange functions as a whole.

The exchange consists of a number of subsystems or modules. Within Figure 9.2, rectangles represent items of hardware, while clouds represent software functions, although in many cases the functions of a particular unit may be carried by a combination of hardware and software modules. It is also worth noting that many of the modules of the local exchange are also found in the trunk exchange. The remaining sections of this chapter describe the major subsystems of the exchange.

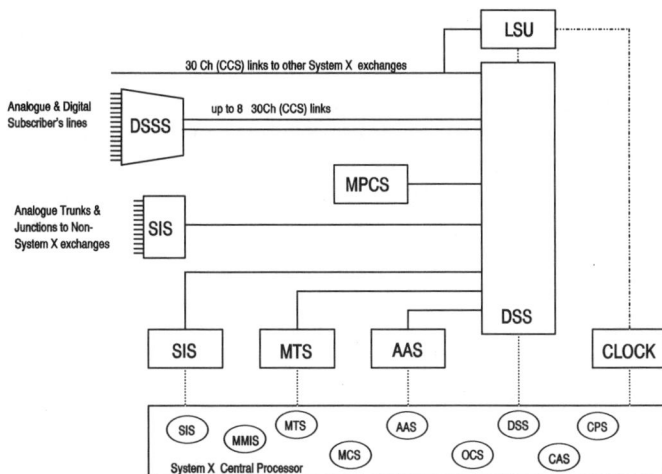

*Figure 9.2  The System X local exchange*

## 9.2.2 The Digital Subscribers' Switching Subsystem (DSSS)

The DSSS consists of hardware units and software functions. The software functions are resident in the Processor Utility in the Local Exchange. The hardware consists of a number of units which together are also known as a Subscribers' Concentrator. This may be located within the same building as the main exchange equipment or used as a Remote Concentrator, as described in Section 9.1.4, in which case it can be located some distance from the exchange. There will normally be several subscribers' concentrators in a local exchange, the exact number being dependent upon the number of subscribers being served by the exchange.

The role of the DSSS is to provide subscriber access to the exchange for a individual subscribers and PABXs. It would be unnecessarily expensive to permit every single subscriber to be able to place or receive a call at the same time, therefore access is limited so that a maximum of between 10 and 20 percent of all subscribers can be involved in calls at any one time.

## 9.2.3 The DSSS Block Diagram

The Subscribers' Concentrator is a processor controlled switching unit which can terminate up to 2048 analogue or digital subscribers' lines and carries out the following functions:

❑ Analogue to PCM Digital Conversion, and vice versa on analogue lines

❑ Detection of signalling from the subscriber (LD, DTMF or digital signalling is catered for)

❏ Supply of supervisory tones and ringing current

❏ Concentration of traffic from Subscribers' Lines on to PCM highways to the main exchange digital switch.

❏ Analogue or digital connection of PABXs, with Direct Dial In (DDI) if required.

The unit is connected to the Digital Switch by up to 16 2Mbit PCM systems and thus has a maximum capacity of 480 circuits available to the subscribers connected to it. Two of the of these 16 PCM Systems connecting the DSSS to the host local exchange will have TS16 reserved for use by the Message Transmission System (MTS) for Common Channel Signalling purposes. The TS16s of the remaining PCM systems are not used. Figure 9.3 shows a simplified block diagram of the Subscribers' Concentrator.

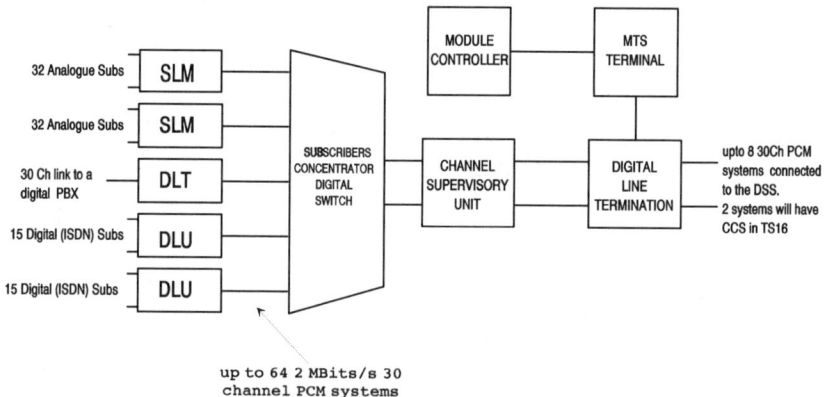

*Figure 9.3 Subscribers' Concentrator Block Diagram*

The Module Controller (MC) includes duplicated commercial 32-bit processors to control the operation of all modules of the concentrator, and communicates with the call processing software of the host exchange using CCITT #7 and the Message Transmission Subsystem terminal.

The 2,048 subscribers' lines are organised into groups of 32 on Subscribers' Line Modules (SLM) each of which is connected to a small dedicated digital time switch within the concentrator by a 32-channel 2Mbit highway.

The digital switch carries out the concentration function by being able to connect any of the 2,048 subscriber channels to any of the available 480 channels connecting the subscribers' concentrator to the host exchange digital switch. Once all the available channels to the host are in use further calls to or from the concentrator will be blocked.

This blocking problem is more likely to occur in business areas as commercial subscribers tend to make and receive more calls per hour than domestic customers. On a fully equipped concentrator placed in a commercial area such as a business park, during busy periods it is very possible for all channels to the local exchange to be in use at any one time and thus any new calls would be blocked. To reduce the risk of blocking occurring in business areas the concentrator is sub-equipped. In other words, not all possible 2,048 lines are actually equipped, thus providing a better grade of service for those lines which are connected.

## 9.2.4 DSSS Subscriber Line Interface

In any digital exchange it is necessary to provide a number of functions relating to the termination of analogue subscriber lines. Figure 9.4 shows how these functions are provided on a Subscribers' Line interface Circuit, one of which is provided for every line terminating in the SLM.

**Battery feed:** provides the necessary DC power feed to the subscribers' line to power the telephone (typically 25-40mA).

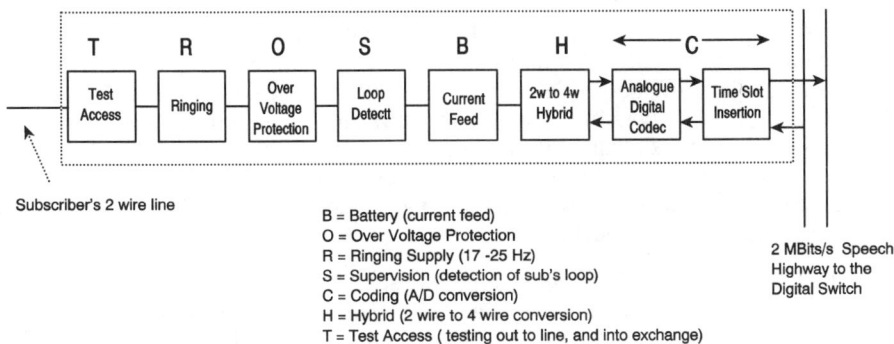

B = Battery (current feed)
O = Over Voltage Protection
R = Ringing Supply (17 -25 Hz)
S = Supervision (detection of sub's loop)
C = Coding (A/D conversion)
H = Hybrid (2 wire to 4 wire conversion)
T = Test Access ( testing out to line, and into exchange)

*Figure 9.4 The BORSCHT functions of the SLIC*

**Over voltage protection:** to protect the exchange equipment from any high voltages on the line due to mains contact or lightning. This is required even though there are protectors mounted on the Main Distribution Frame as these devices may take up to half a second to operate.

**Ringing:** the insertion of ringing current (75V at about 17Hz) to the subscribers' line.

**Supervision:** the detection of the DC loop when the subscriber lifts the handset even when this happens during ringing.

**Coding:** the conversion from analogue to 64 Kbit/s A law encoded PCM and subsequently inserting the PCM signal into the time slot assigned to that SLIC. This also includes the necessary decoding process for received PCM signals. (Strictly, these functions are carried out in the Subscribers' Line Audio Processing Circuit but can be included here for our purposes.)

**Hybrid:** Provision of a 2-wire to 4-wire interface to connect the subscribers' line to the PCM conversion circuitry. This interface should ideally have high attenuation between receive and transmit pathways to prevent received signals being retransmitted as unpleasant echo.

**Test:** Provision of access to the circuit to permit testing towards the exchange, or towards the subscribers telephone. The necessary switching is currently provided by miniature relays.

The initial letters of these seven functions form the acronym BORSCHT, which is now often used to describe the functions of the SLIC for analogue subscribers.

It is worth noting that many public exchanges also have the ability to provide a metering pulse to drive a subscriber's private call meter to enable the subscriber to estimate the cost of calls. To provide this facility it is necessary to inject an analogue signal, typically at either 50 Hz or 16KHz into the subscriber's line.

## 9.2.5 Signalling and Call Handling in the DSSS

Signalling from the subscribers' line, i.e. Loop and dial pulses, are detected by the Line Controller (LC). The LC informs the Module Controller (MC) that a particular subscriber has gone off hook. The module controller must then select a free output channel from the 480 possible channels and arrange for the concentrator switch to connect the subscriber's channel to the selected output channel. The MC informs the host Call Processing Subsystem (CPS) of the new call by sending a *seize* signalling message containing the subscriber's ID and the number of the allocated channel via the message transmission system.

Once the CPS has checked that the calling subscriber is permitted to make outgoing calls, it returns a *send digits* message back to the MC. Dial tone, which is generated digitally, is then sent to the caller's line in the correct TS from the Channel Supervisory Unit (CSU) via the concentrator switch under control of the MC.

The LC informs the MC of the digits dialled by each subscriber. The MC then formats this information into further signalling messages and passes these to the Message Transmission Subsystem Terminal for onward transmission to the host exchange CPS.

The MTS terminal transmits all signalling messages to the host exchange CPS by sending them in one of the two dedicated common signalling channels to the MTS Terminal at the host exchange. The MTS terminal then delivers the message to the host exchange CPS which then uses the information in the signalling messages to carry out the call set up requested.

If the calling line is identified as having DTMF dialling, the MC will allocate an MF receiver to receive the dialled digits, and arrange for the concentrator switch to connect the selected MF receiver to the subscriber's line. The MF Receiver then passes digit information directly to the module controller which itself uses the MTS to transmit the information to the host exchange CPS.

The concentrator is informed of any incoming calls by similar MTS messages sent to its module controller by the CPS. The messages will identify the required subscriber line and the channel on the switch to concentrator PCM highways which has been selected for the call.

Other tones which may then be required, e.g. ring tone to distant end subscribers and so on, are injected into the appropriate PCM channel from the CSU.

The DSSS can also terminate digital lines from individual subscribers and 30-channel PCM systems with common channel signalling from digital PABX. However as these facilities are provided as part of the Integrated Services Digital Network they will be discussed in a later chapter.

# 9.3 Digital Switching Subsystem (DSS)

## 9.3.1 The Role of the DSS

The DSS is found in both System X trunk and local exchanges. In the previous section we described how the Subscribers' Concentrator described switched analogue and digital subscribers' lines onto digital circuits which terminate on the DSS. For a local call the role of the DSS in the local exchange is to provide the required connection between subscribers' lines by connecting the appropriate digital circuits from the concentrator together. For a trunk call the DSS has to connect a subscribers' line to an outgoing, or incoming trunk circuit.

Within a trunk exchange the role of the DSS is to switch incoming trunks to outgoing trunks as part of a trunk call set up.

The DSS terminates a very large number of 2Mbit PCM systems, and switches traffic from any one incoming system to any outgoing system under software control. As well as switching traffic from other System X exchanges, the DSS will also switch local traffic from the subscribers' concentrator, traffic from non System X exchanges via the Signalling Interworking Subsystem and recorded messages from the Automatic Announcement Subsystem.

The DSS also provides a digital cross connection between TS16 of the PCM systems and a channel on a 2Mbit/s link to exchange MTS terminal. In this way signalling traffic received from the Subscribers' Concentrator and the MTS terminal of other System X exchanges is routed to the exchange's own MTS terminal.

In this section we will describe the DSS of the trunk exchange rather than the local exchange as the trunk DSS uses a fully implemented space switch which is not required on the smaller local exchanges. The few slight differences between the trunk DSS and that of the local exchange are not significant to the understanding of the local exchange as a whole.

## 9.3.2 The DSS Block Diagram

The DSS is a digital Time-Space-Time switch which can, in its largest form, terminate up to 3,072 30-speech channel PCM systems (92,160 channels), though in practice a maximum of only 2048 PCM systems (61,440 channels) are actually terminated. The DSS consists of the following units:

❑ Digital Line Termination (DLT). One DLT for every 2Mbit 30-channel system connected to the DSS.

❑ Time Switch. Each time switch can handle 32 2Mbit 30-channel systems, i.e. approximately 1,000 channels. Each time switch is divided into two halves: a receive half and a transmit half.

❑ Space Switch. The space switch is modular and can be configured to handle 32, 64 or 96 time switches.

❑ DSS Control Unit. This unit receives *allocate* instructions from the call processing subsystem and allocates the required time slot interchanges necessary to connect two PCM channels together.

❑ Other units for timing waveform generation, alarm processing, maintenance and so on.

A simplified block diagram of the DSS is shown in Figure 9.5.

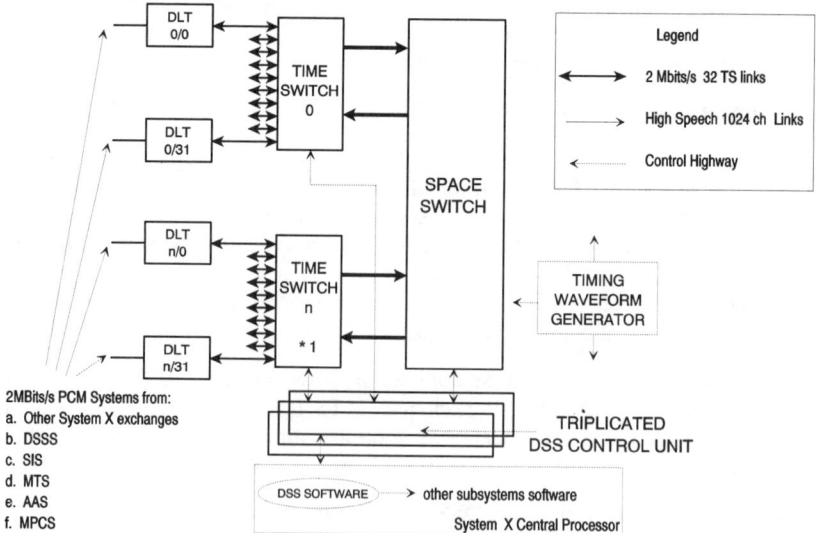

*Figure 9.5 Block Diagram of the digital switching subsystem*

## 9.3.3 Digital Line Terminations (DLT)

Each 2Mbit PCM system consists of two pathways: a receive path and a transmit path, both terminating on the DLT. On the receive side the DLT converts the incoming line signal from its HDB3 format to binary, while the reverse process is carried out on the transmit side.

The DLT monitors the received TS0 word to enable it to carry out frame alignment and generates the frame alignment word for the transmit side.

All the DLTs in the DSS are synchronised so that on every PCM system all TS1s are sent at exactly the same time, followed by all TS2s etc.

However on the receive side, the DSS will not receive all TS0s at the same time due to the different propagation delays between the various exchanges in the network. To present all incoming TS1s to the DSS at the same time the DLT carries out an aligning function.

Each incoming PCM signal is clocked into a shift register known as an aligner, using its own recovered clock signal. The PCM signals are then all clocked out of the shift register in alignment using a single clock signal derived from the exchange synchronisation circuitry.

## 9.3.4 Receive Time Switch (Rx TSW)

Each time switch handles 32 PCM systems, or 1,024 channels. Within the Rx TSW there is a 1,024-location receive speech store which must be written to and read from in a logical order. Figure 9.6 shows the receive half of the time switch.

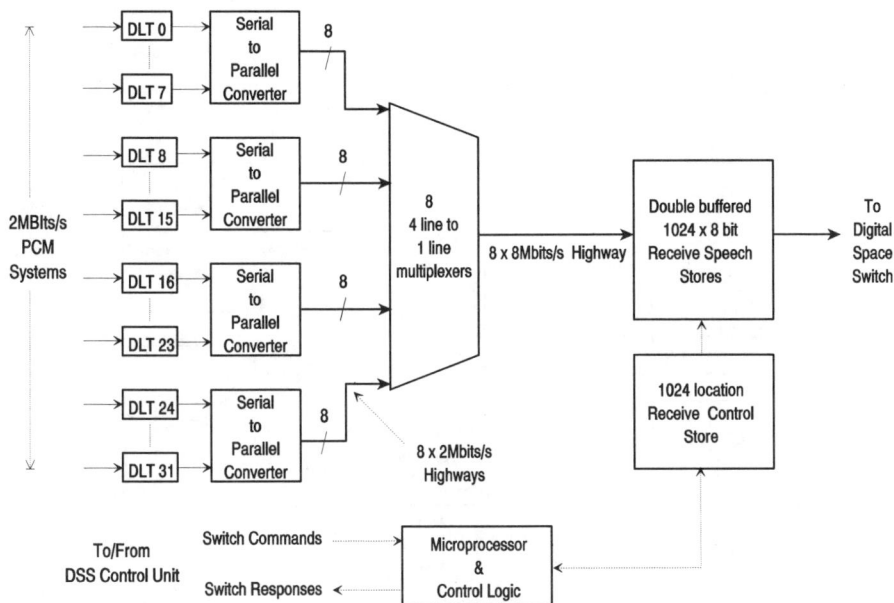

*Figure 9.6 Receive time switch*

Within the Rx TSW there are four DLT interfaces, each of which is connected to the receive side of eight DLTs and performs serial in to parallel out conversion. This conversion is organised so that on the eight wire connection between DLT Interface 1 and the 4 line to 1 line (4:1) multiplex the eight bits which arrived in TS1 on system

1 are presented first, followed by the eight bits of TS1 from system 2, followed by eight bits of TS1 system 3 etc. Similarly Between DLT Interface 2 and the 4:1 Multiplex there will be the eight bits of TS1 system 9, followed by TS1 of system 10 etc.

The 4:1 Multiplex then takes all 1,024 8-bit samples in each frame and delivers them to the correct location in the receive speech store. In fact the receive store is double buffered – in other words there are two identical speech stores. During one multiplex frame one of the stores is being filled, while the other is being read out via the space switch to the transmit halves of the time switches. During the next frame the roles are reversed: the store which was just filled will be read from, while the store which was just read from will be filled with new samples.

## 9.3.5 Transmit Time Switches (Tx TSW)

Figure 9.7 shows the transmit half of a time switch. The transmit time switch carries out the inverse functions of the Rx TSW. The 1024 location transmit speech store is filled from various Rx TSW locations via the space switch. The 8-bit samples from the transmit speech store are read out in PCM system and TS order and passed via the multiplex and DLT i/f to the correct DLT. The DLTs are synchronised so that TS1s for all 32 PCM systems are transmitted at the same time.

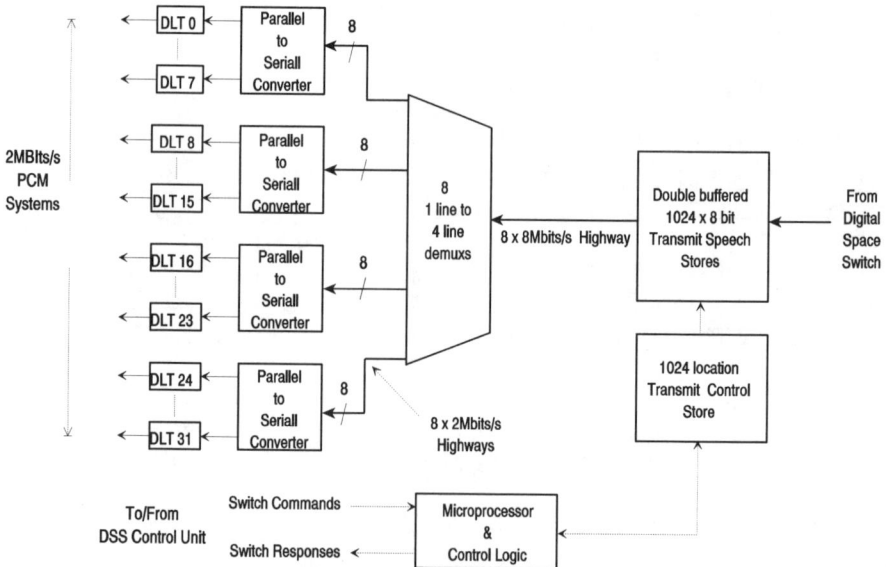

*Figure 9.7 Transmit time switch*

## 9.3.6 The Space Switch

The space switch is based on a digital matrix which can connect the outputs of up to 32 Rx TSWs to the inputs of up to 32 Tx TSWs. In Figure 9.8 only four time switches are shown for simplicity.

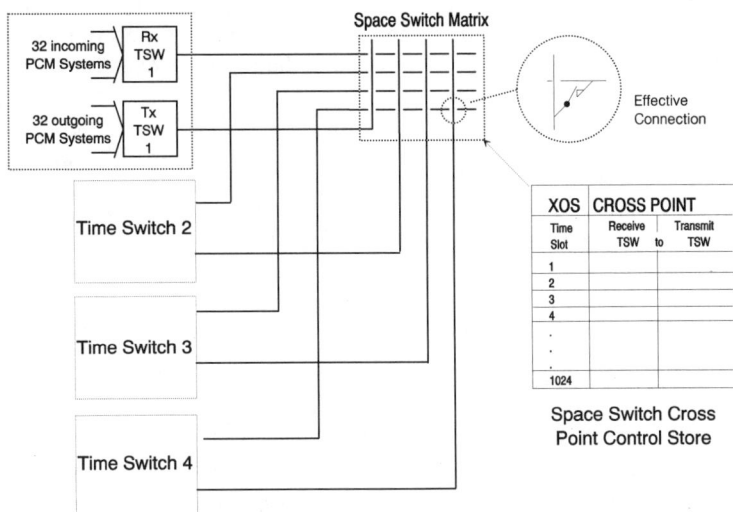

*Figure 9.8 The digital space switch*

Each Rx TSW has a double speech store with 1,024 locations. Each speech store is filled in just one multiplex frame time (125 microseconds), then all 8-bit samples stored must be transferred to the correct Tx TSW speech store locations in an equal time frame as the other receive speech store is filled. The name given to the period of time allocated for each sample to cross the space switch matrix from Rx speech store to Tx speech store is called a Cross Office Slot (XOS). As there must be 1024 XOSs during each 125 microsecond period, a XOS is approximately 0.12 microseconds in duration.

As there are 32 time switches connected to the space switch, during each 0.12 microsecond XOS period, 32 crosspoints in the space switch are operated. Each cross point will connect the output of one receive speech store to the input of one transmit speech store. The actual sample read out from the Rx speech store will be determined by the contents of the Rx TSW control store. This will identify a sample by its PCM system number and its TS number. The identified sample will be read from the speech store and placed on the input to the space switch. The sample then travels across the crosspoint which has been operated and is then stored in the location in the Tx speech store which corresponds to this XOS.

# 9.4 Other Local Exchange Subsystems

## 9.4.1 Signalling Interworking Subsystem (SIS)

When System X exchanges are connected into a network which still includes analogue exchanges there is a requirement for an interface between the various different types of analogue exchange and the System X exchange. In particular analogue speech on the incoming lines must be converted into 30-channel PCM format as this is the only type of system that can be connected to the DSS. There is also a requirement to convert the trunk and junction signalling used in these analogue exchanges to a message format suitable for use in the digital exchange.

This interface is provided by the Signalling Interworking Subsystem (SIS). The SIS consists of six major units as shown in Figure 9.9:

❏ TS16A Unit

❏ DC Analogue Signalling Line Terminations (DC ASLT)

❏ AC Analogue Signalling Line Terminations (AC ASLT)

❏ Voice Frequency Signalling Unit (1VF)

❏ Multi-frequency Signalling Unit (MF Unit)

❏ SIS Application Process Software resident on the Central Processor.

The TS16A unit is connected to the DSS by a 32TS 2Mbit link. TS0 is used for framing, while the remaining 31 TS contain TS16 signalling from various signalling converters within the SIS. Within the DSS semi permanent connections are set up between TS16 of each signalling converter link and one of the TS on the 31 TS link to the TS16A unit. Thus the TS16A unit receives 31 signalling channels carrying the signalling for 930 circuits.

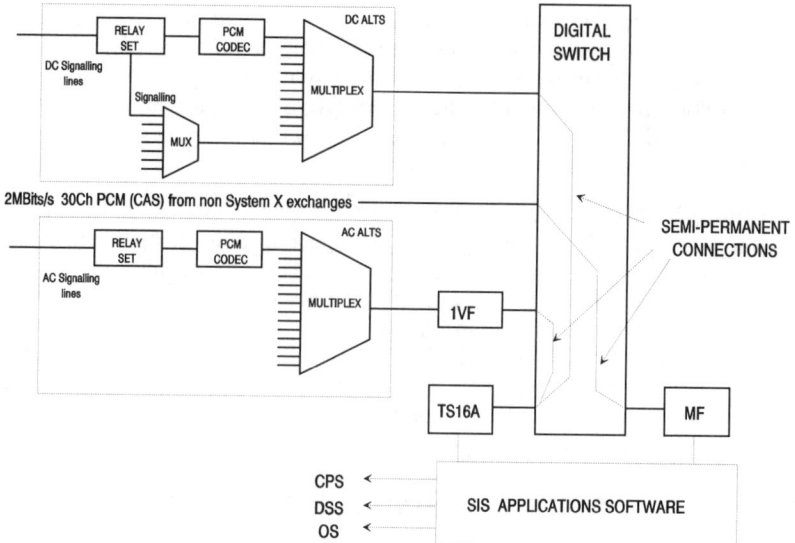

*Figure 9.9 The Signalling Interworking Subsystem block diagram*

DC signalling such as loop disconnect pulses is catered for on a DC ASLT. The Signalling Converter Circuit (SCC) extracts the signalling information from the line and converts it to a suitable 4-bit binary code which is then inserted into TS16. The necessary conversion from analogue speech to PCM speech is carried out in a Codec. Finally a multiplex formats all 30 speech channels and the TS16 signalling channel into a standard 2Mbit system which is connected to the DSS. A semi permanent connection across the DSS connects the TS16 to one to the channels into the TS16A unit.

The TS16A unit validates all incoming signals and checks pulse timing, then formats the signalling information into messages which are passed directly to the SIS software which resides on the central processor utility. The meaning of these messages are decoded by the software and where appropriate other software subsystems, such as call processing, are informed.

AC signalling is similarly handled by the TS16A unit. However a major difference is the method by which the signalling is extracted from the line and inserted into TS16. The AC ASLT does not carry out signalling extraction. VF signalling is encoded as for speech by the PCM codec, but a 1VF unit is placed in the 2Mbit pathway between the AC ASLT and the DSS. The role of the 1VF unit is to examine every speech channel to determine whether or not VF signalling is present. If signalling is present this is extracted and converted to a binary code representing tone on/tone off and inserted into TS16.

Again a semi permanent path across the DSS results in the AC ASLT TS16 being connected into one of the channels into a TS16A unit which will then decode the signalling and present the information in message format to the SIS application software in the processor utility as for DC signalling.

When MF signalling is used an MF Unit is used in place of the TS16A unit to produce the necessary messages to the SIS software. Systems which use MF inter register signalling are converted to 30 channel PCM to terminate them on the DSS. When a seize involving MF signalling is indicated the SIS instructs the DSS to set up a connection between the relevant channel and one of 30 available digital MF receiver/senders in the MF Unit. The MF receiver will detect and validate the MF inter-register signals and format the required information into messages for the SIS application software. Note that the TS16A unit is not used in this case as the MF unit reports directly to the SIS software handler in the processor utility.

In cases where the non-System X exchange is connected by 30-channel PCM (CAS) systems these are terminated directly on the DSS. Semi permanent paths connect the TS16s of such systems to the TS16A which carries out signal validation and message formatting as previously discussed.

As well as converting from incoming signalling in its various forms, all interfaces in the SIS must of course be capable of generating the correct type of signalling in the outgoing direction (i.e. towards the non-System X exchange). Outgoing signalling is generated by the unit concerned on receipt of suitable messages from the SIS application software in TS16.

Another feature of the SIS so far not mentioned is the requirement for test access into the circuits. This is provided by the Automatic Break and Test Access system. The SIS application software can generate the necessary commands to initiate and conduct 2-wire or 4-wire tests of both the line and the exchange equipment.

## 9.4.2 Multi Party Connection Subsystem (MPCS)

The MPCS provides the necessary digital conference bridge to permit multi party calls of up to 10 users to be set up under an individual subscriber's control.

A digital conference involves receiving digital signals from all the parties to such a call, adding the signals to form a composite digital signal containing all the signals and then transmitting the composite signal to all parties to the call.

## 9.4.3 Automatic Announcements Subsystem (AAS)

The AAS system is used in place of tones and recorded announcements where the information that is conveyed is variable in nature. For example each alarm call and advice of call charge and duration requires different information to be passed to the user.

The AAS is connected to the digital switch by 2Mbit 30Ch links. The AAS contains digitally encoded voice messages which are transmitted to subscribers under certain circumstances. These announcements, called voice guidance, range from simple information such as: *"All lines to Bristol are engaged, please try later"*, to instructions on the use of some of the special services. For example during the set up of a reminder call the AAS will send the message: *"Reminder Service, dial the time you want your reminder, followed by square"*. When the subscriber has complied with this instruction, AAS sends the message: *"You have booked a call for 0700 hours"*.

AAS messages are transmitted at the request of the CPS, which arranges for a free channel on the 30ch AAS link to the digital switch to be connected to the subscriber channel concerned. CPS then instructs the AAS to send the relevant message on the selected channel.

To reduce the amount of digital storage required for the large number of messages involved, phrases from which messages are composed are encoded using ADPCM (rather than A law PCM) and stored in PROM. AAS selects the required phrases from the PROM and concatenates them to produce the required message.

## 9.4.4 Message Transmission Subsystem (MTS)

The MTS is the terminal part of the common channel signalling system. The MTS is connected by a number of 2MBit/s links to the digital switch. Each of these links contains 32TS of which TS0 is used for framing, while the remaining 31 are used for the reception and transmission of common channel signalling messages to other exchanges and concentrators.

These other System X exchanges, and the local subscribers' concentrators, will be connected to the digital switch by a large number of 2MBit PCM systems. Signalling information from these sources will be received in TS16 of several of these links, and semi permanent connections in the digital switch connect the various TS16 signalling channels to allocated channels in the 2MBit links to the MTS.

As described in Chapter 8, the MTS system forms part of the CCITT Signalling System No 7 which is primarily used by the CPS to set up calls. However other subsystems are able to use the message transmission facility to pass non call related information to other exchanges.

## 9.4.5 Software Subsystems

The architecture of the System X exchange includes some software only systems. Details of several of these subsystems were given in Chapter 7 in the treatment of digital exchange control, however for completeness of this chapter brief details are also given here.

## 9.4.6 Call Processing Subsystem

This software only subsystem controls the setting up of all calls. When processing calls which originate or terminate at other exchanges CPS must use the common channel signalling system to communicate the necessary information regarding dialled digits and service requested to the CPS at the other exchange.

CPS is not an independent subsystem and will either receive or pass call set up information to several of the other exchange subsystems, particularly DSSS and DSS as these are of course heavily involved in the actual switching of connections required to establish the call. It is also necessary for CPS to pass information to the call accounting and management statistics subsystems.

## 9.4.7 Call Accounting Subsystem

This software-only subsystem receives call details at the start and finish of a call to enable a call charge to be calculated and recorded against the calling subscriber's account. When itemised billing is requested, the charge for the call is calculated as normal and added to the subscriber's record of charge units, but additionally call details such as time and duration of call and number dialled are recorded for further processing by the accounting system.

The Call Accounting subsystem also includes the necessary facilities to permit the network administration to alter the charging tariffs when required.

## 9.4.8 Management Statistics Subsystem (MSS)

The MSS receives, collects and sorts data from other subsystems to produce statistical information required to manage the network effectively. The usual source of this information is the CPS when its attempts a call set up, and typically the information passed to the MSS includes:

❏ Type of call: standard telephony or data call, local or trunk

❏ Identity of the calling party

❏ The route or service requested by the caller

❏ Whether call attempt was successful and if not the reason (engaged, NU, call barred and so on)

MSS can process this and information from other sources to produce statistics such as number of calls, traffic route densities, occupancy of MF receivers and other exchange equipment for use by the network management in making decisions regarding any changes that may be required to the network.

### 9.4.9 Overload Control Subsystem (OCS)

This software system continually monitors the load on the exchange central processor utility and when necessary delays or sheds work of a non urgent nature to ensure that during times of excessive demand for new calls, the exchange can continue to set up calls at the maximum possible rate up without crashing due to overload.

### 9.4.10 Man Machine Interface Subsystem (MMIS)

Very simply, this software subsystem provides facilities to enable changes to the exchange database to be made, and also provides the necessary command interpreter to enable the exchange staff to interrogate exchange equipment, in order to diagnose faults and where necessary to take equipment in and out of service.

### 9.4.11 Maintenance Control Subsystem (MCS)

This software system provides facilities for the co-ordination of handling system faults and the operation of diagnostic and routine maintenance procedures and allows the exchange databases to be altered by the exchange staff via the MMIS when necessary.

The MCS keeps a software record of all faults detected, and can instruct various other subsystems to take equipment out of service automatically. Subsequent actions taken to rectify the fault are recorded against the original fault report to keep track of the current status of the system.

The word *control* included in the title of this subsystem is important. The nature of the System X exchange makes it extremely complex to diagnose faults unless the correct procedures are adopted. MCS contains procedures which prevent maintenance staff carrying out incorrect or inappropriate procedures.

## 9.5 Synchronisation

### 9.5.1 Network Synchronisation and the Local Synchronisation Utility

It is essential that all digital exchanges in a network are operated at the same average clock frequency. Each exchange has its own master clock. Without some form of control the clocks in every exchange will run freely, and slight differences in frequency and phase will occur between the PCM signals generated by an exchange, and those received from other exchanges. These frequency differences would cause bit slip to occur, which would result in a loss of some information (see Chapter 5).

Some loss due to bit slip can be tolerated by voice users, where the loss will be noticed as slight noise. However, when circuits used for data, common channel signalling or facsimile are involved, any information loss will cause errors which can not be tolerated, e.g. loss of a whole page in facsimile transmission which does not include error control. We have already noted that the telephone network will be used to carry significant amounts of data. It is therefore vital that the whole network is synchronised and that all exchange clocks run at exactly the same average frequency.

Digital networks may be operated in either plesiochronous or synchronous modes. In the former case the exchange clocks are operated entirely independently of each other, but each is maintained within very tight frequency limits. This involves very expensive clocks at each exchange and so is rarely found in practice. However a typical use of a plesiochronous link is the international connection between two national networks, each network being synchronised to its own master clock.

In the second case the frequency of all exchange clocks is controlled, or synchronised, so that all clocks run at the same average frequency and that deviations from this mean frequency are kept within strict limits.

## 9.5.2 Network Synchronisation Schemes

Network synchronisation may be despotic in which case a single master reference clock is used to control directly or indirectly every other clock in the network in a hierarchical master slave arrangement, in which a clock at any given level in the network is slaved to a master clock in the level above.

Another method of achieving synchronisation is to use a mutual strategy involving a synchronisation network connecting all exchanges. Each exchange clock is locked to the average rate of all incoming synchronisation links.

## 9.5.3 Synchronisation of the UK Network

Within the UK network a mutual synchronisation strategy based on a 4-level synchronisation hierarchy is used as shown in Figure 9.10. The top level of this hierarchy is a triplicated caesium reference clock. This is connected by synchronisation links to a number of digital trunk exchanges which form the second level of the hierarchy. This level also contains the standby reference clock.

The remaining exchanges in the trunk network form the third level of the synchronisation network and are controlled by those at level two. The fourth level of the hierarchy contains the local exchanges which are controlled by the trunk exchanges to which they are connected. Within levels 2 and 3 there are also bilateral synchronisation links between exchanges in the same level.

The synchronisation links are actually selected 2Mbit PCM links connecting exchanges. Exchanges at levels 2 and 3 monitor the incoming PCM link from exchanges at the same or lower levels and send synchronisation control signals back (eg. *speed up, slow down*) in TS0 of the return PCM link. These control signals are received by the exchange which compares these signals with others received from other exchanges and using a majority vote decision will take the necessary action to adjust the internal clock as required.

The Local Synchronisation Utility (LSU) ensures that the clock in the System X local exchange remains synchronised with the trunk exchanges to which it is connected.

The corresponding hardware in the System X trunk exchange is called the Network Synchronisation Subsystem (NSS).

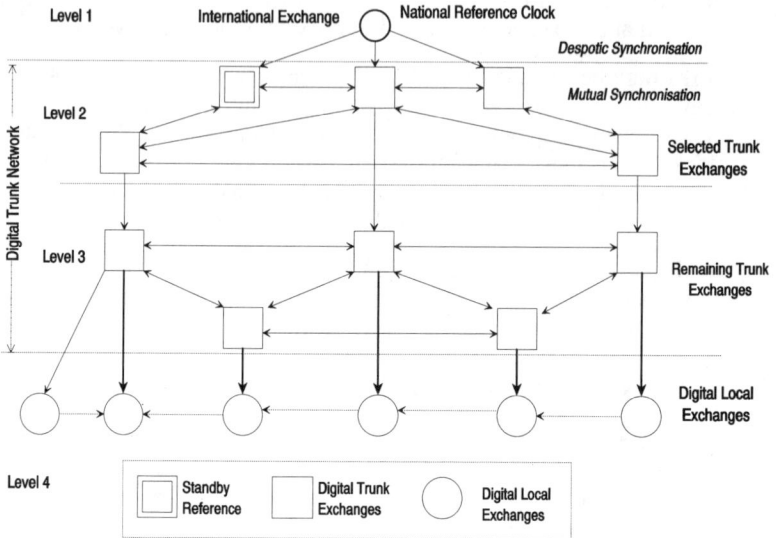

*Figure 9.10 Mutual Synchronisation in the UK Network*

# 9.6 System X Facilities

Today's public digital exchanges provide their subscribers with a wide range of special telephony facilities that were previously available only to users of SPC PABXs. The range of services offered by System X is typical of facilities on other manufacturers' equipment. These System X facilities are known as the star services, probably because they are accessed by dialling the star symbol on the telephone key pad.

It is important to note that although the subscribers' concentrator in the System X exchange can handle normal calls from the old style LD telephones, only MF instruments or the newer digital telephones with the Star and Square buttons will be able to make use of the special facilities.

Because of the nature of SPC it will be quite easy to add new services to those initially offered, to meet subscribers' requirements in the future. The following descriptions of facilities have been taken from GPT literature and should not be taken as meaning that all facilities are available on all BT System X exchanges.

It should also be pointed out that although these type of facilities are available on most manufacturers' exchanges, the names given to them are often different, and there seems to be little standardisation between manufacturers regarding the special facility dialling codes which are used to invoke these services.

## 9.6.1 Abbreviated Dialling and Repeat Last Number

This facility is already currently available on several electronic telephones without using any exchange facilities. Normally up to 10 numbers can be stored by the user in memory devices inside the telephone. However, System X subscribers may store up to 100 frequently used numbers in the exchange database and assign a short dialling code to each stored number. Stored numbers are recalled and dialled using the single or two digit dialling code assigned to each number. This will be particularly useful for longer numbers such as international numbers, or the case where the subscriber often calls the same extension on a PABX using Direct Dial Inward (DDI).

A slight variation on this theme allows the subscriber to repeat the last number dialled irrespective of whether a short code has been programmed for the number or not.

## 9.6.2 Call Diversion

This facility will allow the subscriber to instruct his local exchange to divert all incoming calls to another local number. This permits users who are expecting calls while they are away from their normal telephone to accept these calls on another user's telephone.

Other call diversions can also be programmed by the user. For example in small offices having only a few lines and no PABX, the "Divert on Busy" facility can be used to instruct the local exchange to divert an incoming call to another office number when the incoming call encounters an engaged line. A similar facility exists for Divert on No Reply.

## 9.6.3 Call Restrictions

This facility allows a subscriber to instruct his local exchange to set up only certain types of call. Calls may be restricted to any of, local numbers only, national numbers only or international numbers only, or any combination of the three. The facility is invoked using a special code known only to the subscriber so the facility may be used as and when required. Typically this facility can be used to prevent the telephone being used by children when the parents are away, or to prevent employees from making unofficial private calls.

It is also possible for the subscriber to bar incoming calls should he not wish to be disturbed for a period of time. Emergency (999) calls can still be made irrespective of any call restrictions in operation.

## 9.6.4 Conference Calls

This facility permits subscribers to set up multiparty calls. It can be used to place a current call on hold while a second call is established. The subscriber can then switch between the new and held call, or set up a conference in which all parties can hear and talk to each other.

## 9.6.5 Advice of Call Duration and Charge

This facility will automatically provide the user with the cost of a call and is invoked either before or during the call. When the call is complete the exchange rings the caller back and using the automatic announcement subsystem will give its duration and charge. It is also possible to request itemised billing, in which case details of certain types of calls are recorded and printed on the regular telephone bill for checking by the customer.

## 9.6.6 Call Waiting

This facility provides users with a tone indicating that a second call has been made to the number while it is already engaged on a previous call. The caller is given an automatic voice announcement rather than busy tone. The called user can opt to accept the new call by putting the first call on hold, or he may choose to ignore the second call. In the second case the caller is given a suitable voice announcement and then busy tone.

## 9.6.7 Ring Backs

Two types of ring back are provided. Probably the most used of all System X star services will be the *Ring Back When Free* facility. This allows users who dial a number that is engaged to instruct the System X network to ring back when the engaged line becomes free.

The facility is invoked by dialling a special code when the busy tone is heard. The user then replaces his handset and can carry on with other work. When both the calling and the called number become free, System X rings the caller first, when the caller answers the exchange will ring the called number again. This will obviously save subscribers' time wasted on repeated call attempts.

A similar facility exists to provide a ring back in the event of no reply. In this case the called number is not answered, and the caller continues to hear ring tone. The caller can then invoke the *Call Back When Next Used* facility to cause a ring back to the caller at the end of the next call made on the called user's telephone.

## 9.6.8 Hotline

The System X exchange will automatically route to a predetermined number all calls made from telephones allocated the Hotline facility (also known as fixed destination calling). This facility will typically be used by taxi services who wish to place courtesy telephones in places such as railway stations and supermarkets. Customers simply pick up the phone and press a single button to be connected to the taxi firm. Normal calls to other numbers can not be made from *hotlined* telephones

## 9.6.9 Disabled Subscriber Service – Timed Hotline

This facility can be allocated to subscribers with disabilities which may prevent them from dialling. When a subscriber allocated this facility lifts the handset a timer within the exchange is started. If the subscriber fails to dial before this timer matures it is assumed that he or she is in distress and is automatically connected to a preset number.

### 9.6.10 Other Subscriber Facilities

The description of facilities given in this section has concentrated on those available to analogue subscribers with MF telephones. Other facilities are available to subscribers with the new digital telephones and to PABX users. These will be described in the chapters on PABX and ISDN.

## 9.7 Chapter Summary

This chapter has looked in some detail at the System X network, the facilities offered to its subscribers and the System X Local exchange. While System X is produced by one UK manufacturer, GPT, many of the concepts described have also been adopted by other manufacturers and so the reader studying other networks or equipment should find the underlying principles relevant.

The System X local exchange is based on a modular design in which standard hardware and software modules can be cost effectively built up to provide the amount of subscriber terminations, switching capacity and processing power required in an exchange which is sized for its own unique operating environment. Generally the same modules are used within the System X Trunk exchange, the main exceptions being that the trunk exchange does not have subscribers' concentrators, and tones and announcements are supplied by the Signalling Interworking Subsystem.

While System X was originally developed for the public switched telephone network in the United Kingdom, the manufacturers also market System X for private and public systems around the world.

It is also essential that exchanges from various suppliers can operate together in a network. British Telecom also uses the Ericsson AXE 10 local digital exchange in its network. These exchanges provide the same features as the System X local exchanges and are connected to the network by the same 30-channel PCM links with CCITT #7 common channel signalling.

## 9.8 Further Reading

Several articles on various aspects, from basic principles to testing CCITT #7 links have appeared in the *BT Engineering (formerly the Post Office Engineering) Journal, POEJ* since 1979. Literature on System X, its facilities and subsystems is also available from the manufacturers GPT at their sales offices in Edge Lane, Liverpool.

Penguin also publish, as a paperback, the GPT System X Edition of the *Dictionary of Telecommunications*.

## 9.9 Acknowledgement

The author is indebted to GPT for their assistance and permission to reproduce material and diagrams in this book.

# 10

# Private Digital Telephone Exchange Systems

## 10.1 Introduction to Private Systems

The aim of this chapter is to familiarise the reader with the concepts of private telephone exchanges and key systems. The chapter builds on concepts and information presented earlier in the book and contains some detailed technical information describing the operation of basic functional units of a PBX. This chapter concentrates on the following areas:

❑ PBX User Facilities

❑ Key Systems

❑ PBX Architecture

❑ Major PBX Components

### 10.1.1 The Background to Private Systems

Several years ago two types of private exchange were in service in various types of organisation. Private Manual Branch Exchanges (PMBX), the traditional switchboards, were usually small systems in which all switching was done manually by an operator who plugged in cords, or operated switches to connect callers. Larger organisations were able to afford Private Automatic Branch Exchanges (PABX) which tended to use Strowger, Crossbar or Reed Relay technology. Automatic exchanges tended to have more extensions than manually operated systems, although some very large manual systems were in operation.

Most private exchanges branch from the public telephone network. In other words the organisation rents lines from the local public exchange, and instead of connecting telephones directly to these lines, the exchange lines are connected to the organisation's private telephone exchange.

By definition, a PABX is a telephone exchange, normally owned or leased by a private company to provide internal telecommunications and economical access to the public telephone network. Before the introduction of digital technology, many PABXs were standalone systems providing communications on a single site only, with access to other sites in the organisation via the PSTN. Although some private telephone networks were in existence at this time, these networks were obviously based on analogue exchanges, and employed a variety of inter-exchange channel associated signalling systems.

Not all modern exchanges employ digital switching as it is still more economical to employ analogue switching for small systems. However all use some form of processor control and so it is assumed nowadays that all private exchanges are automatic, and that the single term PBX can be used.

Although the greatest use of the PBX is for voice comms, today's PBX is equally at home handling other services such as circuit switched and even packet switched data. Since the introduction of digital exchanges and more especially common channel signalling systems such as DPNSS, the requirement for private networks has risen dramatically. Such networks offer many facilities for voice and data comms for organisations operating on several sites, even when these sites are in different countries of the world.

## 10.1.2 User Facilities of a Typical PBX

The number of facilities or features provided on a PBX is normally greater than that of a public exchange. This is because when there is a community of interest, such as there is with the users of a PBX, there is a requirement for facilities which would either not exist, or be impractical to implement on a public system.

Table 10.1 lists, in two sections, the facilities that you would expect to find on a modern PBX. Section 1 includes the PBX facilities which are the same as, or similar to, those normally provided on digital public exchanges such System X and the AXE 10, while Section 2 includes some additional features which are normally only found on PBXs.

One of the main drawbacks of many extra facilities is a reluctance on the part of the user to make use of them. Often this reluctance is based on a lack of understanding of the function of the facilities, or lack of knowledge of how to use the facilities in question. Although it is standard practice to provide a small booklet or instruction card with each extension telephone, anything which makes it easier for users, to understand and use these new features should also be considered an advantage. Many PBX manufacturers are now providing further facilities, that may only be used by extensions equipped with a proprietary featurephone normally provided as part of the installation. As these featurephones often include visual displays to assist in the use of the system, they represent a significant move to reduce users' reluctance to make the most of their sophisticated telephone system.

*Table 10.1  List of typical PBX user facilities*

| Section 1 | Section 2 |
| --- | --- |
| Call diversion | Enquiry call |
| Ring back when free | Call transfer |
| Ring back when next used | Call park |
| Conference call | Wait on busy |
| Repeat last number dialled | Night calls |
| Abbreviated dialling | Call intrusion |
| Alarm call | |

As the facilities listed in Section 1 were described in the chapter on System X, only those in Section 2 will be described here.

## Enquiry Call

The enquiry call facility allows users to put an existing call on hold, and make a second call, the enquiry, to another extension, and then return to the first call. At no time are the two calls connected together, so that there is no possibility of the original caller overhearing the enquiry call and vice versa.

## Call Transfer

The call transfer facility is an extension of the enquiry call. It allows an existing call in to one extension of the PBX to be transferred to another. On most systems the facility can be used equally well on outgoing as well as incoming calls, although generally it will be incoming calls which will be transferred to another user who can better deal with the call. In most implementations the incoming call is placed on hold, while the receiving extension dials the number to which the call is to be transferred. If the call is to be accepted, the transferring extension simply replaces his handset, which signals the exchange to connect the incoming call to the new extension. However, if the required number is engaged, or can not accept the transferred call, the extension originally receiving the call can take the call off hold in order to inform the caller of the situation.

## Call Park

Call park permits a user to place or receive a call on one telephone, put the call on hold, replace the handset and then re-answer the call from another extension. This is particularly useful if the user has to go to another location in the building, such as a library or filing room, to get some information pertinent to the call.

## Wait-on-Busy

Normally once the engaged tone has been received the exchange will not connect a caller to a busy extension even if that extension subsequently becomes free before the caller has replaced the handset. The wait on busy facility allows the caller to wait for the busy extension to become free. The caller does not replace his handset but simply waits to be connected. In some cases the exchange will also send a call waiting tone to the busy called extension. When the called extension becomes free, the exchange will immediately ring the now free extension. When the ring is answered, the new call will be automatically connected through.

Different manufacturers use different terms for this facility, the two most common being Wait-on-busy and Camp-on.

## Call Intrusion

The call intrusion facility will normally only be allocated to high priority users such as department managers to allow them to intrude on an existing call when they receive engaged tone having dialled the required extension. This facility would be used instead of wait-on-busy if the caller decided his call was sufficiently important to interrupt the existing one.

Where this facility is implemented there is usually a scheme in which users are allocated a level of priority by the system management. The exchange will then only allow a user to interrupt calls from or to users of a lower priority. This facility is found to be very useful in military systems where the requirement to pass important information or orders often dictates intruding on existing calls.

This list is by no means exhaustive. Some facilities not listed here will be covered later in the chapter, where their inclusion becomes more relevant.

The ability of the exchange to handle various types of equipment other than basic telephones is a feature which is not directly visible to the telephone user, but nevertheless should be available on even a small PBX.

Figure 10.1 shows a small PBX, providing normal telephone service, with other facilities such as:

❑ Music on hold

❑ Connections to a public address system to enable individuals to be paged by the operator

❑ Data communications using modems

❑ Two types of signalling (AC15 and E&M) over leased lines to other PBXs

❑ Connection to the public network exchange lines.

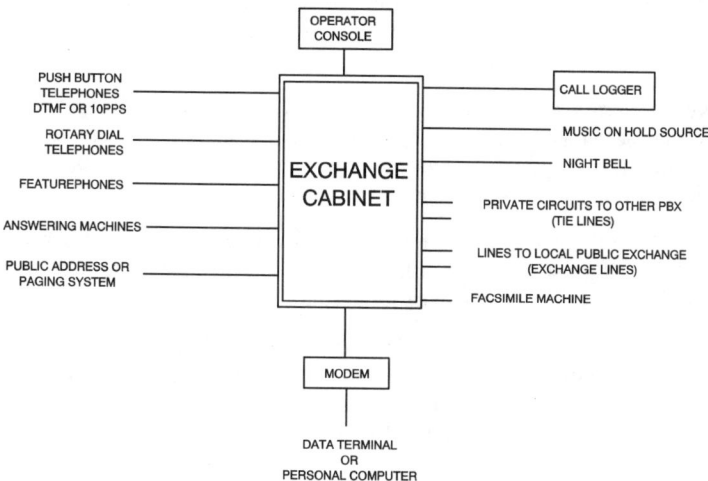

*Figure 10.1   Facilities of a small PBX*

Some necessary basic system management functions are also provided. These include an operator console, which is also used for making changes to the system database, and an RS 232 port which may be connected to some form of recorder or printer to record call logging information.

This typical small PBX can cater for up to 80 extensions and 32 trunk lines. The 32 trunk lines give access to 24 exchange lines to a local public exchange, and eight tie lines, or private wires to other PBXs.

Before going on to look at the features of medium to large PBXs, it would be advisable to consider the alternative to a small PBX, that is the key system. Many PBX manufacturers are offering the key system method of providing a basic telephone service to small businesses, or small departments within a large organisation

## 10.1.3 What is a Key System?

Today's key systems derive their name from an earlier system known as the Key and Lamp unit. This was a very small manually operated switchboard with between 5 and 10 extensions, and typically one or two lines to the public network. Simple key switches enabled a receptionist to connect incoming calls to any of the unit's extensions, and vice versa. A key system is a modern automatic version of this system.

There are two main distinctions between a key system and a PBX. The first is one of capacity, or size. A typical PBX will handle between 50 and 2,500 extensions, while a key system terminates far fewer, typically between 2 and 48.

The second main distinguishing factor is that a key system will often be supplied with proprietary telephones These will have a normal dial keypad but will also include a set of pushbuttons (the keys of the key system) to enable the user to call any other extension by pressing one key. An alternative solution which allows the use of standard telephone instruments is the provision of a key module through which the standard telephone is connected to the key system.

A typical key system will also be able to terminate between 1 and 16 exchange lines. In the very simplest case one exchange line can be shared between a manager and secretary, while the larger key systems can provide access to up to 16 public exchange lines for any extension.

Generally speaking there are not as many user facilities available on a key system as on a PBX, although most of the popular facilities are provided and manufacturers are tending to increase the functionality of their smaller systems.

Two key systems, The Philips Sopho K1 and the Ferranti Rhapsody, will be described to give some appreciation of what a key system is, what it is capable of, and some typical applications. These descriptions should not be taken as a comparison as these key systems are designed for entirely different applications.

## 10.1.4 The Philips Sopho K1 Key System

This system is designed for small offices in which one public exchange line must be shared between two or three telephones, not necessarily located in the same room. One method of providing sharing such as this would be to use simple parallel extension telephones, such as would be provided in a domestic system. Although this is the simplest way of addressing the problem, there are several disadvantages with this approach.

❑ All telephones are permanently connected to the exchange line and thus calls of a confidential nature can not be made without the risk of being overheard.

❑ An incoming call answered by one extension can not easily be transferred to another, as the second extension must receive the call before the first puts down the handset.

❑ No inter-extension communications are provided.

A key system such as the Philips Sopho K1 overcomes these problems and provides a few additional features.

The diagram in Figure 10.2 on the next page shows a basic Philips Sopho K1 system. At the heart of the system is a mains powered wall mounted Central Control Unit (CCU). The CCU is connected to the exchange line and up to three telephones, and carries out all control and switching functions.

As shown, at least one telephone must be connected via a key module. This key module would be located on the desk of the person who normally answers incoming calls and with the CCU provides:

❑ Access to the public telephone system, with indication that the line is in use

❑ Intercomm between extensions

❑ The means of transferring incoming calls to other extensions.

The buttons on the key module are used to send the necessary control signals to the CCU. Note that more buttons are provided than is necessary for this simple system as the key module is also designed to be used on the K2 system which is a six extension plus two exchange lines version.

As shown, the telephone not connected to the key module can not access the exchange line direct. However each telephone could be connected to its own key module to provide direct exchange line access and inter-comm facilities.

In this system the telephones may be standard loop disconnect signalling, push button instruments, although a special type featurephone may also be used to provide facilities such as display number dialled, redial last number and so on.

The list of major advantages below shows that even this simple system provides a vast improvement on the basic parallel extension wiring.

❑ Each extension can have private use of the exchange line

❑ The system provides an intercomm facility between all extensions

❑ Incoming calls can be answered on one extension and held, while an enquiry is made to another extension

❑ Incoming calls can be answered on one extension, held, and easily transferred to another.

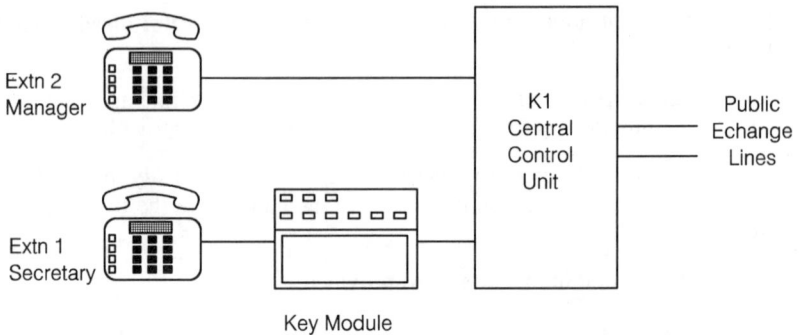

*Figure 10.2  Philips K1 key system. (Reproduced by permission of Philips Business Systems)*

Note that as the CCU is mains powered, the system must be capable of automatically connecting at least one of the telephones to the exchange line in the event of a mains failure, so external communications are provided for emergency use.

This system is ideal for a manager/secretary system, in which case the system can be connected to one extension line of a PBX, or for use in a small business, e.g an estate agents or doctor's surgery, in which case the system would be connected to a local exchange line.

There are several versions of the Sopho Key system, the largest of which handles up to 24 extensions and eight exchange lines.

## 10.1.5 Ferranti Rhapsody

This key system is available in four sizes from eight extensions plus three exchange lines, to 48 extensions plus 16 exchange lines. It is typical of systems designed to meet the needs of the small to medium sized business.

The Rhapsody is similar in concept to the Sopho K1, in that the heart of the system is a small mains powered Central Control Unit. The CCU is more complex as it handles more extensions and exchange lines. However instead of connecting extensions to standard telephones via key modules, each extension is connected to a proprietary featurephone, such as that shown in Figure 10.3. This featurephone is designed to operate on the smallest Rhapsody system. As this diagram shows, it has the normal push button dial pad and three keys at the right which are used to access the three exchange lines. Each of these keys has an integral LED to indicate whether the exchange line is busy or free.

Instruments designed for use on the larger systems will have 16 keys in this area to enable the user to select any of the exchange lines not in use. At the top of the instrument a LCD display provides information such as time of day, date, calling party identification, dialled number, and called number status – engaged, free or absent. Below the display are 10 keys, arranged in two rows of five. Each is associated with a number store which can be individually programmed with one internal and one external number. This provides a personal directory for single touch dialling, which is additional to the 100 commonly used numbers stored within the Rhapsody's CCU.

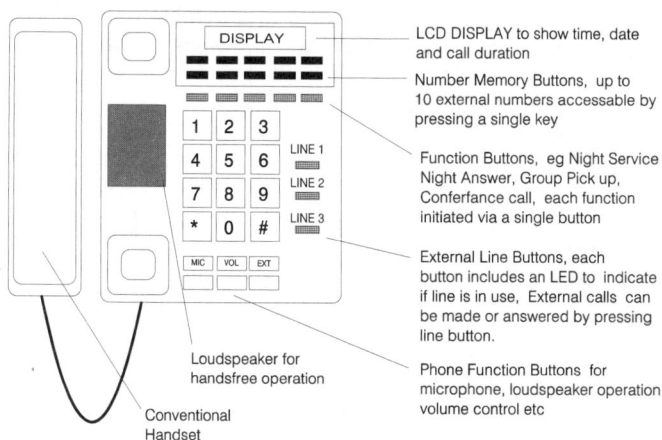

LCD DISPLAY to show time, date and call duration

Number Memory Buttons, up to 10 external numbers accessible by pressing a single key

Function Buttons, eg Night Service Night Answer, Group Pick up, Conferfance call, each function initiated via a single button

External Line Buttons, each button includes an LED to indicate if line is in use, External calls can be made or answered by pressing line button.

Phone Function Buttons for microphone, loudspeaker operation volume control etc

Loudspeaker for handsfree operation

Conventional Handset

*Figure 10.3 A Rhapsody Featurephone with integral display.(Reproduced by permission of Ferranti Business Communications Ltd)*

The next row of five keys provides access to the extra features of the system. Again each key can be individually programmed to invoke a certain feature, thus the phones can be tailored to each user's own requirements.

At the bottom of the phone, six individual keys provide control of the functions such as volume of the internal loudspeaker, redial last number, programming other keys, call hold and so on. In applications where it is preferred to have all incoming calls answered by a receptionist an additional unit may be used with a display phone to produce a simple attendant's console.

A wide range of features is available on sophisticated key system such as the Rhapsody. The list below should considered typical of what can be found on most larger systems.

❏ Selective Call Barring, i.e preventing certain extensions from making trunk, or international calls

❏ Call Hold and Transfer

❏ Conference Calls

❏ Follow Me Call Diversion

❏ Wait on Busy Extension

❏ Wait on Busy Exchange line

❏ Call Logging Information, such as number dialled, call duration and costs for each extension

## 10.1.6 The Use of Key Systems with a PBX

As was suggested earlier, a key system can usually be used in conjunction with a PBX. This method of operation is often referred to as piggy backing the key system on to the PBX, as the key system is connected to a number of PBX extension lines, rather then the public telephone network as shown in Figure 10.4.

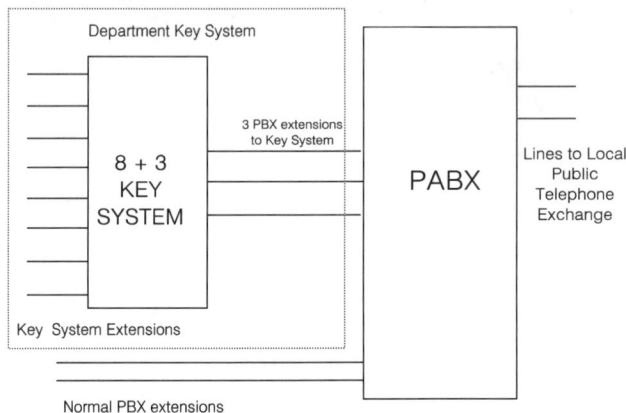

*Figure 10.4   A key system piggy backed on to a PBX*

In this diagram a small department of eight users are connected to a key system to provide their own small departmental system. The key system is connected to three PBX extension lines to enable the department staff to call other users of the PBX, and to gain access to the PSTN. This arrangement is also an ideal way of providing additional telephones to departments when providing extra PBX extensions is not viable for reasons such as the maximum capacity of the PBX has nearly been reached. Note that prudent administrations will usually keep some extensions in reserve for contingency purposes.

## 10.1.7 Summary of Small PBXs and Key Systems

Both types of system offer a cost effective method of providing a small number of users with a private telephone system. Key systems are generally smaller in terms of capacity, and less complex as they provide fewer user facilities than the PBX.

The key system may be connected directly to either a public telephone system or piggy backed onto a PBX. Some manufacturers are now offering the ability to link key systems in different sites, to provide a larger key system network.

# 10.2 PBX Architecture

## 10.2.1 Relationship Between PBX Capacity and Architecture

The architecture of a PBX will to some extent depend upon its size, for example dual processor operation tends to be available only on the larger PBX. Table 10.2 shows that there is a wide variety of PBXs available from several manufacturers, and shows the capacity for each PBX in terms of the number of maximum number of extensions and trunks that may be connected.

Note that the actual switched traffic capacities are not given. Usually the smaller PBX will use a fully non blocking architecture, while the larger systems may incorporate a concentration stage prior to a non blocking switch.

*Table 10.2 Various PBX capacities*

| Manufacturer | Small PBX | Medium PBX | Large PBX |
|---|---|---|---|
| Ferranti | Omni S1S 256 exts 64 trunks | Omni S1 384 exts 92 trunks | Omni S3 2,048 exts 1,024 trunks |
| Philips | Sopho S250 200 exts | Sopho S1000 900 exts | Sopho S2500 20,000 exts |
| Mitel | Sx-50 160 exts 56 trunks | Sx-2000S 600 exts | Sx-2000SG 2,500+ exts |
| GPT | Mini ISDX 80 exts 16 trunks | Small ISDX 240 exts 48 trunks | Large ISDX 2,448 exts 352 trunks |

In this table the figure quoted for trunks is taken to mean the total of both PSTN exchange lines and inter PBX tie lines. In all cases the capacities quoted are the maximum for a particular combination of extension lines and trunks. Some PBX designs allow for an adjustable mix, and will quote the maximum number of ports or channels on the digital switch. As some of the ports on the switch must be used for common equipment such as attendant's console, tone supplies, MF receivers and so on, not all ports can be connected to extensions or trunks. Typically a 256-port switch will handle about 200 extension lines and 30 trunks.

Although some of the figures quoted are for the largest version of a particular exchange, several manufacturers are able to offer a range of sizes for one specific model of PBX. This permits a purchaser to choose the capacity which best suits his requirements at the time, and provide an upgrade path so that should the requirement increase, he does not have to purchase a completely new PBX. Two examples of this approach are the Mitel SX-2000SG and the GPT Large ISDX. The expansion details of these two PBX ranges are shown in Table 10.3.

*Table 10.3 Expansion options for the Mitel SX-2000SG and GPT Large ISDX PBX*

| Equipment | MITEL SX-2000SG | GPT Large ISDX |
|---|---|---|
| 1 cabinet | 750 lines | 432 exts + 64 trunks |
| 2 cabinet | 1,800 lines | 1,104 exts + 160 trunks |
| 3 cabinet | 2,500 lines | 1,776 exts + 256 trunks |
| 4 cabinet | 2,500 + extra trunks | 2,448 exts + 352 trunks |

The architecture of a typical medium sized PBX is shown in Figure 10.5 on the next page. This diagram shows a PABX capable of terminating up to 512 ports and is based on several items of equipment and is not the block diagram of any particular exchange. In this initial description we will not be dealing with the ability of the PBX to handle circuit switched data in an ISDN type system. Such explanations will come later in the chapter.

The main hardware components of the PBX are:

❏ Digital Switch terminating 16 32-channel PCM/TDM systems, and its associated control unit.

❏ Central Processor Unit: Duplicated processors arranged in worker standby configuration, with associated main memory and Winchester Disk Drive backing store for main programs and exchange database.

❏ Line cards for various types of telephone line.

❏ Digital Termination Units for 30 Channel PCM systems with CCS in TS16, e.g. DASS links to System X exchanges, and DPNSS links to other digital PBXs.

❏ Attendant's Console and associated interface card.

❏ Common equipment card with tone supplies, MF signalling receivers and conference bridge.

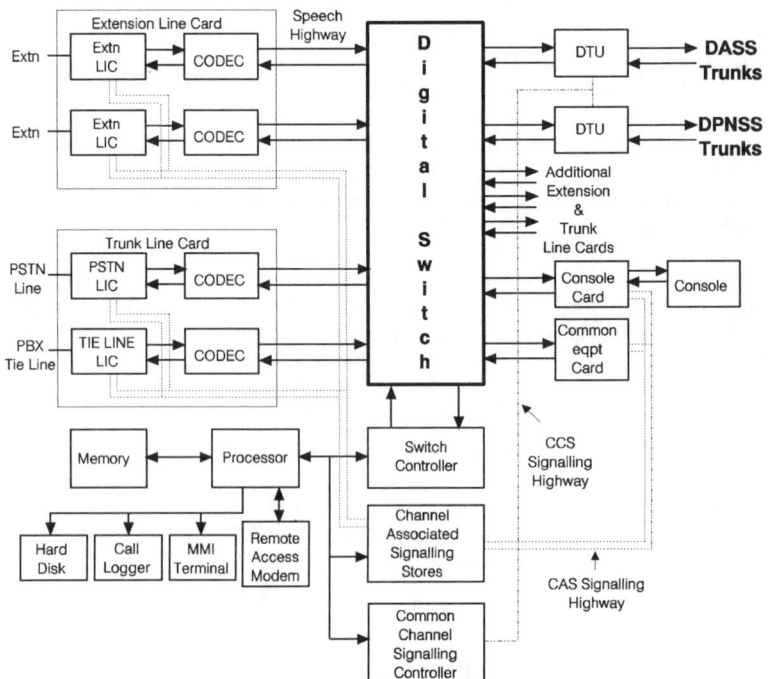

*Figure 10.5  Medium sized PBX system block diagram*

❑ Signalling stores which provide the mechanism for transmitting signalling information between the central processor and the lines cards. The store is divided into two halves. The first is an incoming signalling store for receiving signalling information from line cards. This information can then be accessed by the exchange processor unit. The other half is an outgoing signalling store for receiving signalling commands from the exchange processor which are then transmitted to the line card.

❑ Signalling controller which provides an interface between the central processor and the common signalling channel of the DASS and DPNSS trunk systems.

Four main types of digital highway interconnect the various components of the exchange:

❑ The Digital Speech Highway: 2MBit/s 32-channel Systems interconnecting groups of extensions, and analogue trunks with the digital switch. These highways are also used to connect the console, common equipment and digital trunks to the switch.

❑ Channel Associated Signalling Highway: This is a 256 Kbit/s TDM system with 32 timeslots each of which is eight bits long. The signalling highway is used to transmit signalling words between the various line circuit and the central processor, via the signalling stores.

Each line card is directly associated with a TS on a signalling highway and a location in the signalling store. When the line card has detected a new signalling condition, e.g off hook, or dialled digits, it will transmit a suitably encoded 8-bit word in the designated TS, to the associated location in the signalling store. The processor continually scans the incoming signalling store to detect any new signalling information from the line cards.

In the opposite direction the processor loads signalling commands into the relevant location in the outgoing signalling store. This information is subsequently transmitted in the designated TS to the required line card, which then carries out the required action, e.g. applies ringing current to the line.

❑ Common Channel Signalling Highway: DASS and DPNSS signalling messages pass along these 64KBit/s highways which connect the CCS controller with the TS16 interfaces in the DTUs.

❑ Processor Control Highway: This is a conventional bi-directional bus which is used for the transmission of address signals, control signals and control data to various hardware elements of the PBX.

The PBX will be contained in a cabinet, with several equipment shelves. Each shelf is designed to accept a particular group of printed circuit boards or cards. The description continues with an outline of the function and operation of the more important cards. This description is based upon several PBXs currently available from several manufacturers, rather than a single manufacturer's exchange.

## 10.2.2 The PBX Digital Switch

At the heart of this exchange is a fully non blocking 512 channel digital switch, which could be realised using two switch ICs such as the Siemens PEB 2040, as shown in Figure 10.6.

The switch terminates up to 16 32-channel PCM/TDM systems, each running at 2.048 Mbit/s. For reasons of simplicity in the diagram only six are used, the remaining 10 being available for expansion.

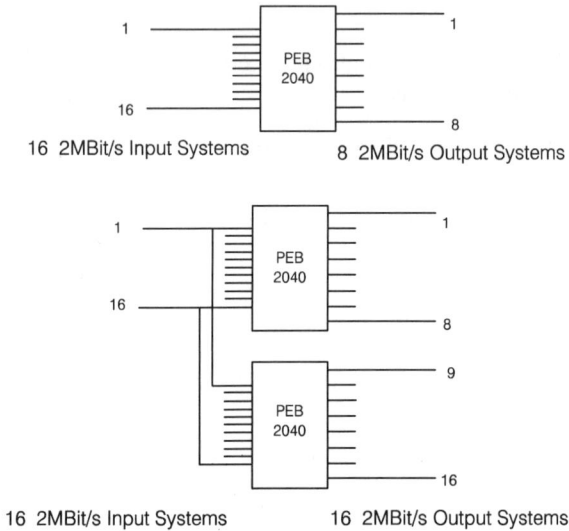

*Figure 10.6 512-channel non-blocking switch using two Siemens PEB 2040 ICs*
*(Reproduced by permission of Siemens)*

Each PCM/TDM system, except those connected to the digital trunk ports is capable of handling 32 PCM voice channels. This may seem to conflict with the 30 Channel PCM systems described in Chapter 5. However it should be noted that since the whole system is mounted in a single cabinet and controlled from a local central timing source there is no requirement for a synchronisation signal to be transmitted in timeslot zero. Also signalling from extensions and trunks is extracted at the line interface and passed to the control unit over a dedicated signalling highway, thus it is not necessary to use timeslot 16 for signalling within the exchange.

But, as the digital trunks will be connected to other exchanges, synchronisation using timeslot zero must be provided. The use of common channel signalling systems such as DASS and DPNSS on these links requires that timeslot 16 must be used, so these links are 30-channel similar to conventional PCM/TDM systems. They are connected to the outside world via a Digital Transmission Unit (DTU) responsible for Frame Alignment Word (FAW) insertion on the transmit side of the link, and monitoring of receive FAW on the other side, to ensure synchronisation in both directions. The DTU is also responsible for the insertion and extraction of signalling information in TS16.

As shown, 2Mbit systems terminating on the switch are organised as conventional 32 timeslot, eight bits per timeslot, serial binary bit streams, and thus 8-bit serial to parallel conversion must be provided in the switch IC.

Other architectures, for example the GPT ISDX, are based on two 8-wire parallel data highways, one for each direction of transmission. Organising the data in this way removes the requirement for serial to parallel conversion in the switch, and although the overall data transfer rate over the bus is still 2.048 MBit/s the actual bit rate on each wire is only 256 KBit/s.

The switch controller contains the necessary logic and processing power to set up and release connections across the switch upon receipt of suitable instructions from the exchange processor.

## 10.2.3 PBX Extension Line Cards

The extension line cards on early digital PBXs, such as the Monarch 120, manufactured by both GEC and Plessey when they were separate companies, were only equipped to handle four extensions per card, with typically eight cards per shelf. The later Monarch 120B was fitted with line cards giving eight extensions per card, and today's ISDX has line cards catering for 16.

Thus since the early 80s there has been a four-fold increase in the number of extensions possible in a given size of cabinet. Our typical PBX has 16-line per extension line cards, and thus two cards are required to fully utilise one 2MBit/s speech highway and one 256 KBit/s signalling highway.

The line card has five terminations:

❏ Two-wire to the extension telephone

❏ 2 MBit/s speech highway to the digital switch

❏ 2 Mbit/s speech highway from the digital switch

❏ 256 Kbit/s signalling highway to the signalling store

❏ 256 KBit/s signalling highway from the signalling store.

Each extension line card is equipped with a single microprocessor which controls many of the operations of the line circuits on the board. It is particularly involved with signalling tasks such as dialled digit to binary data conversion.

For each line circuit on the card, there will be an individual extension line interface circuit which has the same basic functions as the Subscribers' Line Interface Circuit (SLIC) in the System X Digital Subscribers' Switching Subsystem covered in Section 9.2.3. These are the BORSCHT functions:

❏ Battery Feed

❏ Over voltage Protection

❏ Ringing Current Feed

❏ Supervision and signalling

❏ Coding

❏ Hybrid 2w/4w conversion

❏ Test Access

These functions have already been discussed in the chapter on System X.

In our representative PBX, the line circuit deals with speech and signalling separately as shown in the block diagram of a typical extension line card in Figure 10.7.

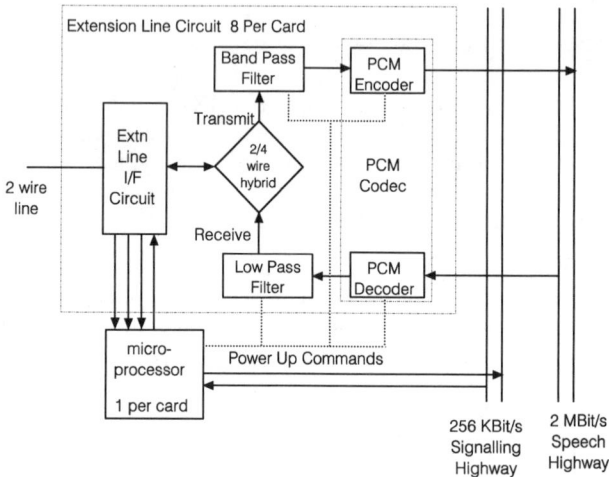

*Figure 10.7 Typical PBX extension line card*

The microprocessor is connected to the 16 SLICs on the card. Each SLIC is able to communicate to the microprocessor, the status of the a and b wires of the telephone line, and the condition of the subscribers' loop, i.e. open or closed. In this way the initial calling loop, and subsequent dialled digits from loop disconnect telephones is detected by the microprocessor. For each line condition the on board microprocessor sends an eight bit signalling word in the timeslot allocated to the extension line concerned along the signalling highway to the signalling stores.

In systems which use earth recall, operating the recall button on the telephone will cause either the a or b leg to connected to earth, whereas in timed break recall systems, a disconnection of the extension loop for 66 millisecs is interpreted as a recall signal. In either case the recall signal is received by the on board processor which sends the relevant signalling word to the signalling stores as before.

The line card must be able to cater for both loop disconnect (decadic) and MF dialling. Loop disconnect pulses are detected by the microprocessor and converted to 8-bit signalling words and passed to the signalling stores over the signalling highway.

As more users migrate to MF telephones, there will be less requirement to provide decadic dialling detection on the line card, and thus scope for cost reduction in this area. Some exchanges, such as the GPT ISDX, use a different approach to handling decadic dialling, in which decadic dialling detectors are part of the common equipment group (see Section 10.2.7) and thus can be provided on an as required basis.

MF dialling tones received on the line card are processed by the circuits on the card in exactly the same way as speech. In this case the microprocessor plays no part in the signalling conversion as this is carried out in an MF receiver circuit as will be explained later.

The microprocessor is also able to control the application of ringing current to the extension by controlling a miniature relay on the board. Generally a single ringing source will supply all line circuits on the PBX, although for reliability reasons the source may be duplicated.

Two to four-wire conversion is achieved by the use of a hybrid circuit. Two-wire to four-wire conversion is necessary to interface the 2-wire telephone line to the Codec, which is by nature a 4-wire device, two wires for transmission, two for reception.

The main purpose of the hybrid in this respect is to provide signal isolation between the receive and transmit paths, and thus prevent any received signal being retransmitted. If this isolation were not provided voice users would experience unwanted oscillations, and on connections that involve several digital switches, or satellite links, unpleasant echoes. The degree of isolation required is of the order of 30 to 40 dB and is achieved using electronics rather than hybrid transformers.

Note that 2-wire circuits which include satellite links, often also include echo suppressors to further reduce the effects of unwanted echo due to imperfections in the hybrids.

The forward speech signal is then passed to a band limiting filter, which is required to restrict the input signal to approximately 3.4 KHz at the upper end of the pass band. At the low frequency end the filter will have around 40dB of attenuation at 50Hz to reduce to negligible levels mains induced hum that will be picked up on the telephone line.

Following band limiting the forward signal is fed to the A/D side of the codec, the output of which is 64 KBit/s PCM encoded speech samples. In this system each codec is allocated a specific timeslot in which to transmit each 8-bit sample. A timing signal from the microprocessor provides the necessary synchronisation to ensure each codec on the card transmits its sample at the correct time.

The backward speech signal is received from the speech switch along the 2MBit speech highway. The codec extracts each 8-bit sample from the correct timeslot, and carries out the D/A conversion. The resulting speech signal is passed to the receive side of the hybrid via a low pass filter, which is required to remove any high frequency components remaining in the audio signal after the D/A conversion.

PBX installations tend to involve a wide range of extension line lengths, typically from as short as 100m, to as long as 2.5kms. To ensure that the input signal to the codec is relatively constant irrespective of line length, the line interface circuit is equipped with an attenuator which is switched in circuit to cater for short lines.

To save power, active circuits on the line card are powered down when not in use. The circuits are powered up when a new call is received from the line, or the microprocessor is informed of a new outgoing call by the exchange processor.

## 10.2.4 Exchange Line Card

The exchange line card will terminate between 4 and 16 2-wire lines from a local public exchange. The card is essentially the same as the extension line card except that the line interface has to cater for a different style of signalling. In this case, the line interface circuit has to be able to detect the presence of an incoming 17Hz high voltage ring from the local exchange, and to interpret this as a new incoming call. In most installations, the call processing software will cause this new call to be routed initially to the attendant's console, where the call can be answered and transferred to the required extension.

Some PBXs can now be fitted with Direct Dial In facilities. However the local public exchange must also be equipped with the relevant line interface cards to be able to offer this service.

Calls from the PBX to the public network are made by generating an outgoing calling loop on the exchange line interface (note that in some cases earth calling is used). There will then be a short but indeterminate delay before the local exchange returns dial tone, and the dialled digits can be pulsed out.

It is essential that the PBX does not cause dialling out to begin before dial tone has been received otherwise one or more digits may not be correctly received by the local exchange. A circuit in the common equipment card is used to detect the presence of dial tone on the line and alert the PBX software that dialling can commence.

Speech processing on the exchange line circuit is the same as for the extension line circuit, except that during dialling out the speech circuits are inhibited to prevent the extension user from hearing loud unwanted clicks in his handset.

## 10.2.5 Inter PBX Tie Line Card

Most manufacturers will supply a range of inter PBX cards to cater for the wide variety of tie line signalling systems in use. It should be noted that some systems use a 4-wire speech path, and catering for these systems often takes up two speech channels per trunk rather than one. Once again the speech elements of the card are similar to those of the extension line card, and as expected the differences lie in the signalling areas. Cards to cater for the following types of inter PBX circuit are normally available:

❏ SSDC 5 (E & M) with 2-wire and 4-wire speech circuit.

❏ SSDC 10 which may also be 2- or 4-wire

❏ Loop Disconnect Bothway Junction (2-wire)

❏ SSAC 13 2- or 4-wire speech with in band signalling using 2280Hz pulses

❏ SSAC 15 4-wire using 2280Hz tone-on idle, i.e. the tone is switched off to indicate signalling information.

Several manufacturers now offer multiple DC and AC cards which can be set up to operate one of several signalling systems. Although the card is more complex, economies are made as the number of different types of board to be manufactured and ranged as spares is reduced.

## 10.2.6 The Console Card

This card is used to connect an attendant's console to the exchange. Although shown here as connecting directly to the digital switch via its own 2 Mbit speech highway, the console line card will usually take a slot which could be occupied by a trunk card and thus will use timeslots on the 2MBit speech highway associated with the trunk circuits. The basic block diagram of the console and its line card are shown in Figure 10.8. Each console card terminates two 4-wire speech circuits to permit the attendant to speak individually to both parties of a call without the other overhearing. Speech passes between the console and the line card in analogue form over these four wire circuits. On the line card each operator speech circuit is associated with its own interface, codec and timeslot on the speech highway.

Console signalling is far more complex than basic telephone signalling. As well as dialled digits and incoming call signals, the console will also be used to call up, display and alter many of the entries in the exchange database. Such tasks range from displaying and changing the time and date, to invoking call diversion, and call barring functions for particular extensions.

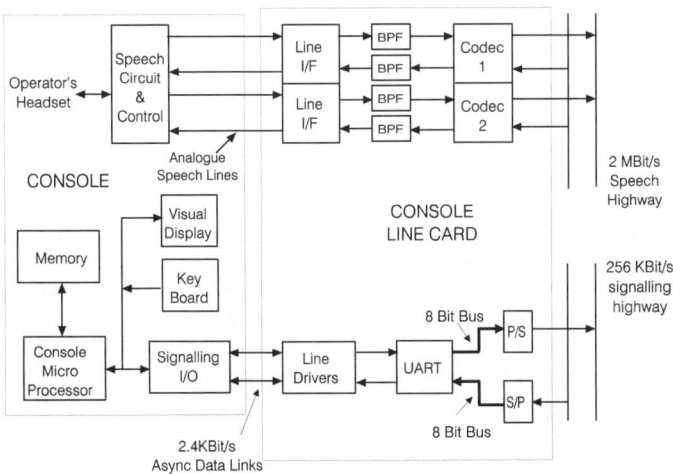

*Figure 10.8 Attendant's console and console line card*

Console signalling passes between the console and its line card as 2.4 Kbit/s asynchronous data, i.e. 8-bit signalling words are enclosed in start, stop and parity bits. The signalling converter on the line card consists of a Universal Asynchronous Transmitter Receiver (UART) which converts the incoming serial data to parallel form, removes the redundant bits and passes the signalling words to a parallel to serial converter which transmits each signalling word in the correct timeslot on the signalling highway. In the reverse direction, signalling words received from the highway are converted into parallel form, loaded into the UART, where the necessary start, stop and parity bits are added prior to transmission at 2.4 KBit/s to the console.

For reasons of security, the 2.4 KBit/s data link between the console and the console line card is duplicated. When the console is idle it transmits an idle signal to the line card, should no signal be received from the console over either of the data links, after a preset period a "console disconnected" alarm is raised. The exchange software will then route all incoming calls to a preselected extension telephone.

## 10.2.7 Common Equipment Card

The common equipment card, unlike any of the cards dealt with so far does not connect to any external equipment. It is connected to the digital switch by its own 2Mbit speech highway and to the signalling stores via its own signalling highway to provide a number of services or facilities which are required during the processing of calls. For example all calls will require dial tone to be sent to the calling extension, while to process calls originating from MF telephones the use of an MF receiver is required.

The following services are provided on the common equipment card:

❏ Tones Generator

❏ MF Signalling receivers

❏ Conference unit

❏ Dial Tone Detector

❏ MF sender

❏ Test Unit

Each of these units will be connected to the speech and signalling highways in different ways as shown in Figure 10.9. For reasons of clarity the test unit is not shown in this diagram, however it is connected to both incoming and outgoing speech highways, and incoming and outgoing signalling highways.

## 10.2.8 Common Equipment Card: Tone Generators

The tone generators are not connected to the signalling highway at all, and as they are used only for the transmission of tones, they are connected only to the outgoing 2Mbit speech highway. The following tones are produced by this card:

❏ Dial Tone

❏ Continuous ring tone

❏ Cadenced Dial Tone

❏ Cadenced Ring Tone

❑ Engaged Tone

❑ Number Unobtainable Tone

❑ Confirmation Tone

Cadenced dial tone is used to indicate to a user that the call diversion feature has been activated for his extension, i.e. incoming calls to this number have been diverted to another extension.

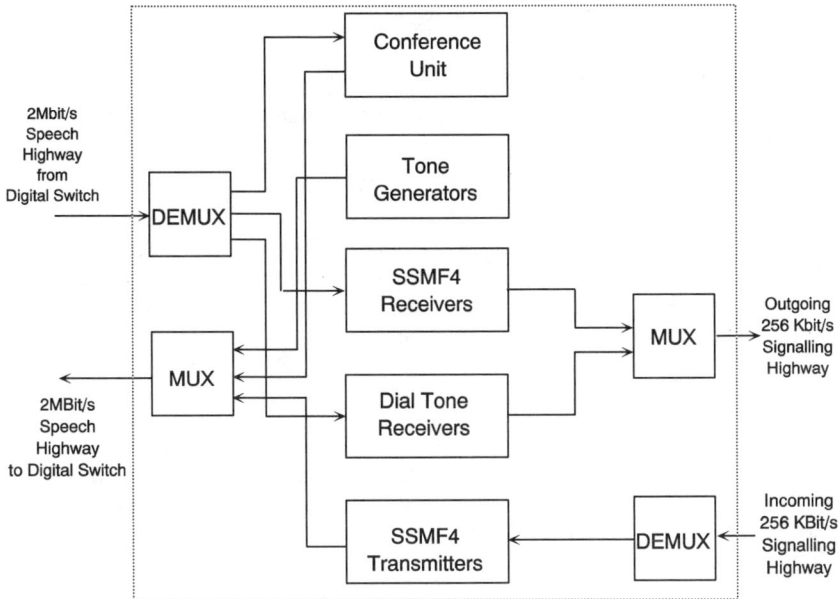

*Figure 10.9 The common equipment card interconnections to the speech and signalling highways*

Confirmation tone is sent to a caller to confirm that his commands to set up facilities such as call diversion have been correctly interpreted and implemented.

These tones are not generated in analogue form in the exchange. Instead the original tones have been digitised and are stored as a succession of PCM samples in a read only memory. Initially, a tone is produced in analogue form, sampled 8000 times per second; the sample values are quantised, and the resulting values stored in binary form in contiguous locations in a ROM. The other tones required are processed in the same way, and stored in the remaining locations in the ROM.

To reproduce the tone it is simply necessary to cyclically read out the binary values stored in the ROM and convert these values back into analogue form. Each tone is allocated a particular timeslot in the 2Mbit speech highway connecting the common equipment card to the digital switch, and the binary values for each tone are constantly being transmitted towards the switch in their respective timeslots.

When, during call processing, it is necessary to send a particular tone to an extension or trunk the exchange processor instructs the switch controller to connect the timeslot containing tone source to a timeslot associated with the extension (or trunk).

A particular advantage of this system is that the tone source can supply several circuits simultaneously without any loading effect, since the time switching mechanism simply copies sample values held in the speech store to the switch output during the required timeslots.

## 10.2.9 Common Equipment Card: SSMF4 Receivers (MF receivers)

The actual number of MF receivers required in any particular PBX will depend upon the number of MF telephones connected, or expected to be connected in the future. The function of an MF receiver is to receive signalling from an MF telephone, decode each tone pair received, and transmit a suitable 8-bit signalling word to the signalling stores. Consequently, as shown in Figure 10.9 the MF receiver needs only to be connected to the incoming side of a speech highway, and the outgoing side of a signalling highway.

When the off hook condition is detected from an extension, the exchange database is checked to determine whether or not the extension is equipped with an MF telephone. If so, a free MF receiver is allocated to the call, and the digital switch controller is instructed to set up the following connections:

❑ Dial tone to extension

❑ Extension to selected MF receiver

Note that this is an asymmetric connection, whereas the connection that would be made between two extensions would be symmetrical. This point is raised to emphasise that for each normal speech connection, two connections are required in the digital switch.

When the user dials the first digit, the MF tone pair will be PCM encoded by the codec on the extension line card, as for speech, and transmitted to the MF receiver via the time switched connection in the digital switch. Thus the input to the MF receiver is actually PCM encoded MF tones mixed with a small dial tone signal due to imperfections in the line circuit hybrid. A filter is therefore required in the MF receiver to prevent this low level dial tone signal interfering with the decoding process. Once the first digit has been properly received and decoded, dial tone is removed by releasing the dial tone to extension connection in the switch, and thus does not cause a problem for the remaining dialled digits. Figure 10.10 is a simplified block diagram of an MF receiver.

Each MF receiver is associated with a specific timeslot of the common equipment card 2MBit/s speech highway. The same numbered timeslot is used on the outgoing signalling highway. The digital signal received in the associated timeslot is extracted and decoded by the Codec, the resulting MF tones being passed to a standard SSMF4 receiver, via a dial tone filter. In the SSMF4 receiver checks are made that:

❑ Two frequencies only are present

❑ There is one frequency from both low and high band groups

❑ The tones are of sufficient amplitude

❑ The tones are present for sufficient time – the recognition period

❑ The inter-digit pause between tone pairs is correct

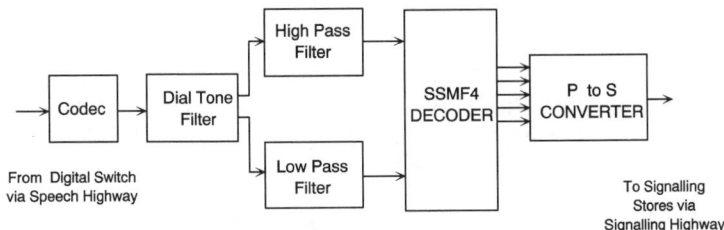

*Figure 10.10 Simplified diagram of an MF receiver*

If all these checks are positive, the signal is considered valid. The value of the dialled digit is then encoded as a 4-bit binary word and passed in parallel to a P/S conversion circuit, along with a fifth bit which is set to indicate that the 4-bit binary number is valid. These five bits form the least significant bits of the eight bit signalling word which is to be transmitted to the signalling stores. The remaining three most significant bits are hard wired, and thus have a constant value.

Table 10.4 shows the MF tone pairs and binary encoded values, for the digits 1 through 0, and the * (star) and # (square) symbols.

*Table 10.4 Coding of dialled digits by the MF receiver*

| Dialled Digit | Tone Pair I/P TO MF RXR | Signalling Byte O/P from MF RXR | | | | | | | |
|---|---|---|---|---|---|---|---|---|---|
| | | A | B | C | D | E | F | G | H |
| 1 | 697 + 1209 | 1 | 0 | 1 | 1 | 0 | 0 | 0 | 1 |
| 2 | 697 + 1336 | 1 | 0 | 1 | 1 | 0 | 0 | 1 | 0 |
| 3 | 697 + 1447 | 1 | 0 | 1 | 1 | 0 | 0 | 1 | 1 |
| 4 | 770 + 1209 | 1 | 0 | 1 | 1 | 0 | 1 | 0 | 0 |
| 5 | 770 + 1336 | 1 | 0 | 1 | 1 | 0 | 1 | 0 | 1 |
| 6 | 770 + 1447 | 1 | 0 | 1 | 1 | 0 | 1 | 1 | 0 |
| 7 | 852 + 1209 | 1 | 0 | 1 | 1 | 0 | 1 | 1 | 1 |
| 8 | 852 + 1336 | 1 | 0 | 1 | 1 | 1 | 0 | 0 | 0 |
| 9 | 852 + 1447 | 1 | 0 | 1 | 1 | 1 | 0 | 0 | 1 |
| * | 941 + 1209 | 1 | 0 | 1 | 1 | 1 | 0 | 1 | 1 |
| 0 | 941 + 1336 | 1 | 0 | 1 | 1 | 1 | 0 | 1 | 0 |
| # | 941 + 1447 | 1 | 0 | 1 | 1 | 1 | 1 | 0 | 0 |
| NONE | MF Rxr idle | 1 | 0 | 1 | 0 | X | X | X | X |

Note that bit D is set to 0 to indicate that the receiver is idle, or the incoming signal is not valid. When a valid signal is present and has been correctly decoded the D bit is set to 1.

## 10.2.10 Common Equipment Card: The Conference Unit

The function of the conference unit in this particular exchange is to connect up to four users together simultaneously, in order that a multiparty call, or voice conference can take place via the PBX. The conference unit is connected to 16 channels on the common equipment card 2Mbit speech highway, thus permitting up to four simultaneous four-party conferences, each using one of four conference bridges.

Each conference is allocated a free conference bridge which produces four different digital output signals. Each output signal is the equivalent of the digital sum of a different combination of three of the four inputs. Figure 10.11 illustrates the simple principle of a four-party conference bridge.

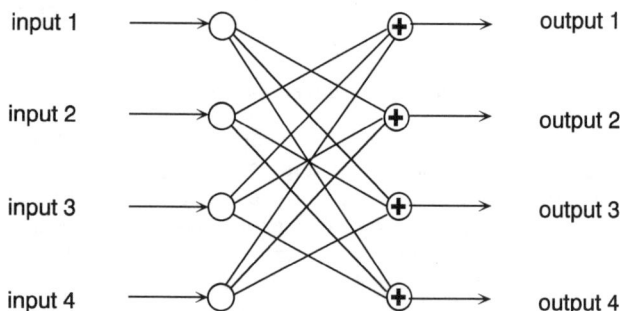

*Figure 10.11 Illustration of the principle of the conference bridge*

When a conference call is established, exchange control will select a free conference bridge. The four inputs to the conference bridge are specified timeslots on the incoming 2Mbit speech highway, thus exchange control must also connect the required extensions to the relevant timeslots on this highway via the digital switch. Similarly exchange control must connect each output timeslot from the conference bridge to one of the required extensions.

The signal sent out in TS 1 is the digital sum of the signals from incoming TS 2, 3 and 4, while the signal sent out in TS 2 is the digital sum of the signals from incoming TS 1, 3 and 4 and so on.

A further complication is that the digital signals to be added are 8-bit A law PCM encoded samples, but it is not possible to simply add A law encoded samples in a binary accumulator and produce a result which is valid.

The conference unit block diagram in Figure 10.12 shows that this problem is overcome by incorporating an A-law to linear converter, so that all A-law encoded input samples are converted to 12-bit linear PCM codewords, which are then added together in a conventional binary accumulator. The output of the accumulator is then converted back to 8-bit A law format before being transmitted out in the required TS.

The block diagram of the conference unit is shown below in Figure 10. Individual conference bridges are not identifiable in this block diagram as the hardware shown includes all four bridges. The explanation below will explain how this arranged.

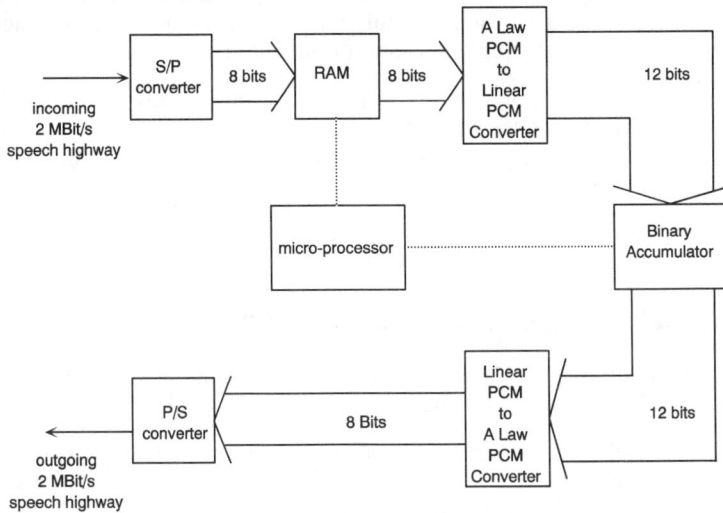

*Figure 10.12 Block diagram of the conference unit*

The conference unit is controlled by its own microprocessor and like the tone generators is operating all the time even if no conference is in progress. Since the inputs and outputs are timeslots (or channels) on the 2MBit speech highway between the unit and digital switch, to set up a conference call when requested, exchange control simply has to arrange for the extension channels to be digitally switched to channels of the conference bridge.

As 16 timeslots (TS1 – 16) on the incoming and outgoing speech highways are reserved for the conference unit. This will only respond to speech samples in these timeslots, and ignore all the others (i.e. TS17 – 32).

As each TS arrives from the digital switch, the 8 bit sample contained in the TS is converted into parallel format and stored in a corresponding location in the Random Access Memory. In this case there are 16 locations in the memory, location 1 reserved for the samples arriving in TS1, location 2 reserved for the samples in TS2, etc. Thus this store is continually being refilled with each frame.

The microprocessor, binary accumulator, A law/linear converters and the RAM are the hardware for all four bridges, bridge 1 consisting of TS1 – 4, and RAM locations 1 – 4, with bridge 2 consisting of TS 5 – 8 and RAM locations 5 – 8, etc. The explanation is for bridge 1, however the other three bridges are operating concurrently in the same way.

Once a four-party conference has been set up, new speech samples will be arriving in TS1 – 4, every frame. At the start of a new outgoing frame from the conference unit, the accumulator is cleared under microprocessor control, subsequently the current sample held in RAM location 2 is converted to linear form and loaded into the accumulator. Then the current sample held at RAM location 3 is converted to linear form and its binary value added to that already in the accumulator. A similar process adds the binary value of RAM location 4 to the accumulator which now holds the digital sum of TS2, 3 and 4 which is required to be sent out in TS1, thus the contents of the accumulator are converted to A law format and transmitted out in TS1.

The accumulator is cleared once more, and a similar process involving RAM location 1, 3 and 4 is used to produce the sample to be transmitted in TS2. This processes continues for 16 timeslots, and in this way a single set of hardware can be configured as four conference bridges.

Each exchange manufacturer offers slightly different conference arrangements, some permitting six or even eight party conference calls.

## 10.2.11 Common Equipment Card: Dial Tone Receiver

The function of the dial tone receiver is to detect incoming dial tone on a public exchange line following selection and seizure of the line by the PBX.

The transmission of dial tone from an exchange is an acknowledgement that a new call request signal has been received, and that the exchange has connected appropriate receivers to the line to detect the digits dialled. In older public exchanges the delay between the line being seized and dial tone being applied may be as much as a second, and in a few instances even longer.

After having dialled the public network access digit(s), the extension user may not wait until he hears the public network dial tone before dialling the national network number required. In this situation, the PBX control software will select a free exchange line circuit (see Section 10.2.4), but it is essential that the exchange line circuit does not commence outward dialling to the local public exchange before dial tone is received as the first digit dialled may be incorrectly detected at the local exchange.

When an outgoing call to the local public exchange is set up, the control software will ensure that the incoming side of the exchange line circuit is routed via the digital switch to the dial tone receiver. When incoming dial tone is detected, the receiver sends an appropriate signalling word to the signalling stores.

When the signalling word has been detected by the control software, the digits of the required national network number are then sent to the relevant exchange line circuit to be pulsed out. Simultaneously the dial tone receiver circuit is released to enable it to be used for the next outgoing call.

The dial tone receiver circuit consists of three main blocks as shown in Figure 10.13. The input signal consists of 8-bit PCM encoded samples from the exchange line card. In the initial few hundreds of milliseconds following seizure of the exchange line, the samples will contain little more than line noise, until dial tone is applied to the line by the local exchange.

Bear in mind that a PCM sample consists of eight bits, one of which indicates the polarity of the sample, while the remaining seven bits indicate the sample amplitude.

```
                    ┌──────────┐      ┌──────────┐
                    │  Level   │─────▶│Persistence│
                    │ Detector │      │  Timer    │
                    └──────────┘      └──────────┘
            ┌──────────┐                          │
            │   S/P    │                     ┌──────────┐
   ────────▶│ Converter│                     │  Check   │────▶
  From      └──────────┘                     │ Circuit  │
  Digital Switch                             └──────────┘
  via                   ┌──────────┐              ▲        To
  2 MBit/s              │ Polarity │              │     Signalling Stores
  Speech Highway        │Alternation│─────────────┘        via
                        │ Detector │                    256 KBit/s
                        └──────────┘                 Signalling Highway
```

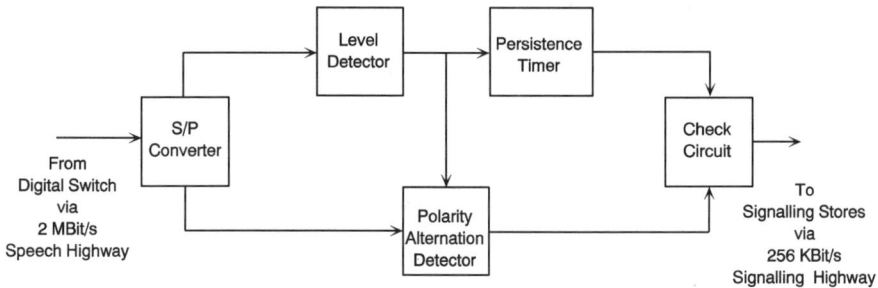

*Figure 10.13 Block diagram of the dial tone receiver*

The polarity alternation detector examines the sign bit, while the amplitude detector examines the most significant bit of the remainder. These detectors determine whether the signal is of the correct frequency and that it has reached the correct level of approximately -20dBm. The output of each of these detectors is a single logic signal which is asserted if the required check is satisfied.

The persistence timer ensures that once the input signal has reached the correct level it remains so for at least 600 mSecs. Once this period has expired the timer output is also asserted.

So long as the check circuit receives asserted signals from the persistence timer and the polarity alternation detector it will output a suitably encoded 8-bit word to the exchange processor via the signalling highway.

As more exchanges in the public network are modernised, the requirement for a dial tone receiver may disappear. Certainly it will not be required for ISDN connections between public and private exchanges. Even where decadic dialling exchange line signalling is used, the delay to dial tone may be small enough to permit a simple delay to be incorporated into the PBX software such that following seizure of an exchange line by the PBX, a short fixed delay occurs prior to the transmission of the first digit.

## 10.2.12 Common Equipment Card: SSMF4 Transmitter

The function of the SSMF4 transmitter is to generate A-law PCM encoded SSMF4 tone pairs for transmission to those public and private exchanges which are designed to accept this type of exchange line signalling. Figure 10.14 shows the block diagram of the SSMF4 sender in which a standard analogue SSMF4 tone generator, similar to that used in MF telephones, produces the required tone pair in response to each digit received in binary form from the exchange processor on the signalling highway. To enable a standard generator to be used, the digit in binary encoded format must first be converted to the same form as that produced by a standard keypad, i.e. four bits representing each row of the keypad, with another four bits representing each column. Thus a total of eight bits is used with only two (i.e. one row, and one column) being set to binary 1 at any one time.

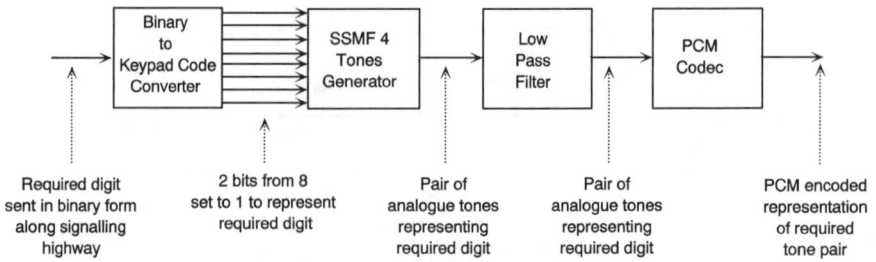

*Figure 10.14 Block diagram of the SSMF4 sender*

The tone pair generation produces some unwanted high frequency components which must be filtered out before the analogue tone pair is converted to A-law PCM format in a standard PCM codec.

When the exchange has to set up a call over a route that requires the SSMF4 sender, the incoming side of the circuit is switched to the dial tone receiver, and the outgoing side is switched to the SSMF4 sender. Once dial tone has been received, the CPU will transmit the required digits along the signalling highway to the SSMF4 sender for transmission. Once all digits have been transmitted by the sender, it is released and the incoming side of the circuit is switched through to the calling extension circuit.

## 10.2.13 Common Equipment Card: Test Unit and Pattern Generator

The test unit was not shown in Figure 10.9. Its purpose is to provide a test unit and receiver to test the digital switch, and a means of testing the signalling stores and highway.

It is not possible to fully the describe the operation of the test unit in this chapter, but a general appreciation of its functions can be gained by describing the process of testing an extension line circuit.

This test involves the use of both speech and signalling elements of the test unit. Figure 10.15 shows that the exchange processor sends a power up command over the signalling highway to the active elements, i.e. codec and filters, of the line circuit under test.

The output of a 400Hz tone generator is connected to the receive side of the relevant line circuit via the digital switch. The transmit side of the line circuit is similarly connected to the test unit.

With the telephone handset on hook, the hybrid (2-wire to 4-wire converter) in the line circuit will be unbalanced and a proportion of the 400Hz tone arriving on the receive leg will leak into the transmit leg, and be transmitted back via the digital switch into the test unit. A circuit in the test unit detects the presence of this incoming tone and transmits a pass signalling word to the exchange processor over the signalling out highway.

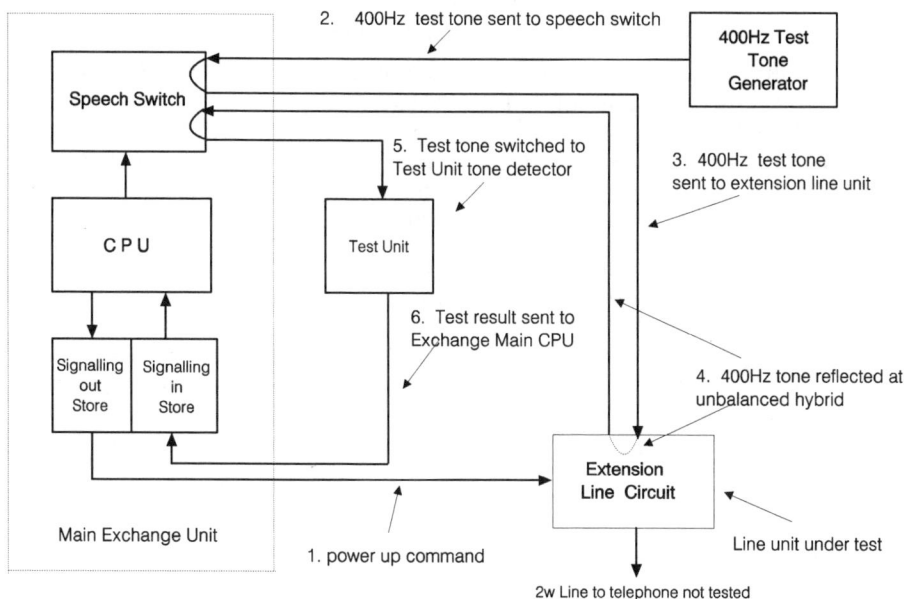

*Figure 10.15 Testing an extension line circuit using the test unit*

This test checks the codec, filters and hybrid circuits of the extension line circuit. If any element is faulty, such that the 400Hz tone is not detected in the test unit, a fail message is sent to the processor, which then initiates action to log the fault, and raise an appropriate alarm on the console, and an external alarm if fitted.

## 10.2.14 Common Equipment Card: Summary

Common equipment comprises those hardware elements, other than the line circuit and the digital switch, which are required by the exchange for call processing. In an analogue common control exchange, much of this would be part of the common control equipment, for example tone generators, MF receivers and senders.

In this case sufficient quantities of the common equipment required can be mounted on a single card, with only one pair of 32 channel speech and signalling highways.

In larger exchanges, due to the greater call handling capacity, more MF receivers (for example) will be required, and so it will not be possible to mount the equipment on a single card. In these cases, several cards will be required to form a common equipment group which may use more than one pair of speech and signalling highways.

## 10.2.15 The Dropback Card

The dropback card is not shown on the typical PBX diagram in Figure 10.5. It is however an important card as its function is to provide access to the public telephone network for designated extensions in the event of a PBX failure. The principle of operation of the dropback card is shown in Figure 10.16.

Prior to installation it is necessary to select those extensions which will require public network access in the event of a PBX failure. The selected extension are not connected directly to the appropriate extension line card but to relay contacts on the dropback card. There will be a limit on the number of extensions that can be given dropback service, and this will depend upon the capacity of the dropback card. Typically up to eight extensions may be catered for.

The required number of exchange lines for use under dropback conditions are also connected to relay contacts on the dropback card. Other contacts on the dropback card are then connected to the relevant extension and exchange line circuits.

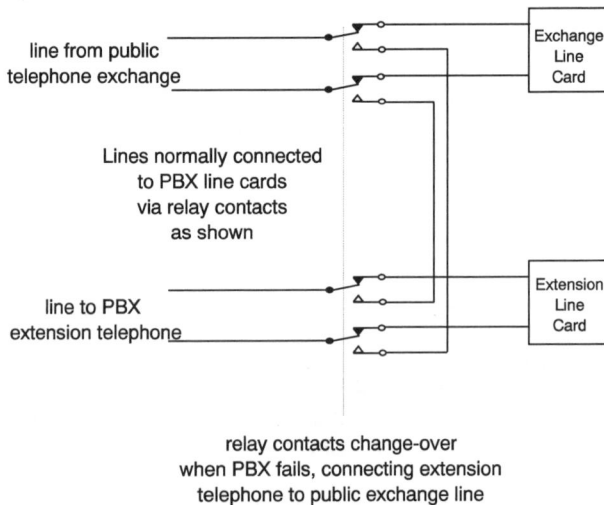

line from public telephone exchange

Exchange Line Card

Lines normally connected to PBX line cards via relay contacts as shown

line to PBX extension telephone

Extension Line Card

relay contacts change-over when PBX fails, connecting extension telephone to public exchange line

*Figure 10.16 Principle of operation of the dropback card*

Under normal operating conditions, the relay contacts are as shown in the diagram, and connect the exchange line to the relevant exchange line circuit, and the designated extension to an extension line circuit.

On PBX failure both sets of relay contacts change over and connect the extension telephone directly to the public exchange line. Since many public exchange lines are still conditioned for decadic dialling the designated extension telephones may have to be capable of this type of signalling. Several approaches are adopted.

The simplest method is to install a decadic dial phone on the designated extensions. This has the disadvantage that most of the time the user has to put up with slow speed dialling, and will probably not have access to all the PBX facilities as the star and square keys are not provided on such instruments.

Probably the most attractive method is to install a dual standard (i.e. decadic/MF switchable) telephone to the designated extensions, and instruct the users to switch to decadic dialling when they become aware that the PBX has failed. Since the PBX dial tone and the public network dial tone will be different this should not be a problem.

Other methods involve providing telephones of each type and instructing users to switch to the relevant telephone, or even unplug the MF telephone and plug in the decadic telephone when dropback occurs.

# 10.3 Basic PBX Call Processing

## 10.3.1 Introduction to Call Processing Descriptions

In order to put the descriptions of the various units within the PBX into context, this section describes two types of call through the PBX. In the first case, a simple extension to extension call is described, this is followed by a brief description of an extension to trunk call. These descriptions include information that is based on software aspects, especially regarding the exchange database.

## 10.3.2 An Extension to Extension Call

To aid the explanation of this example it is necessary to use actual port and channel numbers. These are given below. It is also assumed that the calling extension is conditioned to MF signalling.

The calling extension, extension number 255, is connected to port 1 of the exchange and is assigned to TS 1 of speech highway 1 (TS 1/SPCH 1). The called extension, number 246, is connected to port 42 and is allocated TS 10 of speech highway 2 (TS 10/SPCH 2). Note there is absolutely no correspondence between port numbers and extension numbers. The exchange administration may allocate any of the available extension numbers to any port, the relationship between the two being maintained in the exchange database.

The extension line circuits are also connected to signalling highways as follows:

❏ Extension 255 is allocated TS1 of signalling highway 1 (TS 1/SIG 1).

❏ Extension 246 is allocated TS10 of signalling highway 2 (TS 10/SIG 2).

Signalling words from extension 255 to the control unit will be stored at location 1 of the incoming signalling store, and those from extension 246 will be stored at location 42.

Signalling commands in the opposite direction, i.e control unit to line circuit, is effected by the control unit placing the data in the outgoing signalling store location 1 and 42 respectively.

This call will use the following resources:

❏ Extension Line Circuit 1 (TS 1/SPCH, TS 1 1/SIG 1)

❏ Extension Line Circuit 42 (TS 10/SPCH 2, TS 10/SIG 1)

❏ MF Receiver 3 Common Equipment Port 19 (TS 19/SPCH 6 in, TS 19/SIG 6 out)

❏ Dial Tone (Common Equipment Port 17 TS 17/SPCH 6 out)

❏ Ring Tone (Common Equipment Port 18 TS 18/SPCH 6 out)

Figure 10.17 is a simplified diagram showing the call to be established.

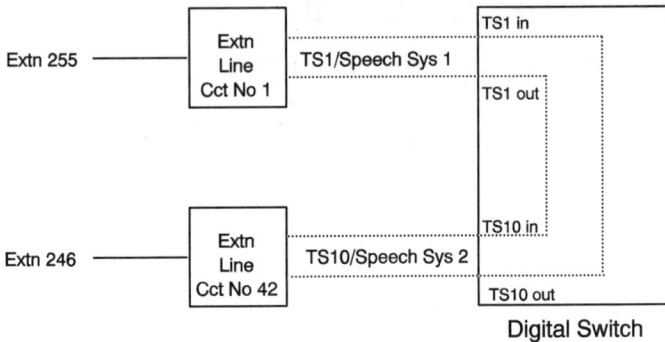

*Figure 10.17 Basic extension to extension call – Extn 255 calls Extn 246*

The first event to be registered is extension 255 going off hook. The extension loop is detected by the extension line circuit, which responds by sending an 8-bit signalling word representing a new off hook signal in TS 1/SIG 1.

The signalling word is stored at location 1 in the incoming signalling store. When this store is next scanned by the exchange control unit, the fact that a new call has been initiated by port 1 is recognised.

The next step is for the control unit and its associated software to access the exchange subscriber database to determine the class of service attributes for this extension, to ensure, for example, that this extension is permitted to make outgoing calls, and if any route restrictions are imposed on the extension (e.g. ability to access the public network). The database will also reveal that the extension is equipped for MF dialling,

When the control unit is ready to process the call, it sends a power up command to the extension line circuit, by placing the appropriate signalling word in location 1 of the outgoing signalling store. This word is then sent out in TS 1/SIG 1 to the extension line circuit, which responds by activating the codec and filter elements of the line circuit.

The control unit must then select a free MF receiver, for example MF Receiver 3, which is allocated TS 19 of the common equipment card incoming speech highway (TS 19/SPCH 6). The control unit then sends a connection command to the digital switch controller to connect TS 1/SPCH 1 (in) to TS 19/SPCH 6 (out).

It is also necessary to send dial tone to the calling extension. Dial tone is supplied by tone generator 1 which is allocated TS 17/SPCH 6, thus the switch controller is instructed, by the control unit to connect TS 17/SPCH 6 (in) to TS 1/SPCH 1 (out).

The current situation, with the calling extension receiving dial tone but yet to dial the first digit, is shown in Figure 10.18.

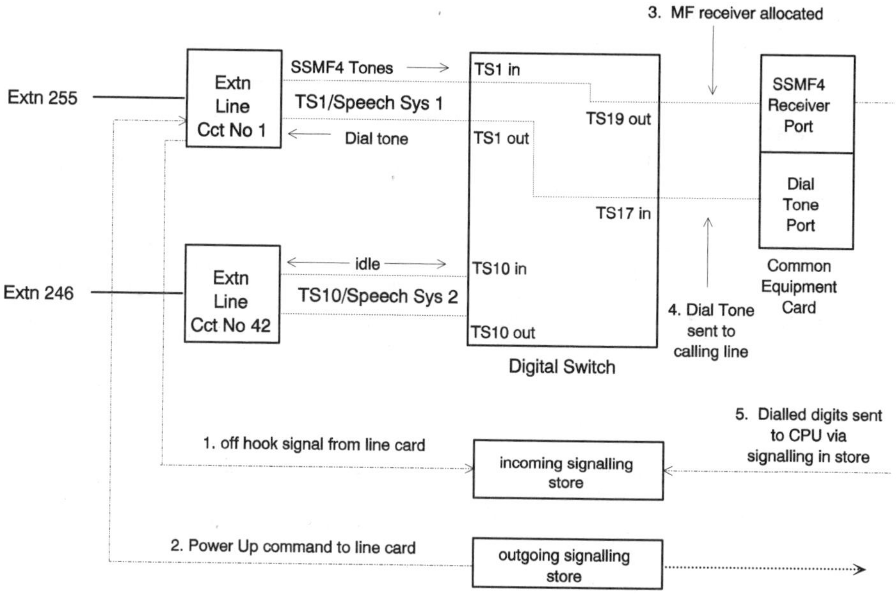

*Figure 10.18 Extension to extension call at the dial tone stage*

The caller then dials the first digit, 2. The MF tones from the telephone are PCM encoded in the extension line circuit codec, and pass in TS 1/SPCH 1 to the digital switch, through the switch, out on TS 19/SPCH 6 to MF receiver 3.

The MF receiver decodes the tone pair and send a binary coded signalling word in TS 19/SIG 6 to the incoming signalling store. As the control unit has allocated MF receiver 3 to this call, it scans its associated location in the incoming signalling store more frequently than previously and registers that the number 2 has been dialled.

The control unit then instructs the switch controller to release the connection between the dial tone port and the calling extension. The digit is checked to ensure it is a valid first digit. If not, number unobtainable (NU) tone is sent to the caller, by connecting the NU tone port to the calling extension port. Assuming the digit is valid, the next two digits are processed in the same way.

When the third digit has been dialled, the control unit can examine the database record to ensure that the called extension is free, and can accept incoming calls. The database also reveals that this extension is connected to port 42.

The control unit then sends an apply ringing command to the called extension line circuit, by putting the appropriate signalling word into location 42 of the outgoing signalling store. This word is then sent in TS 10/SIG 2 to port 42 extension line circuit, where it is decoded. Ringing current is then sent to the called extension.

At this time it is also necessary to send ring tone to the caller. This is achieved by instructing the switch controller to connect the ring tone port to the calling extension port. This situation is shown in Figure 10.19.

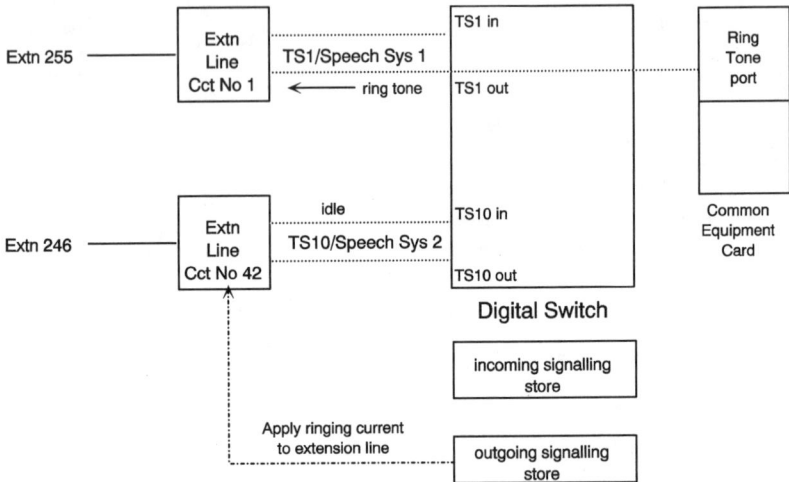

*Figure 10.19 Extension to extension call at the ringing stage*

When the called extension answers, the loop is detected by the extension line circuit which cuts off ringing current, and sends an signalling word representing called extension answer in TS 10/SIG 2. This message is stored at location 42 of the incoming signalling store. When this store is next scanned by the control unit, called extension answer will be detected, and the control unit sends a power up command to the called extension line circuit, and releases the ringing tone to the calling extension. Finally the call is connected through by instructing the switch controller to make the connections:

❑ TS 1/SPCH 1 (in) to TS 10/SPCH 2 (out)

❑ TS 10/SPCH 2 (in) to TS 1/SPCH 1 (out).

The call is now in the speech phase, and will remain connected until either party clears down.

Either user replacing the handset will cause the removal of the extension loop which will be detected by the line circuit which then sends the extension cleared signalling word to the control unit. On receiving this message, the control unit will then instruct the switch controller to release both digital connections made for this call. The control unit will also send power down command to both line circuits via their respective signalling highways.

During the setting up of the call, the control unit has maintained a call record which will eventually include such data as calling extension, called extension, type of call, dialled digits, current phase of the calling (e.g. dialling, ringing, connected etc), time of answer and time of clear.

This data can be extracted from the call records if a suitable call logging device is connected to the PBX. If call logging equipment is not connected, all detail about the call is lost when it is cleared, as the memory space is required to be cleared to be used for another call.

## 10.3.3 Extension to Trunk Call

In this explanation extension 255 is to be connected to a public network number. Call processing in this case will use the following resources:

❑ Extension Line Circuit Port 1 (TS 1/SPCH 1)

❑ Exchange Line Circuit Port 130 (TS 2/SPCH 5)

❑ MF Receiver 3 Port (TS 19/SPCH 6 in, TS 19//SIG 6 out)

❑ Dial Tone Port (TS 17/SPCH 6 out)

❑ Dial Tone Receiver Port (TS 27/SPCH 6 in, TS 27/SIG 6 out)

The call proceeds as far as dial tone in exactly the same way as for the extension to extension call. When the first digit, typically 9, is dialled, the control unit has to determine from the extension class of service record in the exchange database, whether or not this extension is permitted to make outgoing calls, and if so will select a free exchange line circuit. In this case Trunk 2 is selected. The digital switch controller is then instructed to connect the dial tone receiver to the incoming exchange line circuit.

Concurrently the user may already be dialling the digits of the public network number required. As each digit is decoded in turn by the MF receiver, a binary coded word representing each digit is sent to the control unit via the incoming signalling stores. The control unit maintains a record of all digits dialled.

As soon as a positive response has been received from the dial tone detector, each digit will in turn be placed in the outgoing signalling store to be sent to the exchange line circuit. The control unit will also free the dial tone detector and arrange for the incoming exchange line to be switched to the calling extension line circuit.

The exchange line circuit has then to pulse out each digit, in turn, inserting the required inter-digit pauses. As the control unit is unable to determine how many digits are to be dialled, the MF receiver must remain switched to the calling extension line circuit at this time.

Note that at this stage, it is probable that the exchange line circuit is dialling out the first few digits before the user has dialled the last digit of the required number. The is the situation represented by the diagram in Figure 10.20.

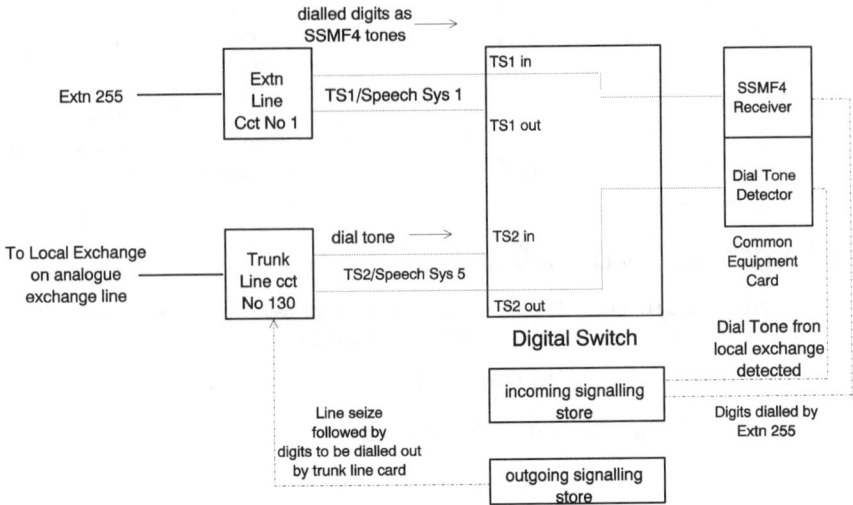

*Figure 10.20 Extension to trunk call – dialling stage*

The public network then processes the next stages of the call. Assuming the call is to be successful, ringing tone is sent back from the public network to the PBX and ultimately to the caller (note ringing tone is not supplied by the PBX).

When the called subscriber answers, the fact is signalled from the public exchange to the exchange line circuit in the PBX, typically by reversing the polarity of the battery ɔotential applied to the exchange line. The exchange line circuit then sends the appropriate signalling word to the control unit. The control unit releases the MF receiver, and the call is finally connected through the digital switch in both directions.

The call is now in the speech phase, and continues until cleared, either by the calling extension, or the called subscriber. If the called subscriber clears, the public network will send a clear signal to the exchange line circuit, which responds by transmitting a signalling word to the control unit, which causes the call to be released.

# 10.4 Chapter Summary

This chapter has introduced the concepts and technology of the PBX by describing how PBXs and their smaller relative the Key System, meet user requirements for internal communications within their own premises, and how such systems also provide connectivity to the PSTN.

A brief analysis of the types of system available was given in order to show that systems are available for all sizes of company, from small businesses such as estate agents to large corporations.

The chapter also provided an introduction to the user facilities generally available to PBX users, but which are not relevant in a public network.

# 11

# *Integrated Services PBX (ISPBX)*

## 11.1 Voice and Data Switching on a Single Exchange

### 11.1.1 Introduction to Data Switching

The traditional method of connecting computers to terminals and other peripheral devices such as printers involved a mix of dedicated point to point wiring and modems where the distance involved precluded a direct connection. In some cases a pair gain was achieved, by using time division multiplexers, and later further gains were achieved by the introduction of statistical multiplexers. In all these cases, the terminal device was effectively permanently connected to the host computer, and thus could only access applications running on this machine.

Many sites had more than one main computer, all running different applications. For example, as a student at college, I had access to a DEC PDP 11, a DEC VAX 370 running the VAX/VMS operating system, and another VAX running Unix. All terminals, and the ports of all computers, were wired to a Digital Distribution Frame (DDF), where the operators were able to physically connect any terminal to any computer simply by plugging a patch cord between the relevant two sockets on the frame.

However the instructor always had to arrange for the terminals in the lecture room he was using to be manually patched to the correct machine at the start of each session. This was a far from satisfactory system as terminals would occasionally appear dead, due frequently to faulty patch cables, and occasionally, I expect, due to a incorrect patch by one of the operators. Even without faults it was a clumsy system which, in a busy business environment, would have given any real computer user a most frustrating time.

Later the system was improved by installing an intelligent data switch. The switch simply replaced the DDF and all terminals and computer were wired to the switch in exactly the same way as they had been wired to the DDF. Logging onto one's favourite computer became a joy by comparison with times past. On switching on the terminal, the data switch sent a log on message indicating what machines were available, and prompting the user to enter which machine was required for the session.

The data switch would then make a digital patch, or connection, between the terminal and a port of the selected computer, if one was available, without the need for phoning for a connection, and the resulting delays from a problem fraught manual patching system. Almost immediately after the name of the required host computer had been entered, a prompt from the selected computer would appear inviting the user to log on for the session. If all ports to the selected computer were in use a suitably worded message would appear on the screen. If necessary the session could be terminated at any time, and the user had the opportunity to log on to one of the other machines. Access security was generally by a password, each user being given a password for those computers he was authorised to use. The data switch was transparent to the password system, allowing a staff user for example to log on to any computer from any terminal.

I should point out that this is a fairly simple description of the data switch used by our college and such devices are normally capable of far more than circuit switching data terminals to computers. The Case Beeline, with integral data switch (DCX) can also function as a Store and Forward, i.e. it uses message switching techniques to provide an electronic mailbox system. In this mode it is also capable of data code and speed conversion, so that messages sent at one baud speed, say 2400 baud, in ASCII code, can be received at a different speed, say 50 baud, in Murray code.

## 11.1.2 Options for Data Switching

Another method of connecting various terminals and devices together, in such a way that a user can have access to a number of facilities, is to use a Local Area Network (LAN). Although no digital switch exists physically it can be assumed to be distributed into the transmission media used for the LAN, as the media is allocated on a time division basis. Data items travel into the LAN from one terminal during their allocated time slot and appear to leave the LAN at the device to which a session has been established. During the next time slot data items can move between any two other terminals connected to the LAN.

Of all methods of data switching currently in use, especially over large geographical areas, it is likely that packet switching is the most popular and its concepts, if not its implementation, best understood. Packet switching offers the ability to construct a wide area data network with the advantages of circuit switching, such as fast call set up and bi-directional communications, while offering some of the advantages of message switching such as the ability to provide speed and code conversion.

However, packet switching offers several distinct advantages over other methods, notably the fact that error correction and detection is built into the system, although in some applications it may be necessary to provide a further correction system for maximum data integrity. The other advantages include the ability to access more than one host computer simultaneously, due to the concept of having a number of logical channels, each of which can be used to access a different host computer, or application. Of course not all users will require this ability, and as in all things in the IT world, the optimum solution for a given communications problem is a matter of providing a system which meets the user requirements most cost effectively, not with the most advanced technology available.

# 11.2 Using an ISPBX

## 11.2.1 The PBX as a Data Switch

This section discusses the use of a PBX as an alternative to point-to-point wiring and local area networks for connecting data terminals together. This should not be viewed as a superior alternative, but as another option which should be considered. In fact, there will be many applications where the PBX is not a viable solution, and systems such as Token Ring LANs, or private packet data networks are best able to meet the users' needs.

Whether discussing the concept of the PBX as a data switch, or the ability of public exchanges such as System X to switch data traffic, there is a convenient common factor linking data from character based terminals and digital telephony, both are based on an 8-bit format. Digital telephony uses 8-bit PCM, while computer terminals tend in the main to use 8-bit ASCII or EBCDIC codes.

Digital switch designs originally produced for PCM encoded voice are then relatively easy to adapt to switch data, thus opening up the possibility of using the PBX in much the same way as the college example above.

Some advantages of this approach to data switching are:

❏ As most organisations must purchase a PBX to meet their telephony requirements, the cost of purchasing extra capacity, and the necessary interface components, to cater for data requirements may well be less than purchasing a dedicated data switch or installing a LAN.

❏ Systems can be designed to make use of existing telephony wiring to carry voice and data traffic simultaneously. This removes some of the requirement to install new cables in existing trunking which may already be nearly full.

❏ By using digital transmission over the telephone wiring, modems are not required and thus the data transfer rate can be far higher than that achievable with conventional analogue lines.

❏ Using the PBX as a voice and data switch may be considered as an entry route into private and public ISDN systems, by initially providing local switching of data resources prior to the provision of a wide area integrated voice and data network.

Chapter 13 is devoted to describing the concepts of Integrated Services Digital Networks more fully, so here we will look only at outline concepts as they are related to private networks. In actual fact there are few conceptual differences between public and private ISDN systems.

## 11.2.2 Integrating Voice and Data, and the ISPBX

The basic ISPBX concept is to integrate a range of voice and data switching facilities on a PBX by providing a switching and control system that is capable of handling both types of traffic. But, as we shall see later on in this section, the full ISPBX concept extends much further than this.

The ISPBX must also, by definition, be capable of providing access to the emerging public ISDN systems. Some of the features, which distinguish the ISPBX from a standard voice only PBX, are as follows:

❑ Ability to connect digital telephones, providing high quality voice transmission

❑ Ability to connect digital devices, e.g. Fax, DTEs, PCs, printers

❑ Simultaneous voice and data calls over single extension line

❑ Ability to access public and private ISDNs, using the appropriate common channel signalling system, i.e DASS or DPNSS (in the UK)

❑ Ability access to a various range of other services e.g. Telex and Packet Switch Networks.

## 11.2.3 A Typical ISPBX

As example of an ISPBX we will briefly examine the Integrated Services Digital Exchange (ISDX) manufactured by GPT. In addition to handling digital telephones and data devices, the ISDX must also be able to handle existing telephones and key systems, as the diagram in Figure 11.1 shows.

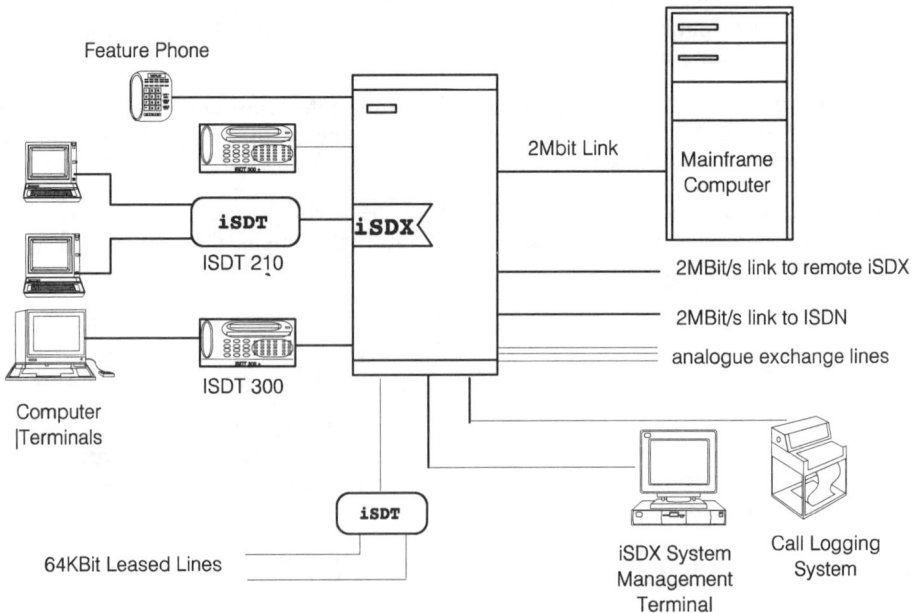

*Figure 11.1 The GPT ISDX system architecture*

This diagram provides a useful basis to describe a typical ISPBX, as it shows a wide variety of terminal equipment, along with the networking features of these exchanges.

On the left hand side of the diagram, three types of telephone are shown. The GPT telephone is a standard instrument with MF dialling and recall features giving the user access to the whole range of normal PBX features. The GPT Featurephone which provides all the functions of the standard telephone, plus other features such as a display of dialled digits and call duration, and programmable keys for storage of personal directory numbers. The standard telephone and the featurephone are both analogue instruments capable of working on most types of exchanges.

## 11.2.4 The Integrated Services Digital Terminal (ISDT)

The third type of telephone is the Integrated Services Digital Terminal 300 (ISDT 300). This instrument has been specifically designed to provide voice and data access to the ISDX. The ISDT is connected to the exchange using the normal telephone wiring via a standard telephone jack. However voice and data are both transmitted over the telephone cable in digital form, rather than analogue.

To prevent any problems being caused by plugging an ISDT into a jack supplied for a standard telephone, the telephone jack and its associated socket are wired so as to use different connectors from those used for normal telephones.

The transmission system employed is a proprietary system which provides two digital channels, each capable of 64 Kbit/s operation, time division multiplexed with a 16 Kbit/s common signalling channel to give a total capacity of 144 Kbit/s, in a form that is strongly based on CCITT ISDN recommendations.

Within the terminal analogue voice signals from the handset are PCM coded/decoded and thus transmitted in digital form to the exchange in one of the two 64KBit/s channels, while a Data Terminal Equipment (DTE) connected at the rear of the ISDT using standard V24/X21 interfaces will use the other as shown in Figure 11.2.

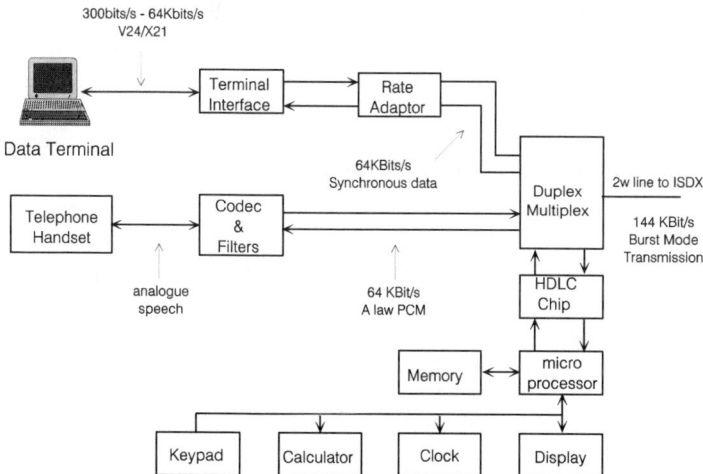

*Figure 11.2 The GPT ISDT 300 showing data equipment connected*

This digital interface can cope with most current types of DTE, At the present time many DTEs are unable to transmit and receive data at a rate as high as 64 Kbit/s, so some form of re-timing, or data rate adaptation, must be incorporated in the ISDT to correctly match the speed of the device to the channel.

Over a switched connection between two ISDTs, the distant ISDT must detect and re-time the incoming 64 Kbit/s signal to reproduce the original 2400 baud sequence. The ISDT can be programmed to use one of two possible rate adaptation methods. The first of these is in accordance with CCITT recommendation V-110 and involves 3 stages of rate adaptation. Thus it is essential that both ISDTs should have knowledge of the original transmission rate so that the received signal can be correctly reconstituted. This can be achieved by the ISDTs making use of the common signalling channel to transmit this and other information regarding the connected DTEs.

The second method of data rate adaptation is multi-sampling in which the incoming data bit stream is sampled at 64 Kbit/s, and a bit with the same value of the sample is transmitted. Thus a binary 1 produced at 2400 baud would be transmitted over the 64 Kbit/s channel as a string of 26 (64K/2400) continuous binary 1s.

The process of multi-sampling can lead to a form of built in error protection. Each bit of the original traffic is represented by the transmission of several bits in 64 Kbit/s channel. Any errors occurring to this 64 Kbit/s stream during switching or transmission are very unlikely to affect more than 50% of the bits transmitted, therefore a majority vote system can be used to predict the polarity of the original traffic should some errors occur. The protection afforded by such a system is not particularly powerful, but is adequate for many applications.

Another function of the ISDT is to strip the redundant start, stop and parity bits from the DTE's asynchronous signal, so that only eight bits of character data are actually transmitted to the exchange. At the other end of the connection the receiving ISDT must reinsert these bits prior to the data being sent to the connected DTE.

It should be noted that the use of the 64 Kbit/s channel does not in anyway increase the throughput of the system above that which the DTEs themselves are capable of.

All signalling between the ISDT and the exchange utilises the 16KBit/s common signalling channel. The system employed is message based and has much in common with the public ISDN signalling system. This has to be the case since it will be possible to access the public ISDN from the ISDT, in the same way that it is possible to access the PSTN from a normal extension.

## 11.2.5 ISDT Variants

Several variants of ISDT are available, one incorporating featurephone facilities, and another, the ISDT 210, which does not include a digital telephone. Instead it is equipped for two identical X21/V24 channels.

Uses of this data only ISDT are shown in Figure 11.3, and include:

❏ Providing a connection between the ISDX and private digital circuits such as Kilostream

❏ Providing connections for two DTEs over one line. Each DTE can be involved in entirely separate data calls

❏ Providing access for two packet mode terminals.

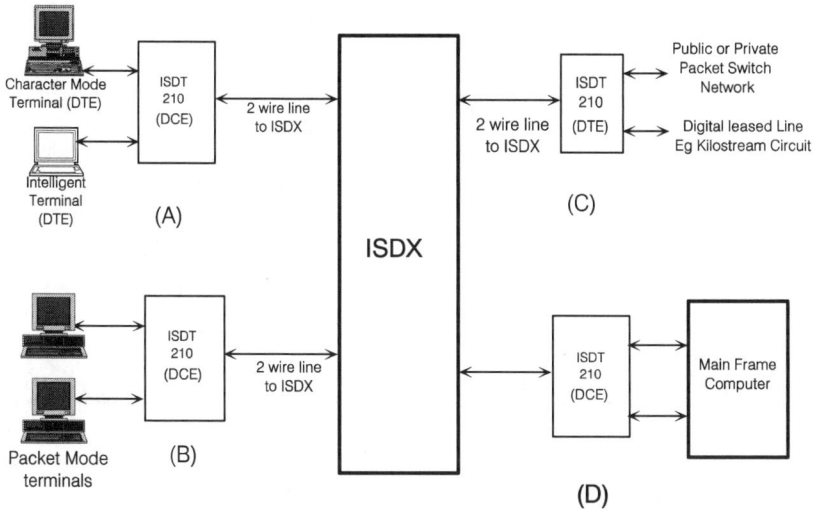

*Figure 11.3 GPT ISDT 210 applications*

## 11.2.6 ISDX Interfaces to Other Equipment

The ISDX System Architecture diagram shows that several other types of equipment can be connected to the ISDX. For example the diagram shows facsimile machines, key systems, and voice messaging equipment.

The diagram also shows a 2Mbit/s connection to a main frame computer. This particular interface conforms to the European Computer Manufacturers' Association (ECMA) standard.

Depending upon the capacity of the ISDX, and its role in a network, one or more ISDX consoles may be connected to provide normal operator facilities. A further option is to use a screen based operator position based on a PC, the advantage of this option being that an integrated software system such as a hotel or directory package can be used.

## 11.2.7 ISPBX Software Considerations

The basic record of each subscriber in the exchange database of an ISPBX must contain more information than the relatively simple data held in a voice only PBX. An important data item which must be stored in the subscriber record is the type of terminal equipment connected. This field will show whether the connected equipment is a basic telephone, an ISDT, a computer terminal or a fax machine. It will also be necessary to store data regarding the types of service to which data users may be connected.

Within the call control software, routines must be provided to ensure only compatible equipments are connected together, e.g. telephone to telephone or ISDT, facsimile to facsimile, computer to printer and so on. These routines will have to check the relevant fields in the subscribers' record to determine the equipment in use and confirm via look up tables whether the equipment is compatible.

### 11.2.8 MMI and System Management

Management of the system is carried out from a dedicated System Management Terminal (SMT). Typically the SMT will be used for making changes to the exchange database when extensions or trunks are added, moved or removed from the system.

The SMT also provides access for instigating various diagnostic checks of the exchange should faults occur, although in most cases faults will have been isolated automatically by in-built software test routines.

## 11.3 Chapter Summary

This short chapter has introduced the concept of Integrated Services Digital Networks, by examining how a single node in such a network, namely the ISPBX, can be used to provide local switching of digital data systems, in addition to its conventional role of switching telephony voice signals.

In order to be able to handle such a wide variety of services, there must be some element of commonalty about these signals. While the original inputs, are very much dissimilar, by arranging that the digital representation of these signals is based on a common 8-bit format it is possible for the switching and transmission systems associated with the ISPBX to handle them all quite transparently.

The exchange database has more information regarding equipment connected to user ports, and software within the control system of the exchange ensures that when a call is established, the two extensions connected have compatible terminal equipment.

## 11.4 Acknowledgment

The assistance of GEC Plessey Telecommunications Ltd in the preparation of this chapter is gratefully acknowledged.

# 12

# *Private Digital Networks*

## 12.1 Introduction to Private Integrated Voice and Data Networks

The aim of this chapter is to provide a gentle introduction to the concepts of integrating voice and data over a single network to produce an Integrated Services Digital Network. In the previous chapter we saw that voice and data could be handled by a single digital switching system. In this chapter we take the idea a bit further to investigate how voice and data can be carried over a single digital network. We will briefly describe how a private ISDN type network can be built from digital exchanges interconnected by common channel signalling transmission links, and to describe some of the features of such a private network.

The concept of private communications networks is not new. As described in earlier chapters, for many years large organisations have operated, or leased private telephony systems based on analogue switches and FDM transmission systems. Prior to the introduction of public packet switched systems, private data networks were the only way to connect computer systems together over wide geographical areas.

Today a private digital network is a switching and transmission system designed to carry voice and data traffic between the various locations occupied by a medium to large sized organisation. In much the same way as an individual PBX provides a cost efficient means of providing communications between users at a single location, so the aim of a private network is to provide an efficient communications system that is tuned to the requirements of the organisation. However the introduction of a private system offers benefits other than cost savings as private networks offer functionality and a degree of security not available on public networks.

Private networks will not be able to satisfy all the communications needs of an organisation. Communications to the organisation's customers and suppliers can only realistically be achieved via the public network, and in many private systems, the public network is used as a fall back, for those occasions when the private system fails or becomes congested.

## 12.2 Private Digital Networks

### 12.2.1 Typical Private Network Architecture

Essentially a digital private network will consist of a number of digital SPC exchanges, typically linked by 2MBit PCM transmission systems. A common channel signalling system is used to provide fast call set up and PBX features network wide.

If the network is to provide voice and data communications between locations, the digital exchanges must be ISPBX types as described in Chapter 11. An example private digital network is shown in Figure 12.1.

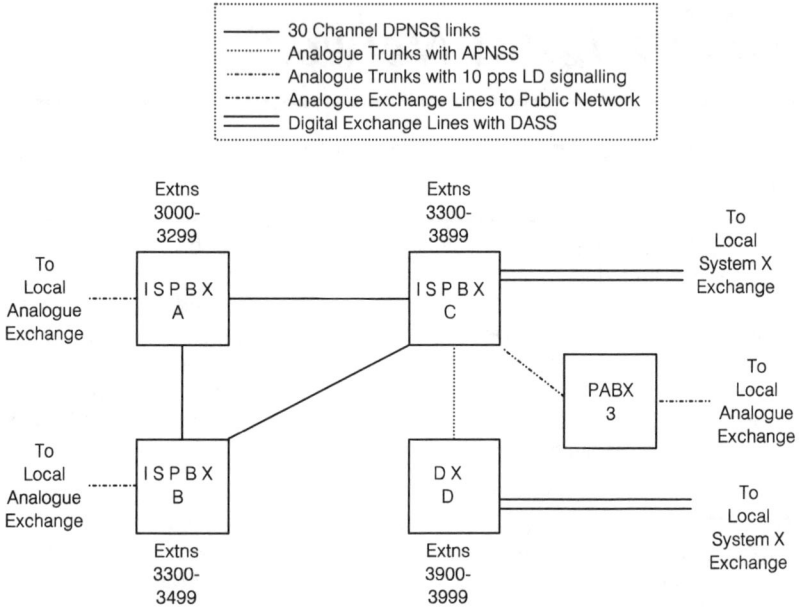

*Figure 12.1  Typical private digital network*

This example is typical of a network undergoing modernisation. The network was originally totally analogue and consisted of five analogue exchanges interconnected by analogue transmission links, and employed a variety of trunk signalling systems, including 10pps Loop Disconnect and AC systems. The modernisation process is almost complete, and the diagram shows that digital exchanges and transmission links predominate, and that there is still a requirement for interworking with older exchanges.

In the modernised network three of the exchanges are ISPBXs which are interconnected by 2Mbit/s digital links, each of which provides for up to 30 trunks and a common signalling channel. A fourth smaller ISPBX is linked into the network by an analogue private circuit to provide five voice only trunks and a common signalling channel.

The fifth exchange in the network is not a digital exchange. Rather it is an old 200-line PABX 3, which uses Strowger technology, and is connected into the network by 10 loop disconnect signalling trunks terminating at ISPBX C. Eventually the 200-line PABX will also be replaced by an ISPBX.

## 12.2.2 The Role of the Digital Signalling System

The signalling system used in the digital portion of the network has been specifically developed for use on private networks by British Telecom and a number of exchange manufacturers. The hope is that the system will be used in its standard form by many manufacturers and thus will eventually allow private exchanges from various manufacturers to be linked together without incompatibility problems.

The system, known as Digital Private Network Signalling System (DPNSS), is based heavily on CCITT signalling system 7 and public ISDN signalling. Like CCITT #7, DPNSS is based on a layered structure:

❑ Layer 1: the physical connection between the exchanges.

❑ Layer 2: based on HDLC frames for transferring signal units, i.e. the DPNSS messages. It is the Layer 2 frame which includes the channel identity.

❑ Layer 3: carries DPNSS messages and procedures.

Although there is generally only one physical connection, the 2MBit/s pathway is treated by DPNSS as 30 traffic and 30 signalling channels. DPNSS treats each channel as a separate link, with signalling channel 1 devoted to traffic channel 1, and so on. A Link Access Protocol(LAP) or procedure is used for each of these channels, thus if traffic on channel 2 is selected for a call, the relevant DPNSS message is passed to the LAP for channel 2, and transmitted in an HDLC frame which contains the channel 2 identity as illustrated in Figure 12.2.

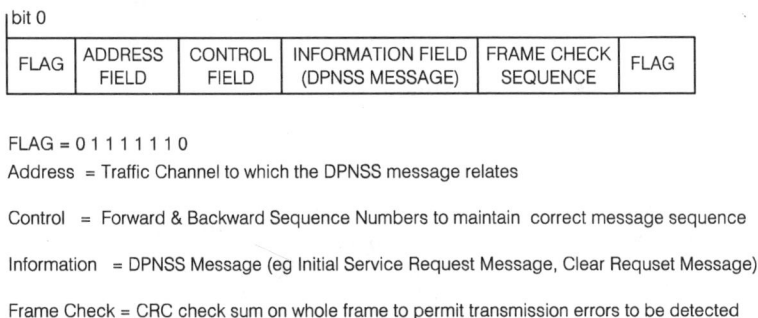

| bit 0 | | | | | |
|-------|-------|-------|-------|-------|-------|
| FLAG | ADDRESS FIELD | CONTROL FIELD | INFORMATION FIELD (DPNSS MESSAGE) | FRAME CHECK SEQUENCE | FLAG |

FLAG = 0 1 1 1 1 1 1 0

Address = Traffic Channel to which the DPNSS message relates

Control = Forward & Backward Sequence Numbers to maintain correct message sequence

Information = DPNSS Message (eg Initial Service Request Message, Clear Requset Message)

Frame Check = CRC check sum on whole frame to permit transmission errors to be detected

*Figure 12.2 DPNSS message transmitted in an HDLC frame*

## 12.2.3 Analogue Private Network Signalling System (APNSS)

Provision has been made for digital exchanges to be connected by analogue transmission systems. Such an arrangement may well be necessary if considerable investment was made in FDM plant just prior to the introduction of digital switches, and the network administration feel that the capital cost of the FDM system can not be written off until it has come to the end of its planned economic life.

Equally where a digital route between two exchanges is simply not available, or can not be cost justified because substantially less than 30 trunk channels are required on the route, digital networking features can still be obtained by using a variation of DPNSS, called Analogue Private Network Signalling System (APNSS).

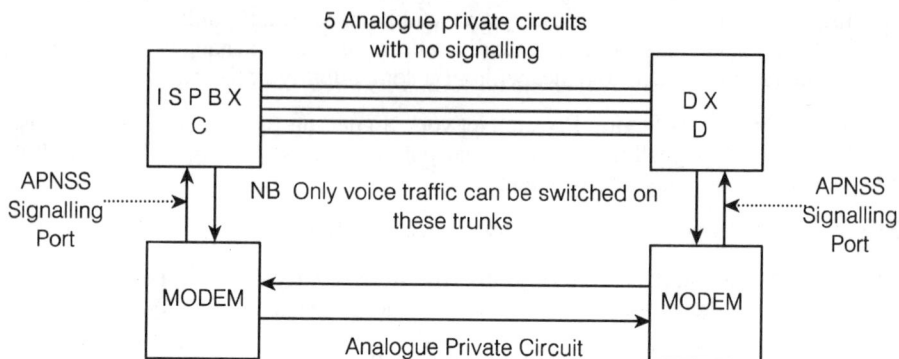

5 Analogue private circuits
with no signalling

I S P B X
C

D X
D

APNSS
Signalling
Port

NB  Only voice traffic can be switched on
these trunks

APNSS
Signalling
Port

MODEM

MODEM

Analogue Private Circuit

*Figure 12.3 An APNSS link*

In the example shown above, five analogue trunks are provided over private wires between the ISPBXs C and D. An additional sixth analogue private circuit complete with data modems at each end provides the common signalling channel between the two exchanges as shown in Figure 12.3. Note that analogue channels can not be used for data traffic, and therefore APNSS controls only telephony calls and requests for supplementary services.

## 12.2.4 What does DPNSS provide?

It would be impossible to list all the functions provided by DPNSS in this book, so a brief explanation covering only the major topics will be given.

DPNSS was developed by a team of exchange manufacturers and British Telecom, and is specified in a document called British Telecom Network Requirement (BTNR) 188. DPNSS must be able to work alongside other signalling systems, including analogue systems such as SSDC5, SSDC10, SSAC15 etc. The interworking, or gateway requirements are detailed in BTNR 189. Some the major features of DPNSS networking are:

❏ Fast call set up to any extension in the network

❏ Universal numbering scheme, allocating each extension a unique network number

❏ Centralised Operator Working, all calls for the operator are routed to a single location, irrespective of their point of origin

❏ PBX features available network wide (e.g. call back when free)

❏ Network Management and Maintenance Capability

The major PBX features specified within DPNSS are listed later.

## 12.2.5 Fast Call Set Up and Efficient Use of Trunk Capacity

Calls from one extension to another on the same ISPBX are rapidly set up by the SPC call processing software, ring tone being extended to the caller, as appropriate, just after the last digit has been dialled. On a network call, the call set up is almost as quick, as the call processing software at each ISPBX involved in setting up the call, exchanges DPNSS messages over the 64 KBit/s common signalling channel as shown in Figure 12.4.

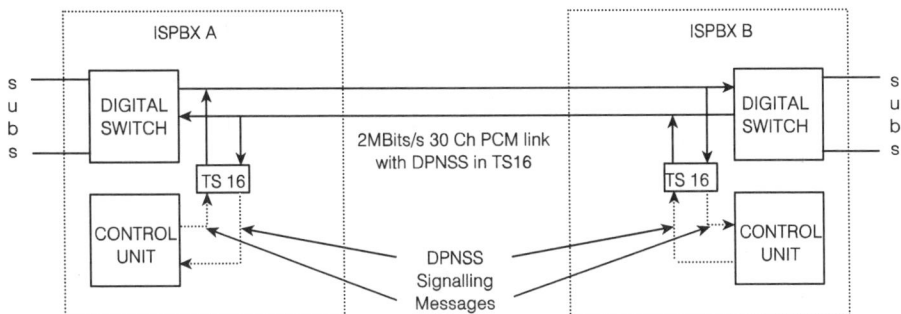

*Figure 12.4 DPNSS signalling between two ISPBXs*

DPNSS is designed to make as efficient use as possible of the 30-channel inter-exchange links. One method of achieving this efficiency is to use fast call set up and clear down to ensure that traffic channels are held for as short a time as possible, when not actually in use for conversation or data traffic.

A further example involves the action taken when a call encounters a busy condition. In earlier networks, calls which encountered a busy trunk route, or a busy extension at the destination exchange, caused trunk channel capacity to be wasted, as the exchange encountering the busy condition returned equipment busy, or engaged tone, to the caller over the speech channel that would have been used for the call.

The caller could have held this channel listening to busy tone for almost any length of time, thus denying other users access to the channel for other calls. DPNSS overcomes this inadequacy by immediately clearing the call on encountering a busy condition, and sending a clear message with the reason for call failure back to the originating exchange. The originating exchange is then responsible for sending a local busy tone to the caller to inform him of the situation. This approach saves tying up the channel originally booked for the call.

# 12.3 The Operation of DPNSS

## 12.3.1 Simple Call Establishment Within a DPNSS Network

The diagram in Figure 12.5 shows the establishment of a simple call, and the sequence of events if the call fails due to a busy or incompatible extension.

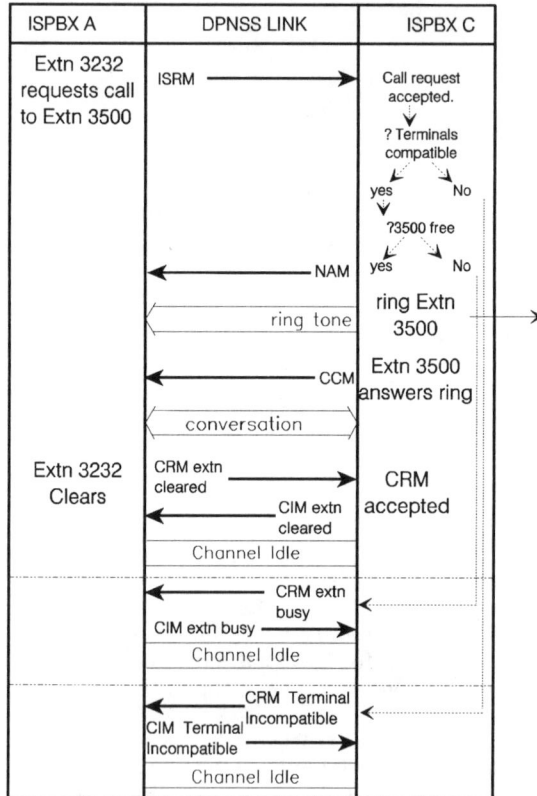

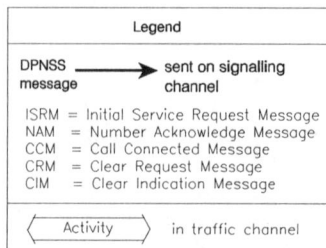

*Figure 12.5 Establishment of a simple call within a DPNSS Network*

Extn 3232 at ISPBX A dials 3500, ISPBX A recognises that this call has to be routed over the DPNSS controlled link to ISPBX C, and sends an Initial Service Request Message (ISRM) to ISPBX C. This ISRM Complete message will containing the following information:

Channel Selected = 2 (Transmitted in address field of HDLC frame)
Message Type = ISRM Complete
Service Indicator Code (SIC) = Telephony      Transmitted in
Originating Line Identity (OLI) = Extn 3232      Information Field
Calling Line Category (CLC) = normal extension      of the HDLC Frame
Destination Address (DA) = Extn 3500

The SIC indicates that this ISRM is for a telephony, rather than data call. Telephony calls will normally use 64 Kbit/s PCM, but if there is not a digital path available to the destination address, these calls can be routed over analogue links. An SIC has also been allocated for specifying a 32 KBit/s ADPCM telephony call. The SIC for a data call will also specify the data rate involved, whether the traffic is synchronous or asynchronous, and if required, how many stop bits are used.

The OLI gives the extension number, or other identification, of the calling line.

The CLC indicates, for example, whether the calling line is an ordinary extension, an operator or an incoming PSTN line.

The DA specifies dialled digits, i.e. the required extension number.

In most cases the DA field of the ISRM will contain all the dialled digits. In this case it is known as an ISRM-complete (ISRM-C). In some circumstances it may be necessary for the originating exchange to send an incomplete ISRM (ISRM-I) containing as little as one digit of the required extension. An ISRM-I must be followed by Subsequent Service Request Messages (SSRM) until all the required digits of the destination address have been transmitted.

On receiving the ISRM, ISPBX C checks that the called line is compatible with the calling line, i.e. both are telephony extensions, or are compatible data extensions, and that the required extension has no call diversion in operation and is free.

## 12.3.2 DPNSS Call Failure

If ISPBX C is unable to set up the call, it will immediately send ISPBX A a Clear Request Message (CRM), which will include the reason for the failure e.g. Extension Busy, Number Unobtainable, Address incomplete, Extension out of service, Incoming calls to the required extension are barred for incompatible extensions. On receiving the CRM, ISPBX A sends a Clear Indication Message (CIM), the channel reserved for the call is then released in both directions.

## 12.3.3 Successful DPNSS Call

If ISPBX C is able to connect the call, it sends a Number Acknowledge Message (NAM), which indicates that all required digits have been received. The NAM will also include the Called Line Category (CLC) and Calling Line Identity (CLI) which are the same type of information as is contained in the OLC and CLC of the ISRM. At this stage ISPBX A and C connect the PCM trunk back through to the calling extension, then ISPBX C sends ringing current to the called extension, and ring tone to the calling extension, via the PCM channel selected for the call.

When the called extension answers, a Call Connected Message (CCM) is transmitted from ISPBX C back to A, which then connects the speech path in the forward direction so that the call can enter the conversation phase. This state will remain until either party clears, at which point the appropriate ISPBX will send a CRM with reason extension cleared, to which the other ISPBX will respond with a CIM, and the PCM channel will be released.

## 12.3.4 Universal Numbering Scheme

One of the objectives of DPNSS is to make a network of exchanges appear functionally as one exchange. This should make the system easier for the users as they do not need to know where a particular extension number is located, and do not need to remember dialling codes etc. Thus the users can make calls to other exchanges in the network as easily as they make calls to extensions of their own exchange.

A universal, or uniform, numbering scheme is one of the ways in which DPNSS achieves this objective. The adoption of a uniform numbering scheme means that each ISPBX is allocated a different set of extension numbers. On any exchange in the network, any extension can be given any of the allocated extension numbers, such that the same extension number does not appear twice in the network. As shown in our typical network diagram the four exchanges have been allocated extension numbers as follows:

|         |                                                     |
|---------|-----------------------------------------------------|
| ISPBX A | Exts 3000 – 3299                                    |
| ISPBX B | Exts 3300 – 3499                                    |
| ISPBX C | Exts 3500 – 3899                                    |
| ISPBX D | Exts 3900 – 3999                                    |
| PABX    | Extns 200-399, (ISPBX users dial 7 plus Extn number) |

Digit translation tables within the software of each ISPBX permit calls to be routed to the correct ISPBX, without the requirement for the user to dial an exchange code. For example, extension 3232 at ISPBX A dials 3205. Analysis of the first two digits reveals that this is a local call for another extension of ISPBX A, therefore the call is then processed at this ISPBX with no DPNSS involvement.

Subsequently extension 3232 dials 3500, the first two digits indicating that this call has to be routed to ISPBX C. Using an ISRM the four digits of the required number are passed to ISPBX C, where the call can continue to be processed. Note that if the direct link between ISPBX A and ISPBX C was fully engaged, i.e. all 30 trunk channels in use, this call could have been routed via ISPBX B, with no perceivable time delay to the user.

## 12.3.5 Central Operator Working

Central operator working allows an organisation to reduce the total number of operators required for a system, and place them in the location, or locations, best suited to their business needs.

In the network shown above each ISPBX is located in a different area, and each has its own connections to the public network. However all calls from the public network into the private network, irrespective of where they originated are routed to the single operator position at ISPBX A. This position is served by a PC based console in which a database containing the directory of the whole network is provided. As well as having the ability to connect the incoming call to the required extension on any ISPBX in the network, the operator is also able to provide assistance when the required extension number is not known by the caller, as complex directory searches by names, department, job title etc are provided in the software. This network directory is of course a function of the operator's PC software, and not a feature of DPNSS.

## 12.3.6 PBX Features Available Network Wide

DPNSS has the ability to make a range of PBX features available on calls to any destination in the network. Prior to the introduction of DPNSS, features such as call back when free, wait on busy and call diversions were available, but only within a local PBX. By transferring suitable messages between ISPBXs, DPNSS permits features such as call back when free to be used across the network.

DPNSS denotes basic telephony and data calls as Simple Calls with all other types of call denoted as Supplementary Services.

BTNR states that all PBX supporting DPNSS must support the simple call. However support of the supplementary services is optional and a matter for negotiation between the vendor and the customer. It is important to note that although particular supplementary services are not provided on some PBX, if these PBX operate with DPNSS they must at least recognise a request for a service and respond in a controlled manner.

The most popular supplementary services are probably Call Back When Free and the various types of call diversion – divert immediate (i.e. all incoming calls), divert on busy, divert on no reply etc, and so these features make good examples for study.

## 12.3.7 Call Back When Free – Initial Phase

This explanation is based upon the previous example in which Extn 3232 at ISPBX A calls Extn 3500 at ISPBX C. The initial call fails due to Extn 3500 being busy. The call commenced with an attempt to establish the simple call as described in Section 12.3.1. Upon receipt of the ISRM-C, ISPBX C determines that Extn 3500 is busy and sends a CRM extension busy message back to ISPBX A. ISPBX A returns busy tone to Extn 3232, and sends a CIM to ISPBX C. The initial (failed) call is now completely cleared.

## 12.3.8 Call Back When Free – Request Phase

The next stage is shown in Figure 12.6. Extn 3232 invokes Call Back When Free (CBWF) by keying the appropriate recall sequence. ISPBX A then sends a further ISRM-C. However this ISRM-C includes a request for the supplementary service Call Back When Free (CBWF-RQ).

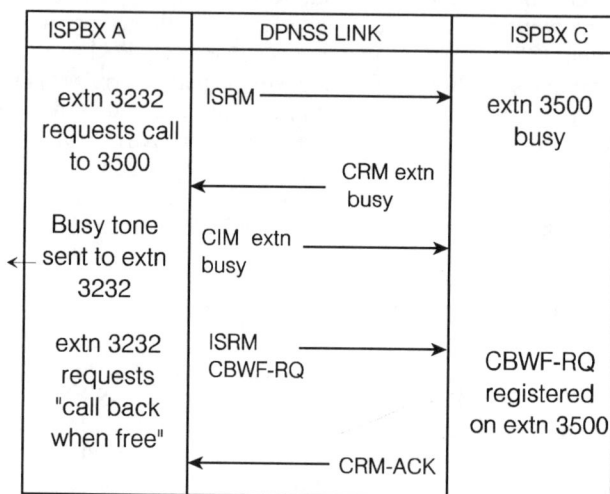

| ISPBX A | DPNSS LINK | ISPBX C |
|---------|------------|---------|
| extn 3232 requests call to 3500 | ISRM ⟶ | extn 3500 busy |
| | ⟵ CRM extn busy | |
| Busy tone sent to extn 3232 | CIM extn busy ⟶ | |
| extn 3232 requests "call back when free" | ISRM CBWF-RQ ⟶ | CBWF-RQ registered on extn 3500 |
| | ⟵ CRM-ACK | |

*Figure 12.6 Call Back When Free – request phase*

Note that as the initial ISRM-C was cleared, this second ISRM-C with the CBWF-RQ must include the required destination address (DA). The ISRM-C contains all the following information:

❏ Message Type: ISRM-C

❏ SIC: telephony

❏ CLC: Ordinary extension

❏ OLI: Extn 3232

❏ CBWF-RQ

❏ DA: Extn 3500

At this stage ISPBX A is not attempting to set up the actual call. The process of sending the supplementary service request is an example of a virtual call within a DPNSS network. Virtual calls do not involve any of the speech channels, and are used for passing signalling or other information which does not require a speech channel to be active.

On receipt of the ISRM at ISPBX C, several options are possible. ISPBX C may not support this facility, but compliance with DPNSS states that in this case, ISPBX C must send a CRM Service Unavailable (CRM-SU) message to ISPBX A to indicate that the ISRM has been understood but the supplementary service is not supported.

There are other options. However the one that is of interest here is the acknowledgement of the CBWF RQ. ISPBX C clears the virtual call by sending a CRM Acknowledge which indicates to ISPBX A that the CBWF has been registered by ISPBX C.

## 12.3.9 Call Back When Free – Free Notification

Figure 12.7 shows the next phase which occurs when Extn 3500 clears the current call. At this point ISPBX C initiates a virtual call to inform ISPBX A that the extension requested in the CBWF RQ has become free. The virtual call is in the form of an ISRM Free Notification (ISRM-FN). ISPBX A responds by sending a CRM Destination Free message, which acknowledges the ISRM-FN and clears the virtual call.

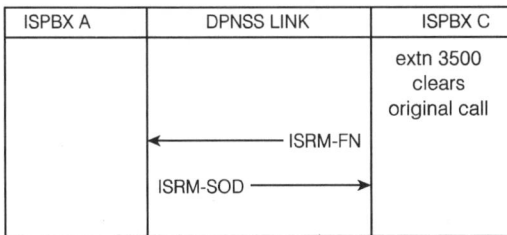

| ISPBX A | DPNSS LINK | ISPBX C |
|---|---|---|
| | | extn 3500 clears original call |
| | ←———— ISRM-FN | |
| ISRM-SOD ————→ | | |

*Figure 12.7 Call Back When Free - free notification*

## 12.3.10 Call Back When Free – Subsequent Call Set Up

The next phase, illustrated in Figure 12.8, is the attempt to set up the call. ISPBX A informs Extn 3232 that the requested call back is being set up, by sending ringing that is easily distinguishable from that used to indicate a normal incoming call. ISPBX A also sends a ISRM CBWF Call Set Up message to ISPBX C. ISPBX C rechecks the state of Extn 3500 to ensure it is still free, and sends a NAM Extn Free message back to ISPBX A. Note that at this point ISPBX C does not send ringing to Extn 3500 as the call will only be set up if Extn 3232 answers the call back ring.

At ISPBX A, Extn 3232 answers the call back ring, ISPBX A sends an end to end message ring out (EEM-RO) to instruct ISPBX C to release ringing to Extn 3500, and return ring tone back to Extn 3232. The call then proceeds, and is subsequently cleared, as for the establishment of a simple call.

This explanation has described the sequence of events for a successful call back, but there are several reasons why a call back could fail. The list below provides some of the reasons.

| ISPBX A | DPNSS LINK | ISPBX C |
|---------|------------|---------|

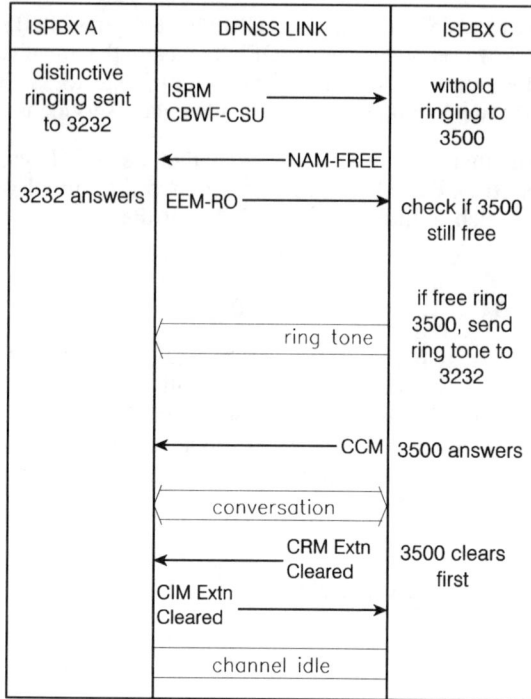

*Figure 12.8 DPNSS Call Back When Free – call set up*

❏ The CBWF-RQ queue for the required extension is full, and no more CBWF-RQs can be accepted for this extension.

❏ The calling Extn cancels the call back request

❏ The calling extension is busy when the free notification arrives

❏ The calling extension does not answer the call back ring

❏ The called extension becomes busy, immediately after clearing the original call.

It is essential that the call processing software and DPNSS include provisions to deal with these and other eventualities which lead to a failed call.

## 12.3.11 Call Diversion Within a DPNSS Network

In this example the initial simple call from Extn 3232 to Extn 3500 can not be set up by ISPBX C as call diversion to Extn 3510 is in operation for all calls to Extn 3500. As shown in Figure 12.9, the call commences with an ISRM from ISPBX A to ISPBX C as for a simple call. ISPBX C responds with a Number Acknowledge Message containing a divert clause, in this case, divert immediate (NAM-DI 3510). Other possible divert clauses include divert on busy, divert on no reply and so on.

There are two points to note here. Firstly, ISPBX C does not attempt to establish the call to the diverted extension, even if it is an extension of the same ISPBX as the originally requested number. Instead the ISPBX returns the NAM, with the extension number to which calls are to be diverted included in its divert clause.

The second point is that the destination exchange does not send a CRM. This contrasts with most cases of call failure, in which the destination ISPBX sends a CRM, with clearing clause indicating the reason for failure. The reason for this slightly different approach will become clear later.

Upon receipt of the NAM-DI 3510 message, ISPBX A sends a Recall Message (RM) to ISPBX C. The RM includes as the Destination Address the number which ISPBX C sent in the NAM-DI message. The call to the diverted number uses the same channel as was originally indicated for the initial simple call, as this has not yet been cleared by a CRM. AT ISPBX C the RM is treated in the same way as an ISRM, and the call progresses as for a simple call.

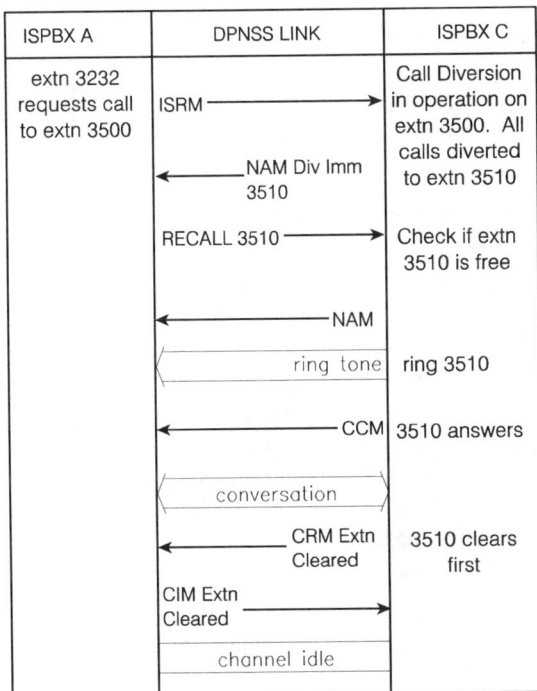

| ISPBX A | DPNSS LINK | ISPBX C |
|---------|-----------|---------|
| extn 3232 requests call to extn 3500 | ISRM ⟶ | Call Diversion in operation on extn 3500. All calls diverted to extn 3510 |
| | ⟵ NAM Div Imm 3510 | |
| | RECALL 3510 ⟶ | Check if extn 3510 is free |
| | ⟵ NAM | |
| | ⟵ ring tone | ring 3510 |
| | ⟵ CCM | 3510 answers |
| | conversation | |
| | ⟵ CRM Extn Cleared | 3510 clears first |
| CIM Extn Cleared ⟶ | | |
| | channel idle | |

*Figure 12.9 Call Diversion – call diverted to an extension on the same destination exchange*

The explanation of call diversion may sound long winded, and you may well ask: *Why is it that ISPBX C does not simply connect the initial call to the diverted extension?* The reason for this is that it is possible within a DNPSS network, to divert calls from an extension on one ISPBX to an extension on another ISPBX in the network. For reasons of conformity DPNSS does not discriminate between calls diverted to another extension on the destination ISPBX, and those which are diverted off the original destination ISPBX to another in the network.

## 12.3.12 Calls diverted to another ISPBX

To explain this, consider a larger DPNSS network than the one in our first example. In this larger network, the exchanges are not fully interconnected and DNPSS signalling messages often have to be routed through one or more ISPBX acting as transit exchanges in the signalling network as shown in Figure 12.10. In this example ISPBX E, with Extns 4000 – 4499 has been introduced between ISPBXs A and C. ISPBX F with extns 4500 – 4999 is connected by a DPNSS link to ISPBX E.

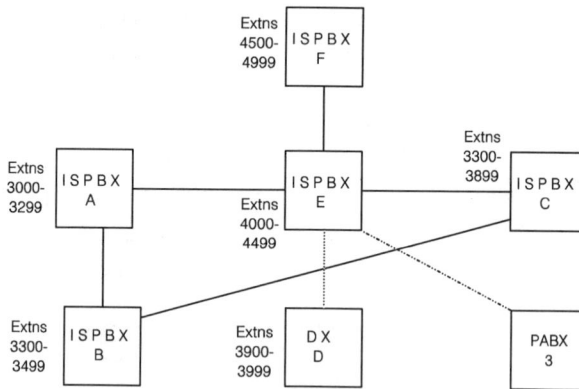

*Figure 12.10 Enlarged DPNSS network to illustrate call diversion to another ISPBX*

When the initial call is made to an extension for which call diversion is in operation, channels are booked for the call on every required link between the originating ISPBX and the destination ISPBX, i.e. links A to E and E to C. When the destination ISPBX, C, sends the NAM call divert message back, it does not use a CRM as this will cause all these booked channels to be released. The concept behind this approach is that as much of the original route as possible is to be used for the recall to the diverted extension.

Using the example in Figure 12.11, assume that Extn 3232 calls 3500, but call diversion to extension number 4500, an extension on ISPBX F, is in operation. ISPBX A checks the routing for number 4500 and determines that the RM must be sent out on the same route that the original ISRM was transmitted. ISPBX E determines that this RM is not to be sent on the same route as the original ISRM, and so ISPBX E must clear the channel booked on the Link E to C by sending a CRM on this link. ISPBX C responds and clears the channel back as far as ISPBX E by sending a CIM.

ISPBX E attempts to establish a new call on the Link E to F, by sending ISPBX F an ISRM quoting 4500 as the destination address. ISPBX F responds as for a simple call by sending a NAM, and extending ringing to the Extn 4500, and ring tone back to the caller. At ISPBX E the speech channels on the links A – E and E – F are connected together through the digital switch.

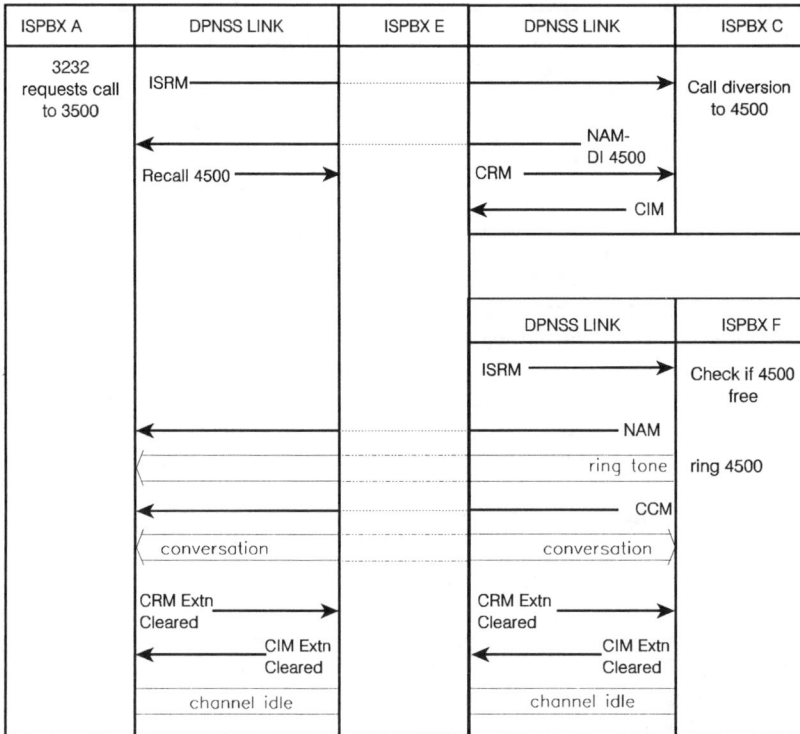

| ISPBX A | DPNSS LINK | ISPBX E | DPNSS LINK | ISPBX C |
|---|---|---|---|---|
| 3232 requests call to 3500 | ISRM ——————→ | | ————————————→ | Call diversion to 4500 |
| | ←———————— | | NAM- DI 4500 | |
| | Recall 4500 ——→ | | CRM ——————→ | |
| | | | ←———— CIM | |

| | | | DPNSS LINK | ISPBX F |
|---|---|---|---|---|
| | | | ISRM ——————→ | Check if 4500 free |
| | ←———— | | ———————— NAM | |
| | ⟨————— | | ring tone | ring 4500 |
| | ←———— | | ———————— CCM | |
| | ⟨ conversation | | conversation ⟩ | |
| | CRM Extn Cleared ——→ | | CRM Extn Cleared ——→ | |
| | ←———— CIM Extn Cleared | | ←———— CIM Extn Cleared | |
| | channel idle | | channel idle | |

*Figure 12.11 Call diversion - call diverted to an ISPBX other than the original destination ISPBX*

The call continues as for a simple call until 3232 clears. The link A – E is cleared first, as ISPBX A sends a CRM Extn cleared, ISPBX responds by sending A a CIM Extn Cleared, and a CRM Extn Cleared to ISPBX. The link E -C is then cleared when ISPBX C sends a corresponding CIM back to ISPBX E. Note that in all cases calls are cleared on a link by link basis, thus releasing channels as quickly as possible.

## 12.3.13 Other Facilities Specified Within DPNSS

One of the aims of DPNSS is to permit interworking between PBX and ISPBX from different manufacturers. It is not mandatory for any PBX manufacturer to implement all the facilities provided by DPNSS on their products, even if DPNSS is provided for inter-exchange signalling, and so an organisation contemplating the purchase of a PBX must ensure that the facilities required are both available on the PBX considered, and via DPNSS. Table 12.1 shows the majority of supplementary services that are specified for DPNSS. This list is not intended to be exhaustive, in fact BTNR 188, issue 5 lists over 30 supplementary services, and the simple telephony and data calls.

---

*Table 12.1 Major Facilities provided within DPNSS*

| | |
|---|---|
| Swap Between Voice and Data | Call Back When Free |
| Executive Intrusion | Call Diversion |
| Call Hold | Three Party Call |
| Call Offer | Call Waiting |
| Route Optimisation | Extension Status |
| Night Service | Centralised Operator |
| Call Back When next Used | Do Not Disturb |
| Traffic Channel Maintenance (#) | Remote Alarm Reporting (#) |
| Time Synchronisation (#) | |

---

The last three items in the above table (marked #) are examples of the ability of DPNSS, like any other common channel signalling system based on CCITT #7, to carry non call related information. In these cases, information regarding network management and maintenance is carried between a system management terminal, which may be located at any node in the network, and all the nodes in the network.

# 12.4 DPNSS Data Calls

## 12.4.1 Circuit Switched Data Within DPNSS

DPNSS can be implemented on a network of digital PBXs where telephony is the only form of traffic. But as DPNSS is designed to operate in an integrated telephony and data network, it provides facilities for connecting compatible data terminals (DTE) through the digital network.

Since an evolving digital network, such as the one in the example in Figure 12.1 will consist of digital and analogue transmission links, protocols within DPNSS ensure the correct type of transmission link is used for data calls. Generally, a wholly digital route must be provided for a data call, although in some cases this route may include analogue links with compatible modems.

Simple data call establishment is slightly more complex than for the simple voice call in that it consists of four phases rather than three. The four phases of a data call are Call Connection, Terminal Synchronisation, Data Communication and Clear.

## 12.4.2 Call Connection Phase

The call commences with an ISRM, in which the two octets forming the SIC indicate the type of service required. A 3-bit Type of Data field in the first octet indicates whether the required service is for voice or data as shown in the coding below:

```
0 0 0 = invalid      0 1 0 = data
0 0 1 = speech       0 1 1 = data
```

Circuit switched data calls may be made between different terminal types so long as they operate in the same mode, and at the same speed. To cater for the wide variety of different traffic speeds possible a 4-bit code is used in the first octet of the SIC.

The following table shows a few of the available speed options for data calls. Note that the actual meaning of the 4-bit code depends upon the value of the 3-bit Type of Data field (shown in Table 12.2) used to indicate data service.

*Table 12.2 Speed options for data calls*

| 4 Bit Code | Type of Data Code | |
| | 0 1 0 | 0 1 1 |
| --- | --- | --- |
| 0 0 0 0 = | 64000 bit/s | 300 bit/s |
| 0 0 0 1 = | 56000 bit/s | 200 bit/s |
| 0 0 1 0 = | 48000 bit/s | 150 bit/s |
| 0 0 1 1 = | 32000 bit/s | invalid |
| | - – - etc – - - | |
| 1 0 0 0 | 9600 bit/s | 75/1200 bit/s |
| 1 0 0 1 | 8000 bit/s | 1200/75 bit/s |
| | - – - etc – - - | |
| 1 1 1 0 = | 1200 bit/s | invalid |
| 1 1 1 1 = | 600 bit/s | invalid |

This table, which is reduced from the full version of BTNR 188, shows that all the common data transmission rates are catered for. A standard data rate adaptation method is used to allow traffic to be transported at 64 Kbit/s irrespective of the original traffic speed. At the far end of the connection, the data is reproduced at the same rate as originally indicated in the SIC. Note that although a standard data rate adaptation method is normally used to permit data to be carried on the 64 Kbit/s channel, there is also provision for end to end speed matching, using flow control procedures to permit a 1200 bit/s terminal to be connected to a 2400 bit/s terminal.

The second octet of the SIC is used to indicate the type of data traffic, and specify certain parameters regarding data format. It is not necessary to reproduce the coding table here, as it is sufficient to understand what information is transmitted in this octet.

Asynchronous or Synchronous traffic
Full or Half Duplex
1, 1.5 or 2 Stop Bits
For Synchronous Traffic                      For Asynchronous Traffic
Byte timing: provided/not provided     Data Format: 5, 7 or 8 bits
Data format: unformatted/packet mode   Flow Control/No Flow control

As in the case of a voice call the destination ISPBX may not be able to connect the call for a variety of reasons e.g. number unobtainable, busy, out of service and so on. There are other reasons why a data call may fail e.g the terminals are incompatible, or the destination DTE is not ready, e.g. not powered up.

### 12.4.3 Terminal Synchronisation Phase

Once the call has been connected (indicated by a CCM from the destination ISPBX) the DTEs are synchronised end to end over the traffic channel. The signalling and traffic channels are transparent to this synchronisation and thus the synchronisation procedure is not defined in DPNSS.

When the user traffic data rate is less than 64 Kbit/s, the rate adaptation scheme can provide an end to end status channel to permit some types of DTE to exchange synchronisation status information.

### 12.4.4 Data Communication Phase

Following synchronisation, the call is in the data communication phase, and the connection should remain transparent to the data traffic until one of the DTEs clears the call.

### 12.4.5 Call Clear

Following a Clear Request signal from the originating DTE, the originating ISPBX will forward a CRM with the clearing clause Call Termination (CRM-CT), and simultaneously transmit an appropriate DCE Clear Confirmation signal back to the DTE.

Upon receipt of the CRM-CT, the destination ISPBX transmits a DCE Clear signal to the destination DTE, and returns a CIM-CT to the originating PBX. The circuit is finally cleared when the destination DTE transmits a DTE Clear Confirmation signal back to its ISPBX.

## 12.5 Permanent data calls

The subject of permanent data circuits in an ISPBX network follows naturally from the preceding section. Permanent calls, or *nailed up* circuits as they are often called, are another alternative to traditional methods of providing point to point circuits.

As mentioned at the beginning of this chapter, many data communications circuits are point to point systems. When such circuits are within one site, e.g. between a user's terminal and the main computer, the circuit may be provided on twisted pair cabling within the building, and may well use modems where long lengths of cabling are involved. The other case occurs when the data communication circuit is between sites. In these cases it is usual practice to use analogue or digital leased lines provided by a public carrier such as BT or Mercury.

The ISPBX can provide an alternative to direct wiring within a building by providing a permanent call between any two compatible DTEs. The concept is similar to the idea of electronically patching terminals to computers outlined in the section on data switching.

The description permanent call refers to the fact that the connection between the two DTEs is via the digital switch within the ISPBX as shown in Figure 12.12.

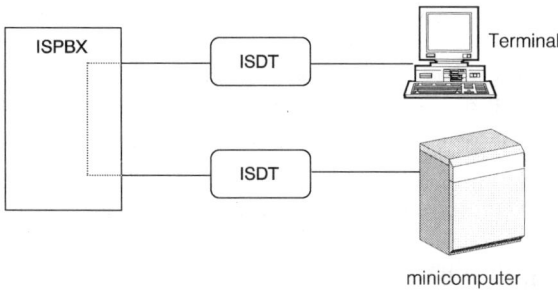

*Figure 12.12 Nailed Up circuit on a single ISPBX*

Nailed up circuits can also be established between compatible DTEs connected to different ISPBXs in the same network. In this case, at both ISPBXs, it is necessary to establish connections between the DTE and a channel on the digital link between the exchanges. The example in Figure 12.13 shows two nailed up circuits. The first is a simple circuit involving the connection of a DTE at ISPBX E to another DTE at ISPBX E. This circuit is connected on channel 30 of the link between E and C.

In the second, slightly more complex case, a terminal connected to an ISDT at ISPBX A is connected over inter ISPBX channels to a mainframe computer. The computer is it itself connected to ISPBX C by a 2Mbit/s 30-channel ECMA link.

The nailed up circuit is established over channel 30 of digital links A -E, channel 29 of link E – C, and channel 1 of the ECMA link.

As channel 30 of link E – C is already in use to provide a nailed up circuit between DTEs at E and C, it can not be used for this second circuit.

Channels used for nailed up circuits are not available for circuit switched voice or data calls, even when the connected DTEs are idle and the channel is not actually passing any traffic.

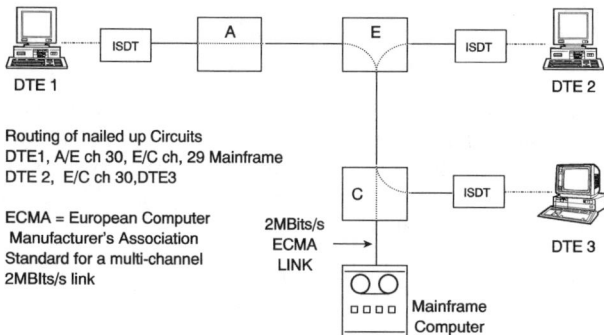

*Figure 12.13 Nailed up circuits across an ISPBX network*

Once established, a nailed up circuit is available for use at any time without the need for call set up by the DTEs, and thus with no DTE to Exchange signalling being required. If the DTEs lose synchronisation for any reason, e.g. power down, end to end synchronisation will again be required prior to data communication, but this is done over the data circuit which will not have been cleared.

When the requirement for the circuit no longer exists, it can be disconnected by entering a suitable MMI command at the relevant system management terminals.

# 12.6 Centrex and Virtual Private Networks

## 12.6.1 Introduction to the Concept of Centrex and Virtual Private Networks

This short section describes briefly how the telephone companies are considering using their new public digital networks to offer ISPBX and private network type services to business users.

The advantages of private exchanges and networks have been discussed in previous sections, and we have seen that it is standard practice to build private networks from a number PBXs networked over suitable transmission links, which invariably have been leased from a public carrier.

Some public network operators are now beginning to offer alternatives to these traditional private systems in the form of Centrex and VPNs. These alternatives offer benefits to certain types of user, however it is unlikely that Centrex or VPN will completely replace true private systems.

## 12.6.2 Centrex

Developed originally in the USA, Centrex is a contraction of Central Office Exchange, i.e. a private exchange within the local exchange, (or using American terminology, the central office).

The concept behind Centrex is to provide a virtual PBX within the hardware of the local public exchange. This is achieved by physically replacing the actual PBX, and its all extensions, by lines to the local public exchange as illustrated in Figure 12.14a.

Although the lines are physically connected to the local exchange hardware, software within the exchange is configured so that the exchange emulates all the functions of a true PBX for these lines. Each user is allocated a Centrex extension number, which can be assigned a class of service giving the features, e.g call back when free, call diversion etc, that would be expected from a real PBX.

Thus Centrex users are given the impression that their telephone is connected to a private exchange which has its own national numbers on the public network exchange as illustrated in Figure 12.14b. As the customers for this type of service come from the business world, the Centrex PBX lines are organised as business groups. The local exchange is capable of providing Centrex service to more than one business group, but each group has its own virtual PBX within the local exchange.

(a) Actual Connectivity

(b) Apparent Connectivity

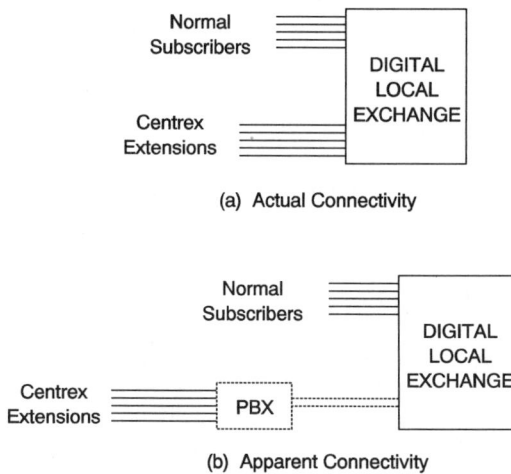

*Figure 12.14 The Centrex concept*

The local exchange is of course also connected to many thousands of normal subscribers. The software must treat these normal subscribers and each Centrex business group as a separate entity, by giving each its own numbering plan, while maintaining a software boundary between normal subscribers and the Centrex groups.

Calls are connected between extensions of a Centrex business group just as if they were extensions of a true PBX, i.e the user simply dials the required extension number. However normal PSTN subscribers can not directly dial a centrex extension simply by dialling the extension number as the exchange software treats the Centrex extensions as a closed user group and prevents unauthorised access to the extensions within the group.

PSTN access to Centrex extensions is made in one of two ways. Firstly the PSTN subscriber dials the national number of the Centrex group. The call is initially connected to an attendant's position, within the business premises. The attendant processes the call just as if it was an incoming call to a real PBX, and transfers the call to the required extension.

In the second case, some extensions within the Centrex group may be configured for direct dial in (DDI), thus PSTN users must dial the extension number prefixed by the DDI national number.

Just as with a true PBX, calls can not be made directly from Centrex group to another without first accessing the public network, except in the case of a virtual private network as described in the next section.

The advantages claimed for Centrex from the user's point of view are:

❑ No requirement to purchase a PBX

❑ No requirement to install and maintain a PBX

❏ Rapid provision of service to temporary sites possible

❏ Service is easily expanded (and contracted)

❏ No requirement to replace PBX when new features are introduced

❏ All interfaces will be to the same standards as normal PSTN and ISDN connections.

Centrex is then an alternative to purchasing, or leasing and maintaining a real PBX.

## 12.6.3 Virtual Private Networks

We have seen examples of a private system in which the connectivity of a PBX includes analogue or digital exchange lines to the PSTN, and tie lines to other PBXs. The tie lines, whether analogue or digital, are usually private circuits which are leased from the same public network operator who provides the exchange lines.

The tie lines will usually be routed to the same local exchange building as the PSTN exchange lines, in order that they can be physically connected to the other lines necessary to complete the route. Thus two different types of interface are required from the PBX to circuits which actually go to the same place.

In a digital environment, DPNSS signalling is used on the lines to the private network, while DASS is used on the lines to the public exchange. VPNs can exploit the opportunity to rationalise the two connections to the local exchange premises by providing access to both public and private networks on the same 30-channel 2Mbit link as illustrated in Figure 12.15. This link carries a mix of DPNSS and DASS signalling messages in TS 16.

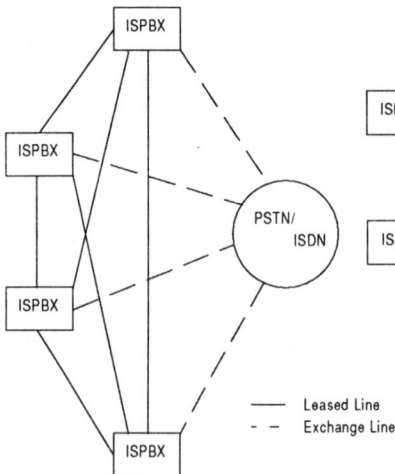

*Figure 12.15 (a)*
*Typical Private Network*

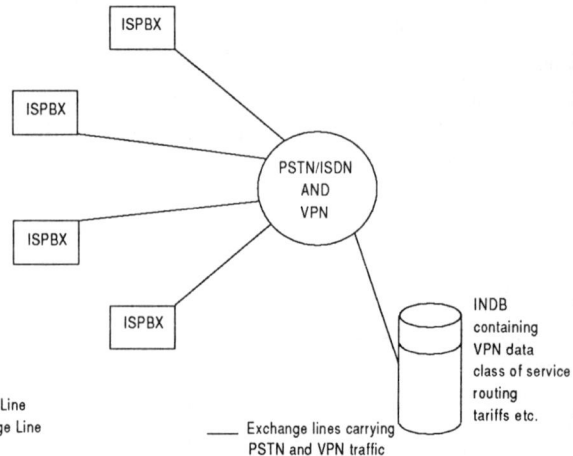

*Figure 12.15 (b)*
*Virtual Private Network*

The private network connectivity is no longer provided on dedicated circuits. Instead capacity is shared on the public network on an as required basis. Within the public network, the CCITT #7 signalling network carries DPNSS messages transparently from the originating local exchange connection to the destination local exchange. The DPNSS messages are then transmitted to the final destination PBX.

To route a VPN call from one PBX to another, the local exchange receiving the DPNSS ISRM message, realises this is a VPN call. To obtain the required routing for the call, the local exchange must access a centralised intelligent network database (INDB) containing the relevant VPN information. From the information received from this central database the local exchange can direct the ISRM through the public network to the required destination.

By providing a single centralised database for VPN routing and other related information, the public network operators do not have to update the software on every local exchange to reflect changes made to any VPN connectivity.

Authorised users of a VPN can alter the apparent connectivity of the VPN by updating the necessary information in the INDB. This information is then available immediately to all exchanges requiring it.

This brief explanation has assumed that the VPN consists of true PBXs linked over virtual private circuits. However it is also possible that VPNs will also include connectivity to Centrex groups.

The benefits of a VPN from a users point of view are based on the fact that the actual network is managed and maintained by the public network operator. Thus the fixed costs of owning and running a private network are not incurred, although the costs of the PBX still exist, unless a full Centrex-VPN is created.

Calls on a traditional private network are not charged by the public network operator, whose revenue is obtained from the annual charge levied for the interconnecting links. At the time of writing, two approaches to VPN tariffing exist. The first is to make a charge for setting up the VPN, then a fixed charge for the service, no charge being made for calls. In this case, the charges are fixed and known in advance. The second approach charges a smaller annual fee for running the service, but calls are charged at a approximately the local call rate irrespective of distance. The cost is thus usage based.

# 12.7 Chapter Summary

Private systems are by their very nature smaller than those employed in public networks. However they offer many features which are not practicable in a public system.

This chapter provided a description of a typical small private system which is currently undergoing modernisation from an all analogue network to one in which digital exchanges and transmission links predominate. The typical network illustrates that there is a requirement for analogue and digital interworking in private networks just as in a public system such as System X, as the change from one system to another can not be accomplished in one step.

Digital private networks within the UK are tending to use DPNSS signalling, although it should be said that other forms of private network signalling are available in Europe. As DPNSS is based on CCITT #7 which was covered in some depth in the chapter on common channel signalling, explanations of the physical and link levels of DPNSS were not required. The space available was devoted to describing how some basic calls and supplementary services, such as call back when free, are set up within a DPNSS network.

The section on ISPBX was completed by a description of permanent call data circuits, the so called "nailed up" circuit. This particular application offers flexibility in terms of connectivity between computers and peripherals, but lacks some of the features, such as speed and format conversion available on a dedicated data switch.

Centrex and VPNs offer an alternative to private systems, which are probably attractive to young, expanding organisations which do not have the required expertise or capital to build and run their own private network, but nevertheless require the advantages of such a system. Large organisations may wish to migrate to a VPN to reduce costs and concentrate effort on their business activities.

# 13

# *Integrated Services Digital Networks*

## 13.1 An Introduction to ISDN

### 13.1.1 What is an ISDN?

In Chapter 4, during the discussion on the evolution of digital networks an Integrated Service Digital Network (ISDN) was defined as an Integrated Digital Network (e.g System X) which provides the subscribers with a digital access, i.e there is a digital connection between the subscribers' equipment and the local digital exchange. This statement rather simplifies the situation, in that it does not begin to provide any information about what ISDN means to the subscriber and the network provider in practice. We saw in the previous chapter that a private ISDN involved the transmission and switching of voice and data communications in a single general purpose digital network.

This chapter will look at some of the many CCITT recommendations regarding public ISDNs, how the ISDN service offered by British Telecom in the UK has developed from a proprietary ISDN standard to one which is in line with the CCITT recommendations, and how ISDN type services can be provided within digital private networks.

### 13.1.2 CCITT Recommendations for ISDN

The CCITT produced the first of the ''I'' series of recommendations for ISDN in the Red Book of 1984. Since then these initial recommendations have been revised in some respects resulting the publication of the current recommendations in the 1988 Blue Book. Whilst any such recommendations are in a state of flux, it is difficult for organisations concerned with implementing them to know exactly how to develop products which will operate in line with the final stable recommendations. It was for this reason that British Telecom's first steps into ISDN used a proprietary standard which was only loosely based on the emerging CCITT recommendations.

However by 1988 the recommendations became sufficiently stable to become the ISDN standards which the telephone administrations and equipment suppliers will use to implement public, and also, in many cases private ISDNs.

To ensure that public ISDNs world-wide can inter-operate successfully the CCITT recommendations are wide ranging and cover far more than defining interface requirements. The following list shows some of the areas that they address:

❑ ISDN architecture and concepts

❑ User to network interfaces

❑ ISDN services

❑ ISDN signalling

❑ Support for non ISDN terminals

❑ ISDN maintenance philosophy

❑ ISDN numbering schemes

❑ ISDN routeing and addressing

❑ Quality of service and performance of ISDN networks

## 13.1.3 The Background to ISDN

The telephone service has been developed over the last 100 years. Initially its sole aim was to provide simple one to one voice communications between subscribers, but we have seen that technology has influenced the telephone network in two ways. Firstly improvements in technology such as the introduction of digital switching, computer control and common channel signalling have meant that the network can offer its users far more facilities than simple one to one voice calls.

Secondly the introduction of new technology in other business areas has resulted in a situation in which the Plain Old Telephone Service (POTS) is carrying a wide variety of data communications traffic. Although it is true to say that the major use of the network is still for voice communications, a growing percentage of the traffic is accounted for by digital traffic, i.e. data communications and facsimile.

## 13.1.4 Traditional Communications Networks

The old telephone network based on analogue switching and transmission was optimised for voice not data comms, and although POTS is capable of carrying these other forms of traffic, it does so with limitations. These limitations are generally caused by the following factors:

❑ The old network is noisy, resulting in bit errors when used for data communications applications.

❑ Call set up times are long, 5 to 25 seconds. For many data calls the call set up time exceeds the holding time, or is at least a significant proportion of the total call time.

❑ Transmission is limited to a duplex 3 KHz bandwidth pathway. Imposed by the nature of the FDM equipment, this bandwidth limits the maximum rate at which data can be transmitted, even on quiet circuits which are not subject to errors.

❑ Routeing of calls is not fixed, and thus variations in transmission performance due to effects such as group delay are experienced on different calls between any two given locations. The problem with providing a dial up data circuit over the switched telephone network is that each call will be routed over a different mix of circuits with indeterminate group delay characteristics. It is therefore difficult to condition the circuit to achieve optimum performance by using line equalisation techniques.

As the demand for data communications increased, the telephone administrations in many countries began to offer dedicated digital networks which overcome the problems outlined above. These services can be categorised as follows:

❑ Point to Point Digital Leased Lines at a variety of bit rates, e.g. 300 Bd, 600 Bd, 1200 Bd, 2400 Bd, 4800 Bd, 9600 Bd, 19200 Bd, 48 KBd, 64 Kbit/s and 2Mbit/s

❑ Circuit Switched Telegraph e.g. TELEX at 50Bd

❑ Packet Switched Data Network at speeds up to 48 KBd

❑ Circuit Switched Data Networks at speeds up to 64 KBit/s, (a dedicated service such as this is not available in the UK)

Thus over recent years the communications infrastructure has evolved in such a way that services are provided on dedicated networks, each with its own subscriber access and interface requirements.

As an example consider the communications requirement of a large organisation, a situation which is depicted in Figure 13.1.

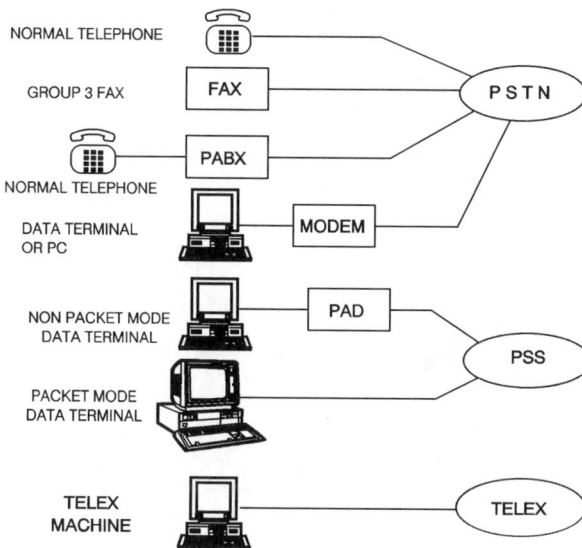

*Figure 13.1: Traditional telecommunications networks*

The communications systems leaving a large office or organisation headquarters will include PABX exchange lines, inter PABX tie lines, datel lines, Kilostream and Megastream circuits, and probably some Packet SwitchStream circuits. Each of these different types of circuit has a different user interface and thus its own network terminating equipment (NTE).

In reality many of these systems are actually carried over the same bearer system within the national network. However to the user it appears that he has individual accesses to several different networks.

From a provider's point of view, the cost of building and maintaining dedicated networks such as these, is so large that it can only be contemplated if the demand for the service is large enough to generate sufficient revenue to make it economic. These high costs therefore prohibit the introduction of new specialised communications services which do not have a large market.

## 13.1.5 Digital Networks and ISDN

In many countries, significant changes are taking place in the telephone network. The analogue systems are being replaced by new digital networks which have been developed to cater for all forms of digital communications. These networks can carry data traffic with the same ease as digital voice.

Figure 13.2 shows the concept of an ISDN, which may be defined as a digital network in which a variety of communications services are carried using common switching, transmission and control plant.

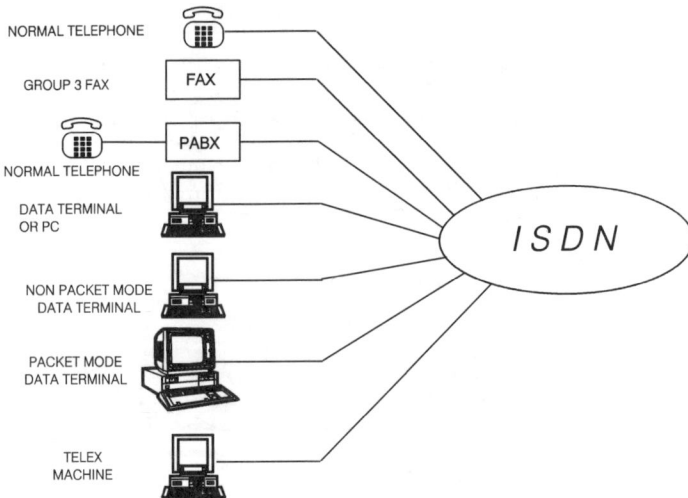

*Figure 13.2: The concept of an ISDN*

The definition of ISDN given above tends to suggest that all an ISDN achieves is producing one network for all services. If this were the case the network provider would gain as he would only be providing and maintaining this single network, but there would be little to be gained from a user's point of view. In fact an ISDN offers far more than this to the users of communications networks.

## 13.1.6 The User's View of ISDN

From a purely technical point of view there are a number of aspects of ISDN which may be harnessed by a user to provide better communications facilities than those provided by traditional networks.

❏ A switched digital circuit is available end to end on a telephony or data call. For voice calls the main advantage here is that there should be a significant improvement in speech quality. As no part of the circuit involves analogue transmission plant, there are none of the group delay problems associated with using switched analogue circuits for data communications.

❏ The capacity of this digital circuit is 64 Kbit/s, far greater than was previously available using modem technology.

❏ There is only one type of access line from his premises to network, and there is a limited set of standard interfaces for this access to the network. This should result in easier connection and reduced unit cost for terminal equipment as manufacturers benefit from the economies of scale.

❏ ISDN offers two simultaneous digital circuits over existing telephone cabling, thus the extra facilities can be provided without the need to install more cable from the local exchange. Each of these circuits can carry digital voice or data and will operate at 64 Kbits/s.

❏ For users of more modern digital PBX, a multi-channel access to ISDN based on standard PCM techniques will be available, this will replace the existing *one exchange line - one cable pair* arrangement.

❏ A common signalling channel between the terminal equipment and the local exchange provides fast call set up and many other features. This signalling channel may also be used to convey end to end signalling information once a call has been established. The signalling channel may also be used for packet switched data in the future.

This is really a list of the technology made available by the introduction of ISDN, which can be used to provide solutions to the users' communications requirements.

An important point which should be made somewhere in a book such as this is that user take up of ISDN will probably depend upon whether or not it offers a cost effective means of providing the necessary communications for the users' information services (in whatever form they may be). So, if you look at the user requirements of an ISDN from a non technical angle, the concept of integrating voice and data can mean more than simply arranging things so that essentially separate voice and data traffic are handled by the same network.

Many businesses today rely on an integration of voice and data at work. Several instances of this type of integration exist, for example a sales desk which conducts most of its business over the telephone. Customers' calls are answered by a sales clerk who operates a VDU, entering the customer's order details into a computer as they are given over the phone. The computer confirms the items are in stock, and prints a delivery notice in the warehouse to enable the dispatch to be made. Subsequently the computer will produce an invoice to be dispatched to the customer, and a copy of this invoice can be called up on to a display unit in the event of any telephone query. This is a case of integrated voice and data at work. Other examples include the financial services market, insurance sales and travel reservations systems.

In all these cases voice and data systems are integrated in the work function, one can not exist without the other. A modern communications system must meet the requirements of this type of operation, and must then integrate voice and data at the user's desk or workstation. In this chapter we will describe how the ISDN helps to achieve this integration.

## 13.1.7 Types of ISDN

At this stage it should be mentioned that there are two types of ISDN currently receiving attention. In this book we will be looking at the narrow band ISDN systems for voice, data, teletext etc. It is this type of ISDN which we have introduced so far in this chapter. There is also research into broad band ISDN services aimed mainly at providing multi-channel video and other data services into consumer premises. This book will not be covering this type of system.

# 13.2 Connecting Users to the ISDN

## 13.2.1 Introduction to Subscriber Access

There will be a large number of subscribers who will not need the end-to-end digital connection provided by the ISDN service. For example, the introduction of ISDN services will not affect normal domestic subscribers wishing to use only telephony services, who will be able to continue to use their older style analogue telephones.

There are two basic types of subscriber who may wish to have ISDN services:

❑ Small business users with single telephones, fax and Telex and some data comms requirements.

❑ Corporate users with a PBX or ISPBX, and a wide variety of data comms.

The CCITT has specified digital access arrangements for each of the two types of ISDN subscribers identified above. These accesses are known as:

❑ **Basic Rate Access:** Digital access for individual subscribers. This is also often referred to as Single Line Access.

❑ **Primary Rate Access:** Digital access using 24-channel or 30-channel PCM techniques, mainly for PBX users. This is often also referred to as Multi-line Access. (Note the recommendations also cater for connecting a PBX to the ISDN using Basic Rate Access. This would be the case if only a few ISDN Exchange Lines are required.)

## 13.2.2 ISDN User-Network Interface Recommendations

CCITT recommendation I. 411 describes a reference model for user-network interfaces to the ISDN. Figure 13.3, on the next page, identifies different types of customer premises equipment in terms of the functions they perform, and shows the interconnection of this equipment in the digital subscribers' line. This diagram represents the reference configuration for the ISDN user-network interfaces.

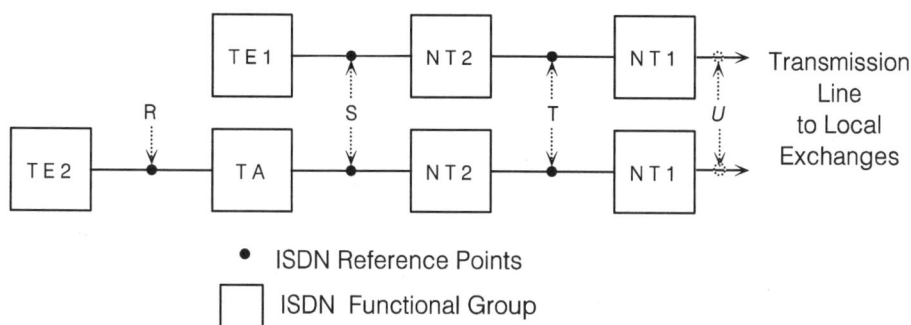

*Figure 13.3: The CCITT ISDN reference points*

## 13.2.3 Definitions of Equipment in the Reference Model.

### Network Termination 1 (NT1)

The main function of this equipment is the physical and electrical termination of the transmission line between the local exchange and the customer's premises. Other functions of the NT1 include maintenance and performance monitoring by providing digital loopback facilities, and the ability to feed DC power from the transmission line to other equipment in the installation.

### Network Termination 2 (NT2)

This may be a PABX, a Local Area Network (LAN) or a terminal controller. The functions associated with an NT2 include protocol handling, multiplexing, switching, concentration and other maintenance functions. Additionally the NT2 provides interface functions to other equipment and the NT1.

### Terminal Equipment (TE)

A TE is a user equipment, typically a telephone or data terminal, the functions of which include physical and procedural interfaces, and maintenance, as well as the general communications function of the device.

### Terminal Equipment 1 (TE1)

A TE1 is a TE as defined above, and will be a digital telephone, data terminal, facsimile terminal or other work station that complies with the ISDN user-network interface recommendations. Generally these will be the more modern equipment which has been specifically developed for ISDN operation.

### Terminal Equipment 2 (TE2)

A TE2 is a TE as defined above, but one which does not conform to ISDN user-network interface recommendations. Generally these will be older types of equipment such as Data Terminals conforming to V or X interface specifications and Group 3 facsimile machines.

Other types of terminal equipment, e.g. certain manufacturers' data terminals, which do not conform to any existing CCITT recommendation are also covered by the term TE2.

### Terminal Adapter (TA)

All types of TE2 will be connected to the ISDN by a corresponding Terminal Adapter which will be responsible for the conversion of the physical and procedural interface of the TE2 to an ISDN standard interface.

## 13.2.4 Definitions of the Reference Model Connection Points

### The T Interface

This interface point has been specified by CCITT as the interface between an NT2 and the NT1. The connection of any PABX, LAN or terminal controller to an NT1 must comply with the physical, electrical and protocol specifications which have been defined for the T interface.

### The S Interface

ISDN Subscriber Terminal Equipment (TE1) will normally be connected to the subscriber side of the NT2 at the S interface. Thus all terminal equipment specifically designed for ISDN must conform to the recommendations for the S interface

To permit the use of non-ISDN Terminal Equipment, i.e TE2s, Terminal Adapters provide an interface between the TE2 and the NT2. The TA will connect to the NT2 at the S interface and so the network interface from a TA must also conform to the recommendations for the S interface point.

### The R Interface

Non ISDN Subscriber Equipment (TE2) will connect to the subscriber side of the TA at the R interface. Specifications for this interface are not included in the I series of recommendations as this interface point will normally use one of the existing data communications interfaces.

Note the use of the term Network Termination (NT) to indicate certain groups of functions, rather than Network Terminating Equipment (NTE). An NTE may actually include one or both of the NTs shown depending upon national regulations governing the network provider. For example some countries do not allow customers to connect their own (approved) equipment to the national network. In these cases the NTE will probably include the S and T interfaces and may well be implemented inside the actual terminal equipment.

### The U Interface

The transmission line from the exchange is terminated by an NT1. The CCITT did not expect that users would be allowed to connect their own equipment directly to the transmission line and so have not specified interface requirements for the NT1 to transmission line interface.

However much literature, especially from suppliers of integrated circuits for use in NT1s does discuss this interface point, and for reasons that are obvious, this point has become known as the U interface.

It is generally true to say that this interface point is not of interest to the European user. It is however of great interest to North American users, and to the European network providers, whether public or private. For these reasons this book will include some material on the U interface.

In European countries the general approach is to consider that the network boundary is at the S interface. Thus equipment on customers' premises, and where necessary terminal adapters, must conform to this standard. The network will be terminated by an NTE which will contain the T and U interfaces.

In contrast, within the USA the Federal Communications Commission (FCC) has stated that the boundary of the public network will be at the U interface. Thus the customer's premises equipment must conform to the network providers specifications for the U interface.

The intention of defining a minimum number of possible interconnections, and thus a limited set of interface standards is to reduce the possibility of a large number of incompatible interfaces being produced by different administrations and suppliers. It must be matter of regret and concern that different interpretations are being used on either side of the Atlantic Ocean, as this will tend to reduce the prospects for manufacturers to produce a standard equipment that can be sold without modification world-wide.

## 13.2.5 Typical ISDN Installations

The diagrams in Figure 13.4 are based on the previous CCITT Reference Points diagram to show how typical small ISDN installations consisting of several items of equipment will conform to the concept of the ISDN reference points.

a. Two digital telephones

b. digital telephone with Group 4 fax terminal

c. Digital Telephone with V24 Data Terminal and V24/ISDN adaptor

d. Digital telephone with X21 Data terminal and X21/ISDN adaptor

Basic rate access

Primary rate access

Primary rate link to another ISPBX

Basic rate access

S* & T* denote the interface at these points may not conform to CCITT recommendations. In such cases the ISPBX manufacturer's own proprietary standard is used.

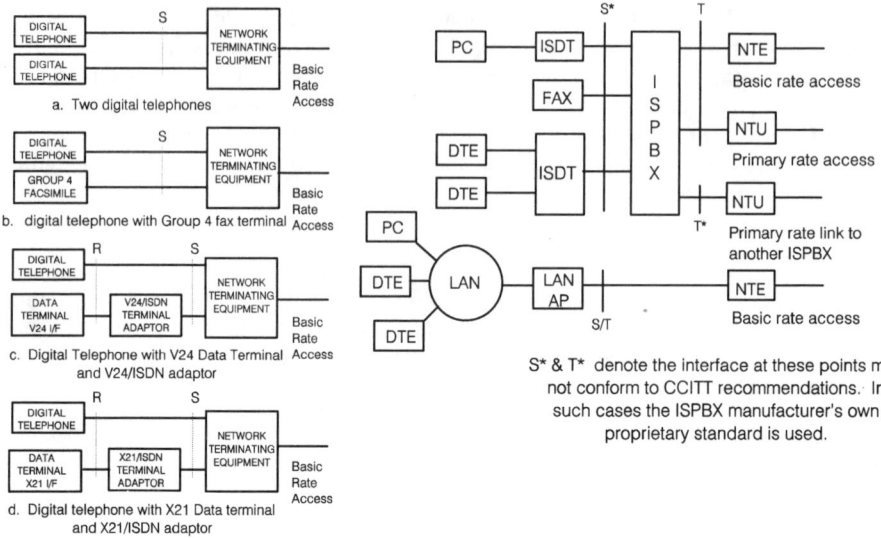

*Figure 13.4: ISDN reference points in typical installations*

## 13.2.6 Structure of the Interface Recommendations

CCITT recommendation I 420 deals with Basic Rate Access, while I 421 deals with Primary Rate. Both I 420 and I 421 are very short, however they refer to several other recommendations to describe the interface between the ISDN terminal equipment and the ISDN network at the S and T interfaces. Following the concept of the OSI 7-layer reference model the structure of these interfaces is layered and maps on to the bottom three layers of the OSI model.

> **Layer 1:** The physical attributes of the interface is covered in recommendations I 430 for Basic Rate Access, and I 431 for Primary Rate Access.
>
> **Layer 2:** Deals with access to the signalling channel, ensuring for example, that problems do not occur when two items of terminal equipment wish to transmit signalling information at the same time. This Access Protocol is covered in recommendations I 440 and I 441.
>
> **Layer 3:** Deals with call connection protocols - basically the messages that will be transmitted between the TE and the exchange to set up circuit switched and packet switched connections through the ISDN. These procedures are defined in recommendations I 450 and I 451.

It is interesting to note that recommendations I 440, 441, 450 and 451 do nothing more than instruct the reader to see other CCITT recommendations in the Q series. This is because these recommendations are specifically involved with the Digital Subscribers' Signalling Systems for use in the ISDN.

The diagrams in Figure 13.5 outline the various recommendations of interest here. Notice that the layers 2 and 3 deal the D channel access and message procedures and that these are common to both Basic Rate Access and Primary Rate Access. However the physical attributes of these are very different and hence covered by different Layer 1 recommendations.

| | I 420 Basic Rate Access | |
|---|---|---|
| Layer 1 | Layer 2 | Layer 3 |
| I 430 | I 440 (Q 920) | I 450 (Q 930) |
| | I 441 (Q 921) | I 451 (Q 931) |
| | I 421 Primary Rate Access | |
| Layer 1 | Layer 2 | Layer 3 |
| I 431 | I 440 (Q 920) | I 450 (Q 930) |
| | I 441 (Q 921) | I 451 (Q 931) |

*Figure 13.5: Relationship between ISDN user-network interface recommendations*

# 13.3 Basic Rate Access

## 13.3.1 The Development of Basic Rate Access

In this section, the arrangements for this digital connection are considered. One of the major factors that influenced the development of basic rate access was the requirement to provide the new digital service as far as possible on the existing telephone cables that formed the local distribution network to subscribers' premises. This local distribution network, designed specifically for analogue use, was already in place and represented a very large proportion of the total capital investment in the telephone system. It would, therefore, have been difficult to justify the cost of installing new cables suitable for digital transmission, even if there was room in the underground ducts in which to put them.

An option may have been to use optical fibre cable and there has been considerable research into the use of optical fibres to subscribers premises. While there are merits in using this technology for wide band services, it will probably be some time before it is economic to lay optical fibre to subscribers' premises for narrow band ISDN.

Basic Rate Access is then, to be provided over the existing 2-wire telephone line. This presented several technical challenges to the designers; firstly a transmission system was required to transmit digital signals at the required data rates over cables up to 5 Kms or so in length (and in some cases longer than this) without the need for additional line plant equipment such as repeaters etc.

A second problem to be overcome was that of transmitting digital signals in opposite directions simultaneously over the same pair of wires without one signal interfering with the other. The main problem was that of unwanted echoes of the transmitted signal from the distant termination and cable joints. These echoes while not a major problem on analogue lines, would have a catastrophic effect on the reception of the wanted digital signals. As it is impossible to remove the echoes from the line, steps have to be taken to overcome their effects.

Solutions to these problems have now been found, with some organisations such as British Telecom adopting a *halfway house* approach in the early days of ISDN in order to get experience in the field and to make its business customer base aware of the concepts of ISDN.

## 13.3.2 Recommendations for Basic Rate Access

The remaining sub sections of Section 13.3 deal with recommendations for the S interface. However for most purposes this description is also valid for the T interface. While it is not the intention of this book to provide an in depth analysis of the CCITT ISDN recommendations, it is important to include some detail as this will assist the reader in the understanding of more detailed literature on this subject.

The CCITT recommendation for Basic Rate Access is contained in I 412 and specifies a 3-channel time division multiplexed digital system, consisting of:

❏ **Two B Channels:** These are 64 Kbit/s digital channels which may be used quite independently. Both B channels may be used for 64 Kbit/s PCM speech, facsimile or data (asynchronous or synchronous) up to a maximum rate of 64 Kbit/s.

❏ **A D Channel:** This is a 16 Kbit/s common signalling channel that will carry message based signalling for both B channels. The D Channel is intended to carry the following types of signalling:
    Subscriber to network signalling and vice versa
    Subscriber to Subscriber signalling
    In the future the D channel will also be used to provide a third digital traffic channel that can access a packet switched service. Packetised data from a subscriber's packet mode terminal, also connected at the S interface, will be interleaved with signalling messages in the D Channel.

Figure 13.6 shows an example of this basic rate access using British Telecom's ISDN 2 service, which was launched in April 1990.

Within the subscribers' premises, a Network Terminating Equipment (NTE) provides the interface between the telephone line and the subscribers' digital terminal equipment. The term, digital terminal equipment, includes digital telephones or other digital devices such as computer terminals, personal computers, facsimile machines and even local area networks. All types of terminal equipment will connect to the NTE using a single standard interface irrespective of the nature of the equipment.

Note that in the case of digital telephones, the A/D conversion (and vice versa) takes place within the telephone and not in the subscriber line circuit in the exchange. As it is essential that calls can take place between subscribers using digital telephones and those still using analogue instruments, the digital telephone must use a 64 Kbit/s PCM encoder/decoder compatible with those already used in the local exchanges for analogue subscribers.

Within the NTE, signalling information is extracted from the digital signals transmitted by the terminal equipment and formatted into signalling messages which can be decoded by the local digital exchange. These signalling messages which are not unlike the signalling messages of CCITT Signalling System 7 are transmitted in the 16 Kbit/s D channel.

The two digital traffic channels and the signalling channel are time division multiplexed to provide a composite digital signal with a total information rate of 144 Kbit/s. The composite signal is transmitted by the NTE along with other information regarding the status of the digital line by the NTE to the local digital exchange.

As Figure 13.6 illustrates there are several arrangements for terminating the 2-wire local line. In the simplest case illustrated at the top of the diagram, the 2-wire line terminates directly on the subscriber's concentrator which will be equipped with a digital line module, rather than an analogue line card.

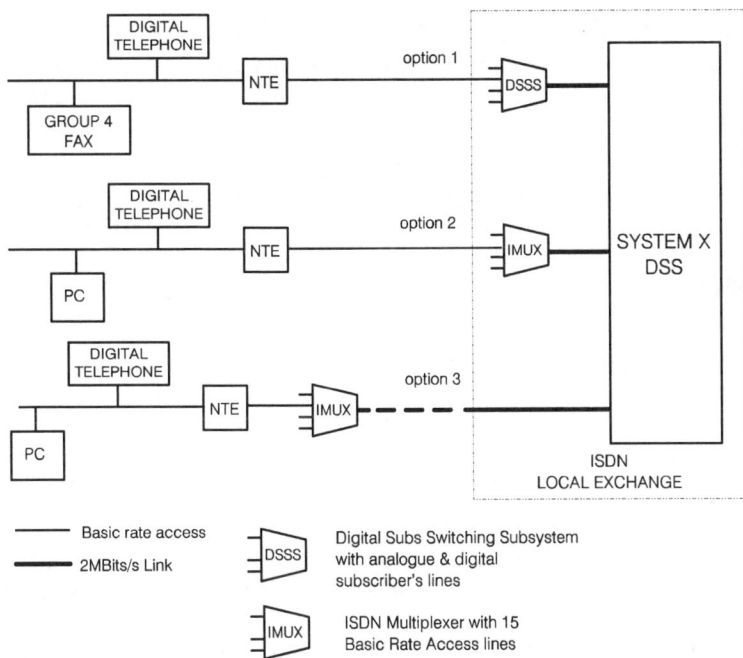

*Figure 13.6 Basic rate access to the ISDN*

The digital line module is then responsible for demultiplexing the two traffic channels and the signalling channel. The B channels are treated like all other traffic channels on the exchange, while the signalling messages are passed to the call processing system, using the Message Transmission Subsystem.

The first option is suitable for a local exchange in a rural area where only a very few subscribers require digital access. In urban and commercial areas where the take up rate for ISDN is expected to be greater the second option involving the use of a separate ISDN multiplexer (IMUX) in the local exchange building will be used. This IMUX has been specifically designed to handle up to 15 digital subscriber lines, and interfaces to the local exchange using multi-line protocols, i.e. 30 B channels and 1 D channel at 2 Mbits/s. In this case the ISDN multiplexer is equipped with the relevant digital line cards.

The third option, in which the ISDN multiplexer is remoted from the host local exchange, is suitable when ISDN access is required by a group of different subscribers who are geographically close together. This would be the case, say, in a shopping arcade, business park or industrial estate.

This third option would also be suitable if ISDN access is required to a group of subscribers whose own local exchange has not yet been replaced by a digital exchange. The subscribers' lines are connected to the ISDN multiplexer which is located in the local exchange building. However the 2 Mbit/s side of the ISDN multiplex is routed to another convenient digital local exchange.

The reason that the two 64 Kbit/s traffic channels are known as the B channels, while the signalling channel is referred to as the D channel is not known by the author. However the use of this terminology has led to the term *2B + D* frequently being used by manufacturers and others to describe ISDN Basic Rate access.

As stated earlier, the 144 Kbit/s 2B + D signal is transmitted over the same 2-wire cable that previously carried 3 KHz bandwidth analogue telephony signals, or data rates of up to 9.6 Kbit/s when used in conjunction with suitable data modems.

## 13.3.3 Basic Rate Access to Packet Switching Services

Although the total data transmission rate is 144 Kbits/s, strictly speaking, the subscriber has at present only two 64 Kbits/s circuits, or 128 Kbit/s total capacity available to be used for carrying traffic. It is intended by the CCITT that in the future it will be possible to access a packet switching function within the ISDN. This packet access will use the spare capacity of the D channel, which will normally only be carrying signalling messages during the set up and release phases of a call.

Calculations on the use of the D channel for Basic Rate Access signalling indicate that under worst case conditions, (i.e. high calling rate and short call holding times), the D channel will only have a 2% occupancy for signalling messages. Even though these calculations do not take account of any user-to-user signalling that may take place after call set up, the overall occupancy of the D Channel for signalling will still be much less than 5%.

Access to packet services will permit this form of connection to be used for those certain types of data traffic for which circuit switching is inappropriate. If, in the future, this packet switching access option is taken, the diagram in Figure 13.6 will probably be modified to that of Figure 13.7.

When packet data is to be transmitted in the D channel, a few restrictions will need to be imposed to ensure that the transmission of signalling messages takes priority over packet data. Packets will be restricted to a maximum of 256 octets to ensure that a new signalling message to be transmitted does not suffer an undue delay if a data packet is already in the process of being transmitted.

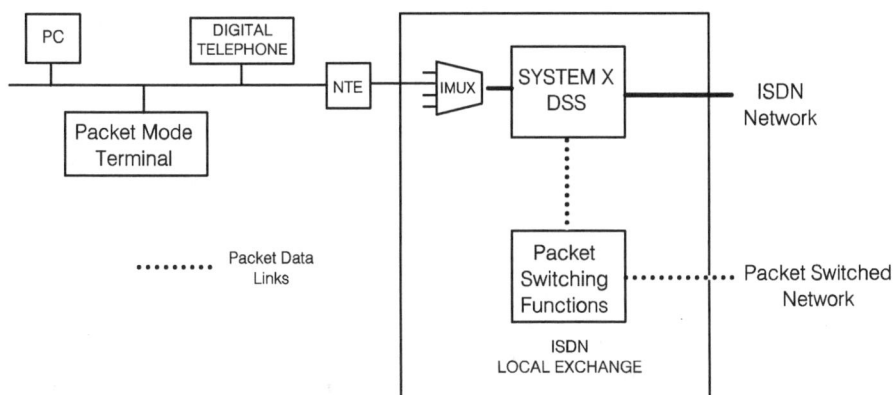

*Figure 13.7: Basic rate access with access to packet switching services*

In the description above, each B channel is associated with a specific terminal equipment, but in a typical office environment there will often be more than two items of terminal equipment which will require access to the ISDN. For example, a typical office will be equipped with telephones and systems such as WP, PC and Fax. The CCITT took this fact into account when developing the standard user interface to the ISDN, i.e. the connection of the terminal equipment to the subscriber side of the NTE, which effectively is the boundary of the network.

## 13.3.4 Basic Rate Physical Interface (Recommendation I 430)

This interface may be a simple point to point arrangement when only a single terminal equipment is required. Alternatively a point-to-multipoint arrangement can be used, in which so long as certain precautions regarding cable lengths are taken, up to eight items of terminal equipment may be connected to an 8-wire passive bus. The utilisation of the eight wires is:

❑ Two wires transmit TE to NT

❑ Two wires transmit NT to TE

❑ DC Power will normally be fed on the phantom of these four wires

❑ Four wires for optional additional power feeding.

Figure 13.8 shows the reference configuration for signal transmission and power feeding at the S and T interfaces. Within the UK, the standard ISDN connector is an 8 way RJ45 socket similar (in appearance) to the standard telephone line jack.

Note that the use of leads c, d, e and f is mandatory as these four wires carry signal transmission in the transmit and receive directions. However as the use of the additional power feeding leads a, b, g and h is optional, there may be some problems when terminal equipment is moved from one network to another.

In the case of the NTE and digital telephones it is likely that under normal circumstances these will be powered from a local mains supply. However should the mains supply fail an emergency mode of operation will permit the NT and the telephone to draw power from the network so that a basic telephony service can still be maintained.

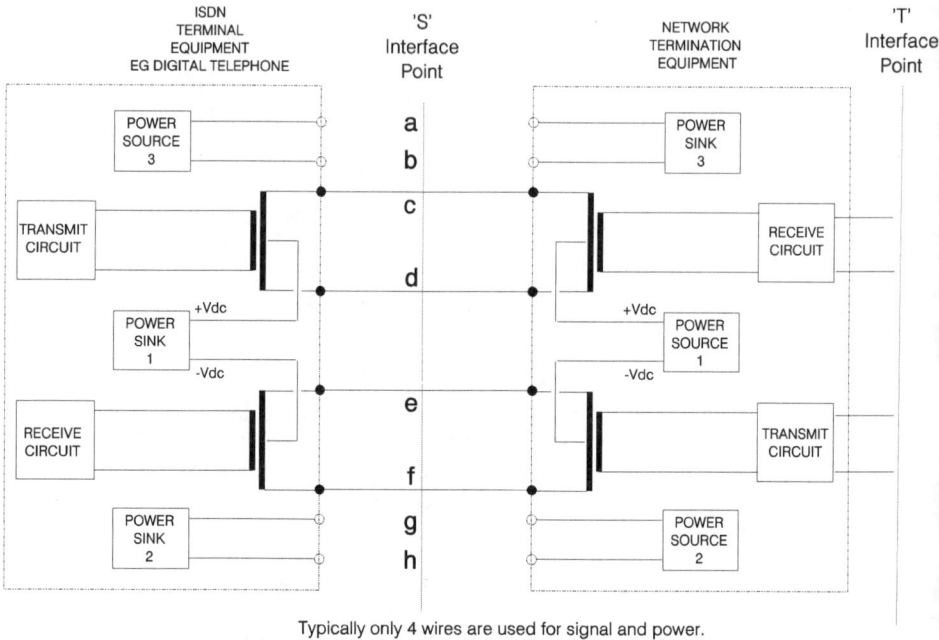

Typically only 4 wires are used for signal and power.

*Figure 13.8: The 8-wire passive bus*

## 13.3.5 Activation and Deactivation of Terminals

Procedures are included to permit the network to power down TEs and NTs which are not in use, and thus reduce overall power consumption. A deactivation message is sent to the NT and TE which will enter a low power state in which only essential circuits such as memories and monitoring circuits, are powered up.

When a TE or NT is required to be woken up, an activate message is transmitted, which will cause all the functions of the equipment to be restored. There may well be some instances in which the installation is kept powered up continuously.

## 13.3.6 The S interface Passive Bus

Figure 13.9 shows that up to eight terminals may be connected directly to this passive bus. A terminal adapter can be provided to permit connection of existing terminal equipment (VDUs, Data Terminals using V and X series interfaces) which are not designed for the ISDN interface. The use of a terminal adapter will permit subscribers to migrate to ISDN without the need to replace all their existing data terminals in one operation. In the future, when this equipment comes to the end of its economic life it can be replaced by devices with standard ISDN interfaces.

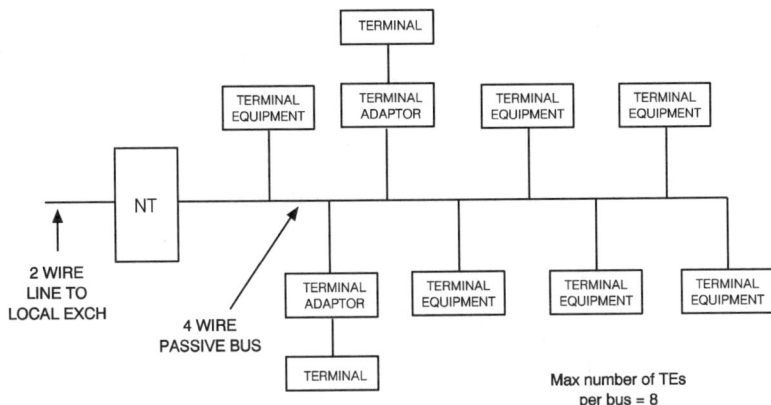

*Figure 13.9: The ISDN Passive Bus Arrangements*

## 13.3.7 Point-to-Multipoint Configurations

The topology of a point-to-multipoint passive bus has to be carefully considered to ensure that the terminal equipment and the NT can function together correctly. Section 4 of I 430 and Annex A (A.2.1.2) contain descriptions of wiring layouts, and explain that the limits on cable length are normally due to pulse propagation delays rather than the cable attenuation at the data rate involved. Two types of passive bus are described:

❏ The Short Passive Bus (up to 200m)

❏ The Extended Passive Bus (500m - 1000m)

On a point-to-multipoint passive bus, terminal equipment may be connected at random to any point on the bus. This should cause no problem as far as transmission in the direction NT to TE is concerned, but in the opposite direction there is a problem due to the multiple access of up to eight terminals. Therefore the NT has to cater for signals arriving from different TEs on the same pair of wires.

As you might expect TEs derive their timing from the network via the NT, by extracting a timing signal from the received signal at the S interface. Thus all TEs should be operating in synchronism. However, as each TE is at a different location on the passive bus, the signals from the NT to the various TEs will suffer different propagation delays and the best that can be achieved is that all the TEs will be operating at the same clock rate but can not be in synchronism.

In the direction TE to NT, the signals from each TE will also be subject to different delays dependent upon their location on the bus. Figure 13.10 illustrates the problem to be overcome, by considering the transmission of a single pulse from the NT to TE, and the subsequent possible transmission of another pulse from one of two TEs situated at opposite ends of the bus, i.e. TE1 adjacent to the NT, and TE2 at the extreme end of the bus. The point to note here is that although only one of the two TEs will actually transmit during the period shown, the NT must be able to detect the pulse irrespective of which TE it came from.

At time t0, the NT starts to transmit the pulse. As TE1 is adjacent to the NT, the leading edge of the pulse is received by TE1 very shortly afterwards at time t1. The reception of this leading edge is used by TE1 as a synchronising signal, and shortly afterwards, at time t3, TE1 may transmit a similar pulse to the NT.

At time t2, TE2 receives the leading edge of the pulse transmitted by the NT, and shortly afterwards, at t4, TE2 may transmit a pulse to the NT.

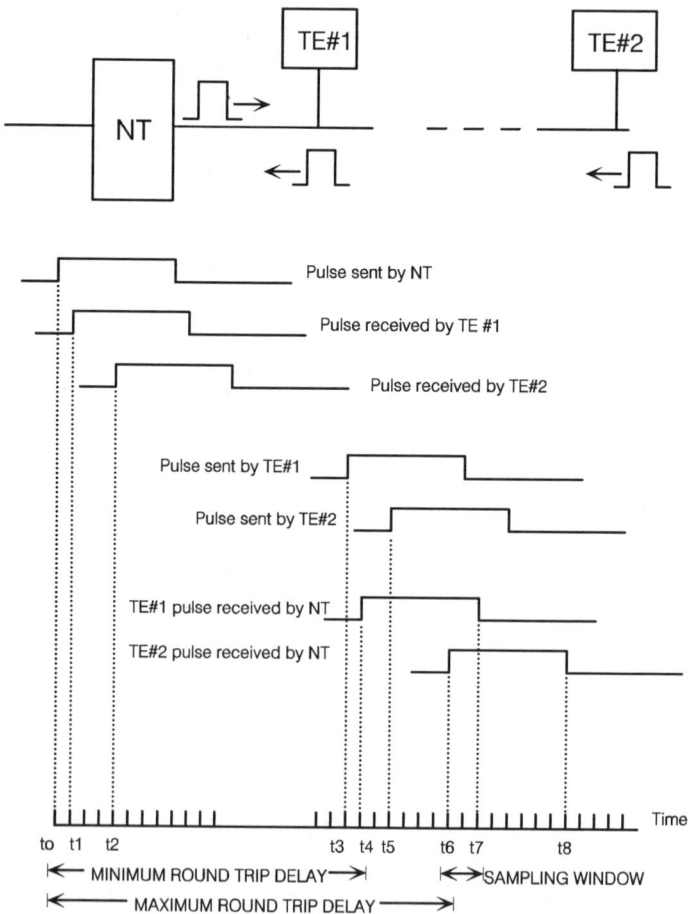

*Figure 13.10 Illustration of the problem caused by different round trip delays on the passive bus*

Due to the differing propagation delays, a pulse from TE1 would not arrive at the NT at exactly the same time as a pulse that had been transmitted by TE2. As the diagram shows, if there is some period of overlap during which a pulse from either TE is present it is possible to detect a pulse from any TE on the bus. There is, however, only a limited period during which the NT can sample the signal, and it should be noted that this sampling window is considerably less than the nominal pulse width.

If the NT samples during the period t5 to t6, it will be able to correctly detect a pulse transmitted by TE1, but will not be able to detect the pulse if it was transmitted by TE2, as this pulse will not yet have arrived at the NT. Similarly, if the NT samples during the period t7 to t8, a pulse from TE2 may be correctly detected, but the pulse from TE1 will no longer be present, and thus will not be detected.

Since the locations of TE1 and TE2 represent the extreme cases, the period between t6 and t7 represents the time of overlap between possible received pulses at the NT. The period t6 to t7 thus specifies the window during which the NT can sample and be able to detect a pulse from any TE on the bus.

The period of time between a pulse being transmitted by the NT and a subsequent pulse being received from a TE is defined as the round trip delay. By specifying the period t0 to t5 as the minimum round trip delay, and the period t0 to t6 as the maximum, a suitable sampling window can be specified.

Increasing the cable length would further increase the maximum round trip delay and reduce the sampling window, and thus make the design of the NT sampling circuit more complex. If the cable were so long that the difference between maximum and minimum round trip delays approached or exceeded the nominal pulse width, the NT would not be able to detect pulse from all TEs connected to the bus. By restricting the difference in delay to about 80% of the pulse width, it is possible to ensure that the NT is able to cater for these variable delays.

## 13.3.8 The Short Passive Bus

In the case of the short passive bus, illustrated in Figure 13.11, a limit of 10 - 14 uSecs round trip delay is imposed. As the speed of propagation on the cable is known, the maximum length of the bus can be calculated from the maximum round trip delay figure.

The minimum delay is enforced by a built in two bit delay at the TE (see Section 13.3.12). The maximum delay of 14 uSecs is calculated for a TE located at the distant end of the bus. These figures also include allowances for timing jitter deviations. Using typical cable this limits the maximum distance for any TE to 100 - 200 metres from the NT. As previously stated up to eight TEs may be connected to any point of this short passive bus.

## 13.3.9 The Extended Passive Bus

It is possible to use a topology in which distances of up to 1,000 metres are possible. This configuration, known as the extended passive bus, restricts the locations of TE to a group at the distant end of the cable as shown in Figure 13.12.

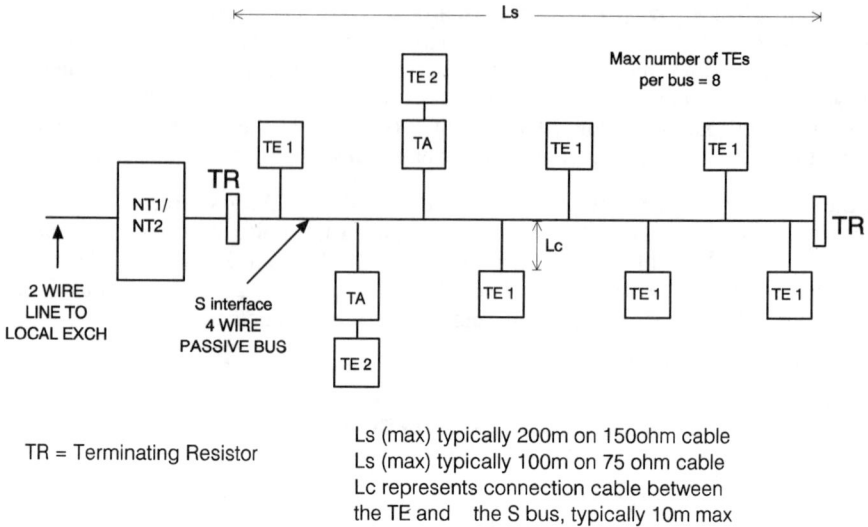

TR = Terminating Resistor

Ls (max) typically 200m on 150ohm cable
Ls (max) typically 100m on 75 ohm cable
Lc represents connection cable between
the TE and    the S bus, typically 10m max

*Figure 13.11 Short passive bus*

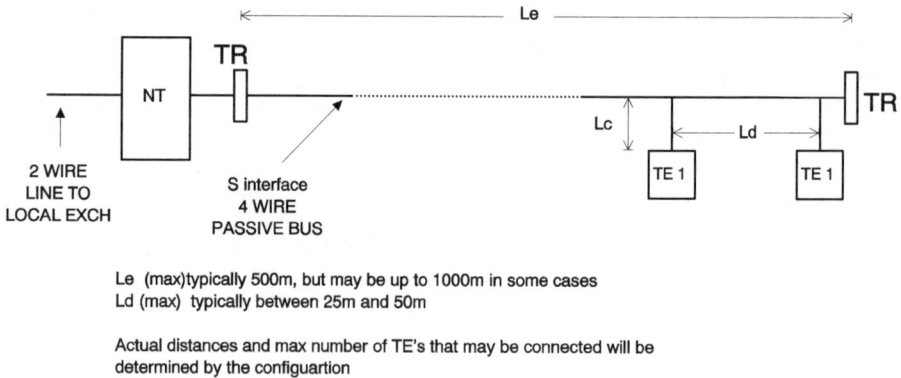

Le  (max)typically 500m, but may be up to 1000m in some cases
Ld (max)  typically between 25m and 50m

Actual distances and max number of TE's that may be connected will be
determined by the configuartion

*Figure 13.12 Extended passive bus*

This arrangement is to ensure that the maximum difference in round trip delay for the closest and the farthest TE does not exceed 2 uSecs.

Using typical cable, a reach of at least 500 metres is possible, with maximum spacing between TEs of between 25 and 50 metres. It may however not be possible to connect 8 TEs in this configuration and the network provider must determine the combination of total length, spacing between TEs and the maximum numbers of TEs to be connected

## 13.3.10 Point-to-Point Configuration

In this configuration, shown below in Figure 13.13, only 1 TE will be connected to the NT, thus the problems of varying round trip delays do not occur. The limiting factor here then is a combination of cable attenuation at the data rate involved, i.e. 192 Kbit/s and a maximum round trip delay.

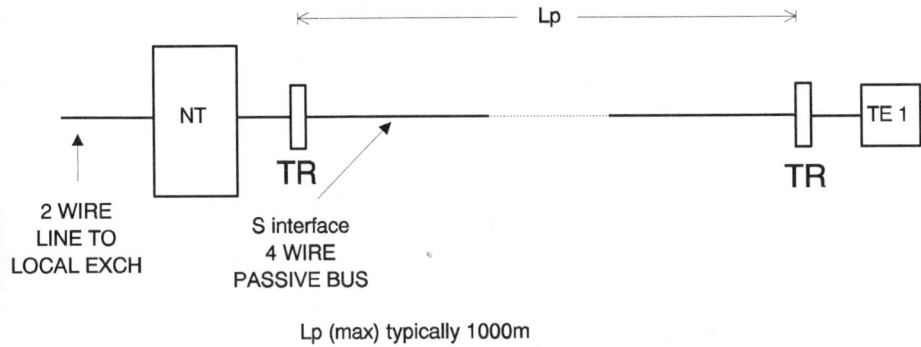

*Figure 13.13 Point-to-point operation*

Point-to-point operation should give up to 1,000 metres separation between the NT and TE. To achieve this objective the recommendation states that the maximum cable attenuation at 96 KHz should not exceed 6dB, and the round trip delay should be between 10 and 42 uSecs.

## 13.3.11 Line Coding at the S Interface

A pseudo-tenary line code is used for both directions of transmission at the S interface. As illustrated in Figure 13.14 a binary 1 is represented by no line signal (zero volts). During this period the output impedance of the transmitting device should be high, i.e. above 2,500 ohms.

Binary 0s are transmitted as pulses of alternating polarity. The output impedance of the device while transmitting a zero should be low, i.e. less than 20 ohms. There are exceptions to this coding rule, which will described in the following section.

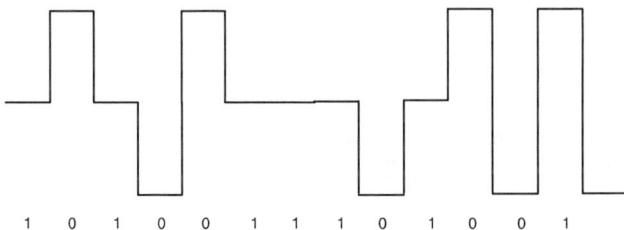

*Figure 13.14 Pseudo-tenary line coding at the S interface*

## 13.3.12 Frame Structure at the S Interface

As Basic Rate Access only provides for two B channels. It is only possible for two of the eight terminals to be given access to a B channel at one time. The procedures for assigning a B channel to a terminal are described in the signalling recommendations, and will be covered later in this chapter. To simplify the explanation of the TDM frame structure at the S interface, consider for the time being that two terminals A and B have been assigned to B channels 1 and 2 respectively.

At the S interface, a multiplexed frame structure of 48 bits, transmitted at 192KBit/s is used in both directions of transmission, i.e NT to TE, and TE to NT. There are however subtle differences between the frame structures used for each direction of transmission.

Both frame structures are shown in Figure 13.15. Note that there is a nominal time delay of two bits (approximately 10 uSecs) between the commencement of the frame transmitted by the NT and that transmitted by the TE. The perceived time relationship between frame structures will depend upon the distance between the NT and the TE, and the position on the bus from which the frames are being observed. The physical separation between NT and TE causes the time delay between the NT transmitting its first bit and that bit being received by the TE.

The TE delays the start of its frame for a 2-bit period after having received the first frame bit from the NT. There is then the subsequent delay for this bit to travel to the NT. If the observation point is at the TE the delay perceived should be a 2-bit period. If however the observation point is at the NT, the delay will be equal to the 2-bit offset, plus the propagation time in both directions. This explanation may help to throw some light on the reasons for the restrictions on cable lengths in the point-to-multipoint passive bus.

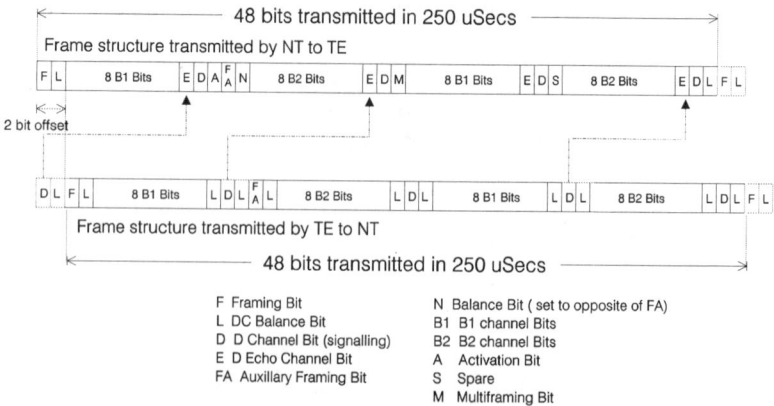

*Figure 13.15 Frame structures at the S interface*

The NT will derive timing from the network and use this timing to control its transmission to the TE. The TE will extract bit timing, octet timing and framing from the signal transmitted by the NT and use this timing to control its transmissions to the NT.

## 13.3.13 Frame Alignment

The first bit of the frame is the framing bit (F). This is always a binary 0, and may therefore be either a positive or negative pulse. It is immediately followed by a balance bit (L), which is a pulse of the same polarity as the framing bit. This framing condition of two consecutive pulses of the same polarity violates the coding rule outlined in Section 13.3.10 and as it is repeated every frame provides a mechanism for rapid realignment should frame alignment be lost.

The framing procedure is further improved by the addition of an auxiliary framing bit, Fa, and its associated balance bit at bit positions 14 and 15 in the frame. In the direction NT to TE, the auxiliary balance bit is labelled L, while in the opposite direction it is labelled N.

The coding of these auxiliary framing bits guarantees that, following a line code violation due to bits Fa and L, a subsequent violation will occur 14 bits or less later. Frame alignment is assumed when three such consecutive pairs of violations have occurred.

Loss of frame alignment can be assumed if 96 bits, equivalent to two frames, have been received without a pair of framing violations. A slight variation to this rule can be used in the direction TE to NT, but in this case 144 bits, equivalent to three frames, have to be received without a pair of violations.

## 13.3.14 Bit Positions of B1 and B2 Channels

Within each 48-bit frame two octets from both B channels are transmitted. The first B1 octet follows bit Fa and L. The positions of the second B1 octet, and both B2 octets can be seen from the diagram in Figure 13.15.

In the direction TE to NT a balance bit (L) immediately follows each B channel octet. Except for the framing balance bits, all balance bits are set to binary 1, i.e. no line signal, if the number of binary 0s following the last balance bit is even. A balance bit is set to zero, and observes the alternating polarity rule, if the number of binary 0's following the last balance bit is odd. This procedure is especially important when power feeding across the interface as it ensures that the average DC voltage due to signals is zero.

In the direction NT to TE, the situation is different in that the whole frame is DC balanced by the last L bit of the frame.

## 13.3.15 Bit Positions of the D Channel

Within each frame four D channels bits (D) are transmitted. In the direction TE to NT, each D bit, is immediately followed by its own balance bit (L). D channels bits in the direction NT to TE are not individually balanced.

## 13.3.16 Accessing the D Channel

To transmit signalling a TE must first access the D Channel, perhaps in contention with other TEs which require access at the same time. Having successfully gained sole access of the channel, signalling information is transmitted by the TE in the form of messages, (or frames), similar to those used in CCITT 7.

Each TE may only transmit one frame at a time. At the end of each transmitted frame the TE releases its access and then must regain access to the D channel before sending the next frame. This procedure ensures that all terminal equipment has fair access irrespective of its position on the passive bus.

All TEs that are active, transmit during the D bit time slots, however if a TE is not in the process of transmitting signalling (or packet data), it must transmit a continuous stream of binary 1s in these timeslots.

## 13.3.17 The D Echo Channel Bit (E)

As up to eight terminals may be connected to the S interface, a procedure is required to ensure that only one TE can access the D channel at any one time. Without such procedures, two or more terminals may attempt to transmit signalling information at the same time, with the result that no intelligible information is received at the distant end.

The D echo Channel bit is transmitted only in the direction NT to TE to provide a mechanism for controlling the access of terminal equipment to the D Channel when contention may arise. However the procedure is also used in the point-to-point configuration.

Prior to accessing the D channel, a TE must first monitor the D Echo channel to ensure that the D channel is not already in use by another terminal. If all active terminals connected to the bus are not in the process of signalling they will be transmitting binary 1s, (i.e. in the 0 volts output, high impedance state) during each D channel time slot. The NT will then receive a binary 1 in these time slots.

However, if any other terminal has access to the channel, and is in the process of signalling, the signalling frames it transmits in the D channel will of course include binary 0s (i.e. low impedance, with positive and negative pulses). Irrespective of the number of terminals which are simultaneously transmitting 1s, if any one terminal transmits a zero, then the signal received at the NT will be a zero.

The NT sets each D Echo bit to the binary value of the last D channel bit it received, and thus reflects, or echoes the data it receives in the D channel back to all connected TEs.

Figure 13.16 is a very much simplified diagram to illustrate, in concept, how a TE gains exclusive access to the D channel to transmit a frame of signalling information. A similar procedure is also used if the TE has data other than signalling to transmit, however the description that follows is generally valid for both cases.

START STATE

TE is transmitting 1's in each D channel Bit Time Slot

'START' FRAME TO TRANSMIT — ( 1 )

SET C = 0
SET n = 1

MONITOR D ECHO

CONTINUE SENDING 1'S IN D CHANNEL

?
BINARY 1 RECEIVED IN D ECHO — NO

YES

INCREMENT COUNTER C

CONTINUE SENDING 1'S IN D CHANNEL

?
IS C = 9

TRANSMIT nth BIT OF FRAME

MONITOR NEXT D ECHO BIT

?
D ECHO = nTH BIT — COLLISION !! ABORT FRAME

YES

?
LAST FRAME BIT E SENT — NO — INCREMENT COUNTER n

YES

?
MORE FRAMES TO SEND — YES

NO — ( 1 )

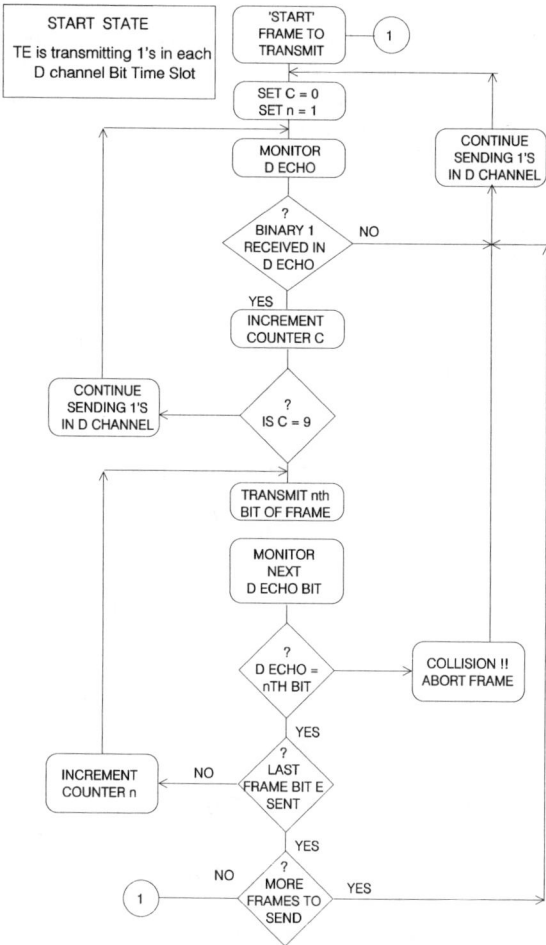

*Figure 13.16 Simplified flow chart to illustrate D Channel access procedure*

A TE that has signalling or packet data to transmit is already transmitting binary 1s in the D channel bit timeslots. A counter (C) is set to zero and a pointer N is set to point at the first bit of the frame to be transmitted.

The TE monitors the next D echo bit received from the NT and if this bit is a binary 1 the counter C is incremented by 1. As the counter has not reached nine, the TE continues to transmit 1s in the D timeslots and monitors the next D echo bit from the NT, this process continues until nine consecutive binary 1s have been received.

If at any time before counter C reaches nine, a binary zero is received in the echo channel, it can be assumed that the D channel is in use and the counter must be restarted from zero.

As it may be possible for a signalling frame to naturally include a string of more than nine continuous binary 1s, another protocol is required to ensure that the count of nine bits in the D echo channel is due to the fact that no other TE has access to the D channel, and not the transmission of a frame with a lengthy string of 1s by a TE which already has access to the channel.

Zero bit stuffing is used to ensure that the maximum number of continuous 1s that can be transmitted in a frame is five, except in the cases of flags with the pattern 0 1 1 1 1 1 0.

Once the counter reaches nine, the TE can transmit the first bit of its current frame in the D channel, continuing to monitor the echo channel to ensure that next echoed bit is identical to the transmitted bit. So long as a match occurs the TE may transmit the next bit of the current frame, and the frame pointer N is incremented for each bit transmitted. This process continues until either the end of the frame is reached or an echo bit is of the opposite value to that transmitted.

If an echo channel bit differs from the last transmitted D channel bit, it must be assumed that two terminals have attempted to access the D Channel simultaneously. Both TEs must stop the transmission of the current frame, and recommence transmitting continuous 1s. The counters will be reset to 0, and both terminals will revert to monitoring the D Echo channel. The procedure ensures that the TEs will regain access to the D channel independently and both will retransmit the frames that were aborted.

A priority mechanism is included to ensure that signalling messages take priority over other types of data, e.g. packet data. A TE with packet data to transmit must allow its counter C to reach a value of 13 before accessing the D Channel.

# 13.4 Testing Digital Subscriber Circuits

## 13.4.1 Digital Test Loop Backs at the S Interface

With the introduction of digital transmission on subscribers' lines it is now possible, in fact it is essential, for a mechanism to be provided for testing the transmission path from the exchange as far as the terminal equipment. The network maintenance functions of an ISDN will include fault location procedures which will be able to determine whether a fault exists on the line, network termination or terminal equipment. In many cases these fault procedures can be carried out automatically by the exchange software before the user is aware that a problem exists.

In general terms the principle is illustrated in Figure 13.17. Either manually or under software control a digital loopback is applied at a specified point in the transmission system so that the signal being received at that point is retransmitted back to the originating equipment.

A known test signal can then be applied to the looped section and the resultant received signal analysed to detect whether any degradation or corruption has taken place. If the received signal from one loopback is good, a fault can be isolated to a particular section of the system by systematically moving the location of the loopback.

The I 600 series of recommendations includes details of the maintenance philosophy but as the procedures involved have an impact on the design of TE and NT equipment, some interesting information regarding test loopbacks is also contained in Appendix 1 to I 430.

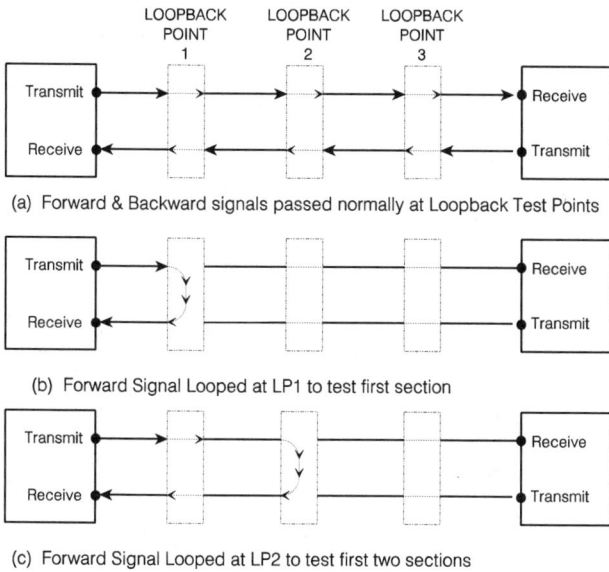

(a) Forward & Backward signals passed normally at Loopback Test Points

(b) Forward Signal Looped at LP1 to test first section

(c) Forward Signal Looped at LP2 to test first two sections

*Figure 13.17 Principle of loopback testing*

Within an ISDN installation there are several obvious points where a loopback would be useful in isolating faults in certain equipment. These points have been identified by either letters or numerals. Letters refer to loopbacks which will operate towards the terminal equipment, while numerals refer to loopbacks which operate towards the exchange.

The diagram in Figure 13.18 includes two test loops that are recommended by the CCITT. These are shown in solid lines and are known as Loopback 2 in the NT1 functional unit, and Loopback 3 in the NT2 unit. Both are complete loopbacks as described in Section 13.4.3.

The dashed lines represent loopbacks which are optional and may not necessarily be implemented in all equipment.

Loopback 2 should be as close to the T interface as possible and loops back toward the exchange. This loopback loops both B channels and the D channel for testing, and is controlled by the exchange using messages sent over the line transmission system.

Loopback 3 should be as close to the S interface as possible, and also loops back toward the exchange. This loopback is controlled from the NT2 by local maintenance procedures or by the transmission of suitable messages in either a B channel or the D channel.

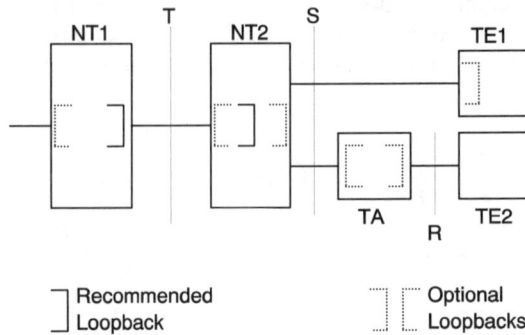

*Figure 13.18 Location of test loopbacks at the user-network interface*

## 13.4.2 Categories of Loopback

Three different types of loopback are specified: Complete, Partial and Logical. Each type may be further categorised as being either transparent or non-transparent. The definitions and diagrams which follow should make the distinctions between these different types and categories clear.

## 13.4.3 Complete Loopback

This is a Layer 1 mechanism, which means that the whole of the received bit stream is retransmitted back to the exchange without any alteration. The method of providing this loopback is not specified, however it is suggested that it may be implemented by some form of electronic gate between send and receive paths within the NT providing the loopback.

## 13.4.4 Partial Loopback

This is also a Layer 1 mechanism, in which only the bits associated with a specified channel, or channels, are looped back toward the exchange. Bits from unspecified channels are passed on unaltered in the forward direction, while bits from the TEs associated with unlooped channels are multiplexed as normal with the bits from the looped channels. A partial loopback permits one B channel only to be tested while the other B channel remains in service.

## 13.4.5 Logical Loopback

This type of loopback may be implemented at Layers 1, 2 or 3. Logical loopbacks operate only on selected information received in a channel or channels. This information may be modified as part of the loopback procedure. e.g. to indicate that certain information has been received and acted on, some of the received bits may be altered prior to retransmission.

## 13.4.6 Transparent Loopback

Figure 13.19 shows that a transparent loopback involves two actions, firstly the backward signal that would normally be transmitted by the NT is inhibited at point X, to prevent it interfering with the looped signal. Secondly the forward signal continues as normal, without modification, but is also looped back toward the exchange.

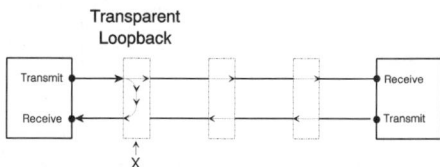

Complete Transparent Loopback: The forward signal received at the loopback point is returned unchanged in the backward direction, and also transmitted on in the forward direction

*Figure 13.19 Transparent loopback*

## 13.4.7 Non-transparent Loopback

Figure 13.20 illustrates that this category is similar to the transparent loopback in as much as the forward signal is looped back without modification. The main difference is that the signal passed forward is subject to some modification as indicated by the inclusion of a unit L which may either inhibit or alter the forward signal.

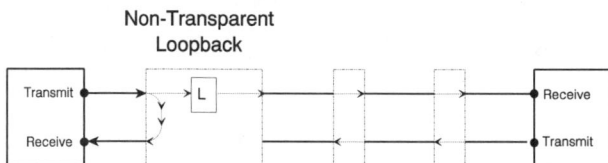

Non-Transparent Complete Loopback: The forward signal received at the loopback point is returned unchanged in the backward direction. The signal transmitted on in the forward direction has been altered or inhibited by unit L

*Figure 13.20 Non-transparent loopback*

It should be noted that in these examples only complete loopbacks have been described. It is also possible to have partial transparent and non-transparent loopbacks.

## 13.4.8 Optional Loopbacks

Several optional loopbacks are shown in Figure 13.17. In this text it will suffice to draw your attention to a couple of these.

Optional loopback 4 will be in a ISDN TE or TA and operate only on a B channel and towards the exchange. This can be a partial, transparent or non-transparent loopback. If implemented it may be controlled by the NT2 or local exchange using Layer 3 messages.

Optional loopback A provides a local self test of the TE or TA. If this optional loopback is implemented within a TE or TA, any transmission from the TE or TA (while in loopback mode) towards the network interface should be inhibited. Loopback B2 similarly provides a check of TE or TA and associated NT2, and also should inhibit any transmission past the loopback point toward the network interface.

More specific details of all the optional loopbacks are contained in two tables in the Appendix to I 430.

# 13.5 ISDN Test Equipment

## 13.5.1 The Requirement for ISDN Test Kit

Suppliers of ISDN equipment and those responsible for installing it must be able to check that the individual items and the installation as a whole conforms to the relevant regulations governing connection to ISDN lines.

Kit is becoming available for the testing of ISDN Terminal Equipment, Terminal Adapters and ISDN customer premises installations. Such test kit requires a high degree of sophistication to permit the installation to be fully tested in both directions of transmission. The following are examples of tests that are required during the initial installation and subsequent maintenance of an ISDN installation:

❑ Terminal Equipment tests (voice/data transmission and signalling)

❑ User Interface (S and T reference points) tests

❑ Local line tests (inside and outside the customer's premises)

❑ Protocol tests for the D channel

Terminal equipment such as digital telephones must be tested to ensure conformity with relevant regulations regarding analogue to digital conversion and vice versa, frequency response, linearity and distortion. The digital telephone must also comply with the protocols for accessing and using the D channel to establish calls.

At the S and T interfaces, the passive bus installation has to be tested to ensure that supply voltages, impedances, return loss, signal balance and propagation delay are acceptable. Additionally in most installations it will also be necessary to confirm that the polarity of the individual wires in the passive bus is in accordance with the relevant specifications, when DC power is being supplied over the passive bus.

Local Line tests will include measurements of the characteristics of the NT and transmission media. It will be necessary to carry out Bit Error Rate (BER) tests on the whole installation.

D Channel protocol tests are required to ensure that call set up and release protocols are correct. This is often a prerequisite for other tests, many of which involve setting up a call for a specified TE to a specified B channel to permit further testing of the B channel and associated TE.

It is also necessary to check the protocols involved for setting up of calls to non voice terminals such as fax and teletext terminals.

A wide range of ISDN test equipment is available from several manufacturers. As examples of such equipment, the following three items from the Wandel & Goltermann company are used to assess various aspects of the operation of the 2B + D user interface :

❏ IBT-1 ISDN bit error tester and test telephone

❏ DA -20 Data Analyser for monitoring D Channel Signalling

❏ ILS-1 ISDN Passive Bus simulator

These three items will be briefly described in the following sections.

## 13.5.2 ISDN Bit Error Rate Tester

This test set would normally be used in pairs to evaluate the quality of a switched ISDN digital end-to-end connection.

The IBT-1 shown in Figure 13.21 consists of the following main items:

❏ ISDN test telephone, with handset and push button keypad

❏ Pseudo Random Bit Sequence (PRBS) generator

❏ PRBS analyser

❏ Keyboard, Liquid Crystal display and thermal printer

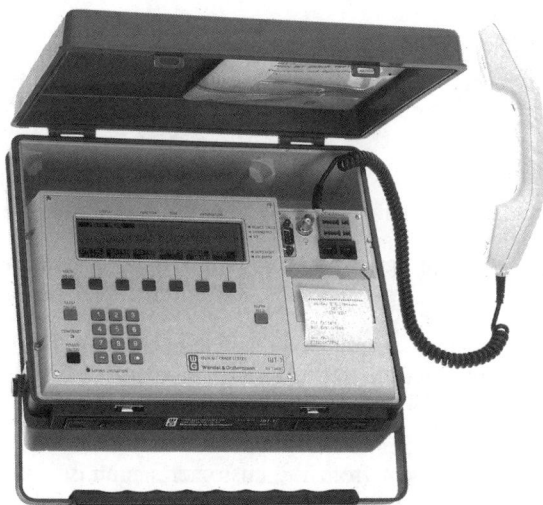

*Figure 13.21 Wandel & Goltermann ISDN bit error tester*

The IBT-1 will connect to any access point on the ISDN passive bus within the customer's premises. In most modes of operation only one of the available B channels is tested, thus the operation of other TEs is not prohibited during testing as these are able to access the remaining B channel when required.

The IBT-1 operates in two modes, Telephone mode and Bit Error Rate Test (BERT) mode. In Telephone mode the equipment functions as an ISDN digital telephone and can be used to give a fast go/no go check of any access point on the installation. This mode is also used prior to BERT testing to permit the technicians carrying out the test to confirm or alter equipment parameters etc.

In BERT mode four basic types of test are possible, and are illustrated in Figure 13.22.

## End-to-end measurement

This is the classic BER test. An IBT-1 is connected to each of the two connection points A and B. Using Telephone mode a connection between the two IBTs is established via the ISDN network. This connection may involve one or more digital exchanges. Both IBTs then transmit a test sequence which may be a standard PRBS or a repetition of a 16-bit word programmed into the machine by the technician.

As the diagram shows the test pattern is transmitted in only one of the B channels, at the distant end the test sequence is received and evaluated, the results being displayed on the screen and printed out if required. Typical evaluations to be made during testing would include; Bit Error Count, Bit Error Ratio (BER), and Error Free Seconds. The IBT will also display and record alarm states and periods when the BER exceeds a specified threshold.

## Test sequence loopback

The IBT at A transmits a test sequence in the forward direction of one B channel. The test sequence is received and evaluated by the IBT at B, and retransmitted back to A on the backward direction of the same B channel.

## End-to-end measurement with result retransmission

The IBT at A transmits a test sequence to B. The sequence is received and evaluated by the IBT at B, which then transmits the evaluated results to B in the backward direction of the same channel. The results can then be displayed and printed at the location which originated the test.

Because the backward channel is used for result transmission, only one direction of transmission can be tested at a time in this mode.

## Self-call BER measurement

This mode will check the system from the customer's premises as far as the ISDN local exchange and back. It therefore provides a BER check of the equipment, local installation and line in cases where a high BER has been measured during an end-to-end measurement, and it is necessary to determine which portion of the whole circuit is causing the problem.

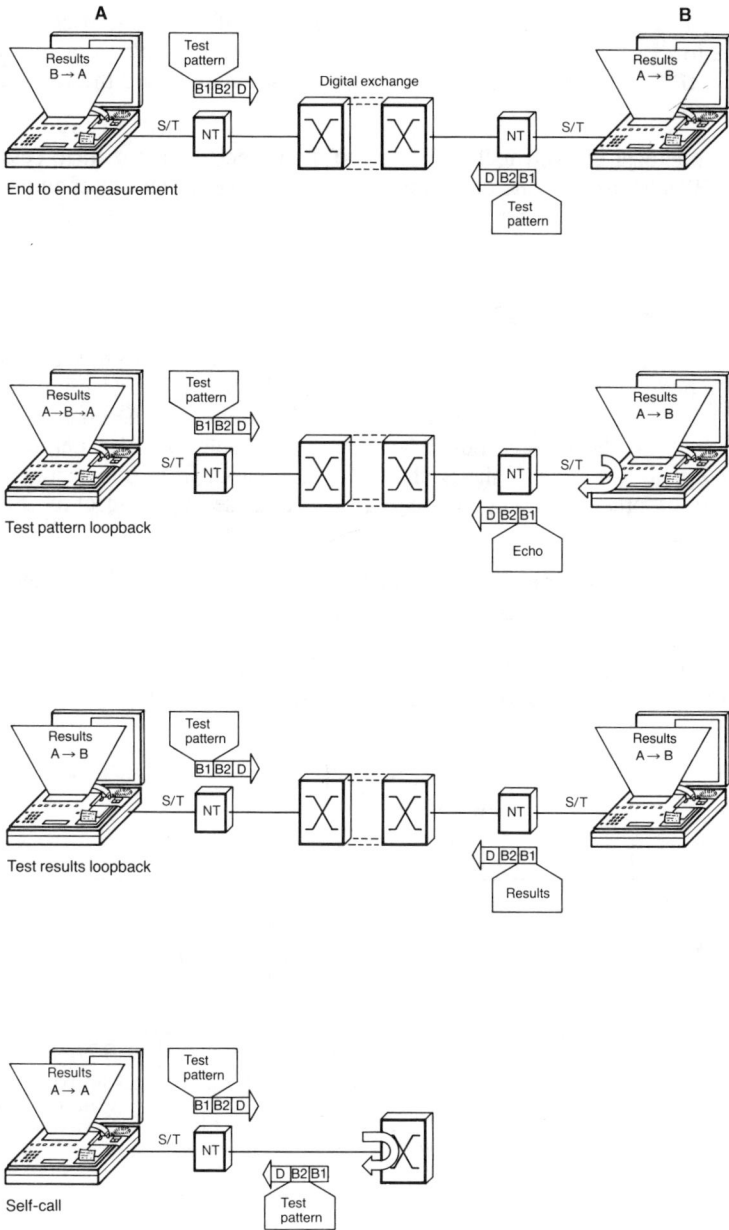

A

Results
B → A

Test
pattern

B1 B2 D

Digital exchange

S/T

NT

X X

NT

S/T

B

Results
A → B

D B2 B1

Test
pattern

End to end measurement

Results
A → B → A

Test
pattern

B1 B2 D

S/T

NT

X X

NT

S/T

Results
A → B

D B2 B1

Echo

Test pattern loopback

Results
A → B

Test
pattern

B1 B2 D

S/T

NT

X X

NT

S/T

Results
A → B

D B2 B1

Results

Test results loopback

Results
A → A

Test
pattern

B1 B2 D

S/T

NT

X

D B2 B1

Test
pattern

Self-call

**Bit error rate test modes**

*Figure 13.22 Test functions of the IBT-1*

Only one IBT is required. The IBT sets up a call to itself using one B channel for the link to the local exchange, and the other B channel for the link from the exchange back to the IBT. A test sequence can then be transmitted around the local exchange loop to test the local circuits.

In all modes except the self call mode, the IBT can be programmed to select either the B1 or B2 channels for the test, or the IBT can use normal D channel protocol signalling thereby having no control over which B channel is selected by the exchange.

## 13.5.3 DA-20 Data Analyser

This equipment is similar in many respects to the protocol analysers used to check packet switched networks. The DA-20 may be fitted with an interface module for the testing of basic rate access D channel protocols by monitoring D channel signals in both directions on the passive bus. The interface module can also demultiplex and extract the traffic from one of the B channels. The output from this channel can then be monitored by other test equipment e.g. a PCM analyser, by connecting it to DA-20 via an external connector.

The DA-20 can be programmed to select only signalling messages referring to a given TE or B channel, or to select only certain types of signalling messages. The selected signalling units are stored in memory so that the results can be evaluated off-line. The use of the DA-20 to monitor the signalling on the S bus of single ISDN lines from a public exchange or ISPBX is illustrated in Figure 13.23.

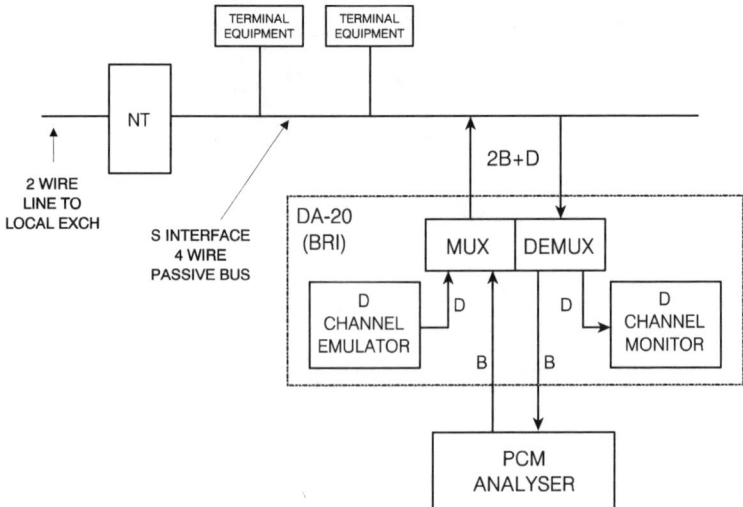

*Figure 13.23: DA-20 used to monitor signalling on a basic rate S bus*

As the DA-20 is able to simulate and evaluate D channel signalling in both directions on the passive bus, the DA-20 can emulate the signalling functions of a TE to set up a call, and subsequently demultiplex and extract the relevant B channel data and output this to other test equipment (e.g. PCM analyser) via another external connector.

Figure 13.24 shows that with a further interface module, the DA-20 can be used to monitor D channel signalling on Primary Rate Access systems, e.g. 30-channel access in Europe with D channel signalling in TS16, or 23-channel systems used North America and Japan. Note that the DA-20 is also capable of monitoring and evaluating CCITT #7 signalling on inter-exchange trunks.

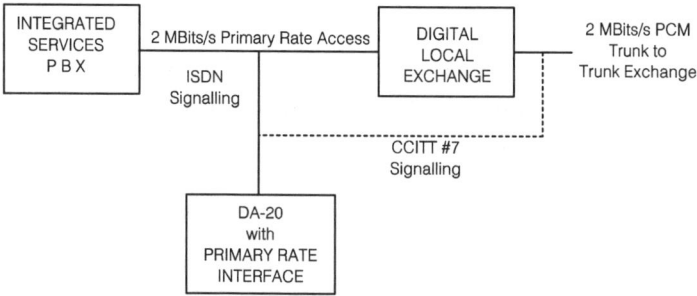

*Figure 13.24: DA-20 used to monitor signalling on primary rate ISDN lines and inter-exchange links*

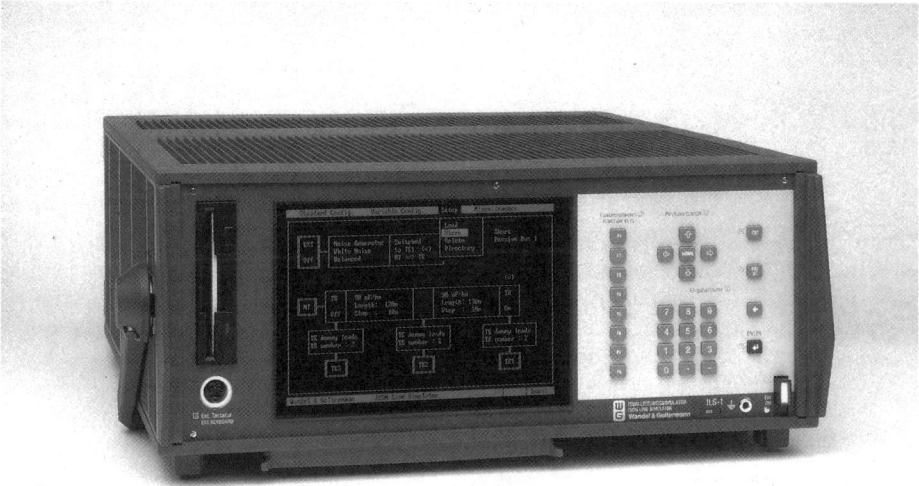

*Figure 13.25: Wandel & Golterman ILS-1 passive bus simulator*

The ILS-1 shown in Figure 13.25 is equipment which has been designed to simulate and measure the physical and electrical characteristics of a basic rate S interface passive bus. The device, which is based around its own internal PC-AT computer, includes individual hardware modules to simulate the following:

❑ Terminal equipment (each with seven metres of connection cable). This module also permits actual TEs to be connected to the simulated bus

❑ Up to 150m and 1,200m of PVC or PE cable (programmable in 10m sections)

❑ External interference (a noise generator capable of producing white noise, noise bursts, impulsive noise and so on).

As there are various cabling options for the passive bus, e.g. point-to-point and point-to-multipoint, the device is able to simulate the following types of installation:

❑ Point-to-point

❑ Point-to-Multipoint Short Passive Bus, with NT at one end

❑ Point-to-Multipoint Short Passive Bus, with NT between the ends

❑ Extended Passive Bus

As we have already seen in an earlier section, control of access to the signalling channel is dependent upon signal propagation times. The ILS-1 permits measurements of propagation delay to be made on actual installations, or for proposed installations to be simulated to determine whether a particular installation is feasible.

An additional hardware module permits connection of the simulator to an NT so that a proposed installation with various TEs can be tested prior to cabling.

The concept is to simulate the installation cabling and connect the actual TEs that will be used, to the simulator to check that the TEs can access both B and D channels correctly.

# 13.6 Digital Transmission in the Subscriber Loop

## 13.6.1 Current Digital Transmission Techniques for ISDN

Basic Rate Access requires duplex transmission, that is digital transmission from subscriber to exchange, and vice versa simultaneously. The telecommunications authorities and IC manufacturers have been developing systems for bi-directional digital transmission over the subscriber's 2-wire line. Two main techniques have evolved, albeit with implementation variations between manufacturers. These techniques are known as:

❑ Burst mode

❑ Echo cancellation.

Both systems are currently found in practice. Burst mode is relatively simple, and therefore cheaper to implement. Echo cancellation is considerably more complex, and hence more expensive, however the maximum range achievable with echo cancelling systems is up to three times that achievable with Burst Mode, and it is for this reason that in the near future echo cancelling systems will be the preferred solution in public ISDN systems.

The next section briefly describes the early BT Pilot ISDN, which although now no longer offered to customers, serves to show the evolution of the British ISDN service and gives an operational example of Burst Mode transmission. Subsequent sections describe some of the ICs available for both techniques.

## 13.6.2 Burst Mode Transmission in the BT Pilot ISDN

In an earlier section mention was made of a halfway house approach to ISDN by British Telecom. BT's early ISDN implementation, known as the Pilot ISDN, but marketed as Integrated Digital Access (IDA) was based on System X local exchanges in Birmingham and Manchester, with two further exchanges located in London.

This early system was developed ahead of the firm CCITT recommendations for ISDN standards, but did provide similar voice and data functions to those discussed earlier, although the data capacity of the second B channel was limited to 8 Kbits/s rather than 64.

Specifically the Pilot ISDN provided:

❏ A 64 Kbit/s B channel for voice or data

❏ An 8 Kbit/s B' channel for data only

❏ An 8 Kbit/s D channel for signalling.

Several versions of NTE appeared for use on the Pilot ISDN. NTE 1 incorporated a digital telephone which was to use the B channel. The NTE 1 had a connector to which a data device could be connected to access either the B or B' channel using X21 bis or X21 leased line variant protocols.

NTE 3 was a small wall mounted cabinet which incorporated up to six interface cards for analogue telephones, fax machines, VDUs, teleprinters and Local Area Networks. As only two devices could be operational at one time, access was determined by a control unit in the NTE rather than the terminal equipment itself. A remote selection unit was provided with the NTE 3 to provide dial functions for those devices, e.g. VDUs, without a telephone keypad.

After a while NTE1 and NTE3 were withdrawn and replaced by an NTE 4. This provided two X21 interfaces to which the user could connect any equipment with an X21 capability. Digital telephones with X21 interfaces became available, as did ISDN cards for PCs and adapter modules to support V24/X21 bis terminals.

All the NTEs used on the Pilot ISDN incorporated a multiplexer to produce a composite digital signal from the B, B' and D channels. The output from this multiplexer was 80 Kbits/s (64K + 8K + 8K). Burst Mode transmission (or Time Compression Multiplex) allowed apparent 80 Kbit/s duplex communications on telephone lines up to about 2.5 to 3 Kms in length.

The Burst Mode technique used in the Pilot system involved multiplexing data from the two B channels and the D channel, and transmitting the resulting multiplex frames in individual short bursts.

Burst mode transceivers were placed in the NTE and in the exchange. In the transceiver, B, B' and D channel data was multiplexed as shown in Figure 13.8 and then transmitted to line in 22 bit bursts (20 data bits and two marker bits) at an instantaneous data rate of 256 Kbit/s using a WAL2 line code (after Walsh, See *Data Communications and Networks*, edited by Dr RL Brewster, published by The IEE).

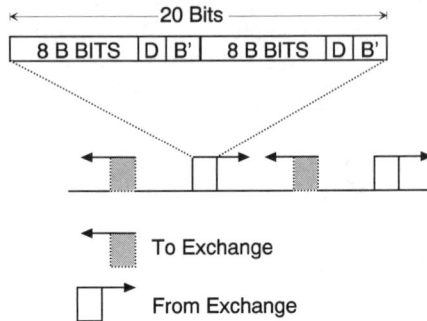

*Figure 13.26 Burst mode transmission in the BT Pilot ISDN*

Each burst can be considered as a multiplex frame with an original duration of 250 mSecs. At 256 Kbit/s, the 22-bit frame is transmitted in just under 86 uSecs. Thus there is slightly less than 250mSecs between transmitted bursts. During this time the transceiver operates in the opposite direction to receive a similar burst from the transceiver at the other end of the line.

Because the actual line rate is 256 Kbit/s, the line signal attenuation is high and this limits the length of line over which it can be used to about 2.5 to 3 Kms. However as many customers, especially in business areas, are located less than 3 Kms from their local exchange, this technique was satisfactory for a large majority of potential users of the services. BT were able to make arrangements to cater for those users who were outside the 3 Kms from a digital local exchange by the use of remote multiplexer units.

Because of the short range available with Burst Mode, it is unlikely to be used in future public ISDN systems. However it will be used in private ISDNs which offer full 144 Kbit/s 2B + D systems over existing 2-wire PABX extension telephone cabling. The following section describes a typical Burst Mode transceiver for such applications.

## 13.6.3 Siemens PEB 2095 Burst Mode Transceiver

This IC was developed for ISPBX applications in which the vast majority of extension lines are under 2 Kms in length. For such lines, Burst mode provides a relatively cheap and simple transmission system. The transceiver is used with other ICs to implement the 144 Kbit/s (2B + D) system, although extra bits are included for synchronisation purposes and the actual line rate is 384 Kbit/s.

The block diagram of the IC is shown in Figure 13.27. Separate ICs carry out the functions of interfacing with terminal equipment and present a 144 Kbit/s serial data I/O interface to the IC buffer. AMI line coding, scrambling, and AGC are used with adaptive line equalisation in the receive section to improve transmission performance. Other features of this device include the ability to power down when not in use to reduce power consumption. The device is also capable of clock and frame recovery to enable the subscribers' equipment to be synchronised to the PABX.

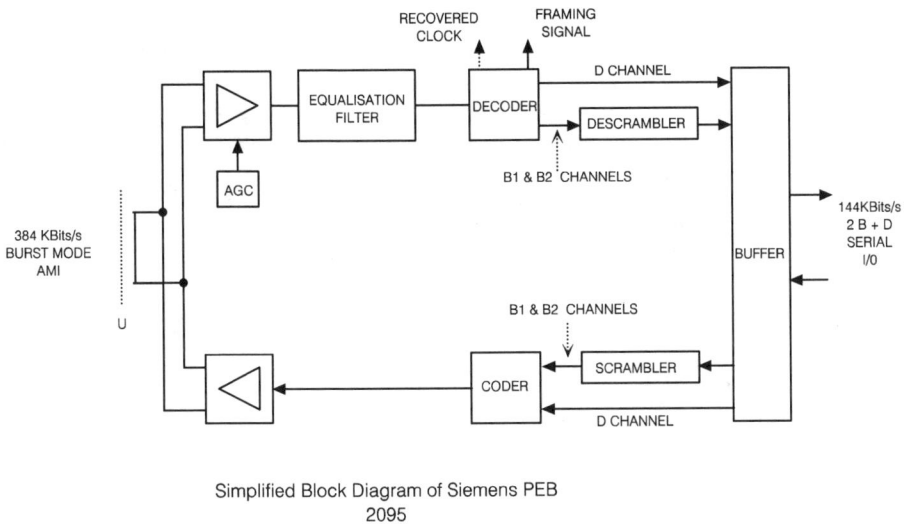

Simplified Block Diagram of Siemens PEB
2095

*Figure 13.27 Siemens PEB 2095 burst mode transceiver (Reproduced by permission of Siemens)*

## 13.6.4 Line Coding

The reach of any transmission system is dependent primarily upon the output power of the transmitter and the sensitivity of the receiver to detect signals in the presence of noise. These two parameters specify the maximum transmission line attenuation that is acceptable. Should this maximum attenuation figure be exceeded, insufficient power will be available at the receiver and the system as a whole will not operate correctly. In this book we are considering the transmission of digital signals over the local loop, and in this case it is fair to assume that signal attenuation is directly related to the frequency of the transmitted signal. That is, given a fixed length of cable signals containing high frequencies transmitted on that cable will suffer more attenuation than signals containing lower frequencies.

Echo cancelling systems are able to achieve greater reach than burst mode systems because the actual line transmission rate used in an echo cancelling system is at least as low as the information rate, and thus the signal contains lower frequency components than a burst mode system operating at the same information rate.

For example in the burst mode system described in the previous section the user data rate of 144 Kbit/s was transmitted (with additional information) at 384 Kbit/s. Thus the reach of this system is limited by cable attenuation at frequencies of 192 KHz.

It is not possible to increase the output power of the transmitter as this would produce unacceptable crosstalk effects into adjacent pairs in the cable. (See *A review of copper pair local loop transmission systems* by P F Adams and J W Cook; BT Tech J April 89.)

The fundamental frequency content of a digital signal may be reduced by the use of line transmission codes which reduce the actual baud rate, without reducing the information rate. For example the baud rate of a binary signal (with two voltage levels) equals its bit rate. But it is practical to have a system in which each transmitted element may have one of four possible voltage levels and thus reduce the baud rate to half the information rate. In such a system one signal element can be used to transmit the information contained in two binary pulses, as shown in Figure 13.28.

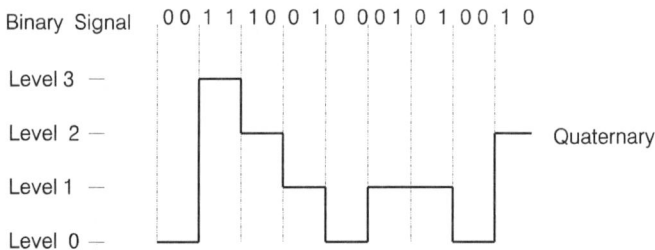

*Figure 13.28 Example of a 2-binary to 1 Quad (2B1Q) line code*

In this diagram, the binary pulse stream is divided into pairs of adjacent bits. Each possible pair combination is indicated by a voltage level which is transmitted to line. If the transmitted signal is analysed it will be found to contain a fundamental frequency at half the bit rate, and thus this digital signal will suffer less attenuation when transmitted over the local line than a pure binary signal carrying the same information.

2B1Q is not the only type of line code available and several are being investigated at research establishments. There are many factors which need to be determined for a suitable line code. Frequency content is just one of the parameters of interest. Another important aspect is error performance in the presence of typical line noise (i.e. BER for given signal to noise ratio). In many cases the investigations involve using computer models to predict the performance of particular line code under various different operating conditions. Modelling is preferable to the development of actual circuits to produce a trial code, as in this way, only those codes which the computer simulations show are likely to produce the best results can be targeted for practical tests.

Reducing the fundamental frequency of a digital signal is not the only reason for using a line code other than binary. The following are factors which may also be taken into account:

❏ The need to remove any DC component from the signal

❏ The need to be able to extract timing information from the signal

❏ The need to reduce crosstalk into adjacent pairs in a cable.

The table below gives examples of types of line code being considered for use in ISDN digital loops and the actual line rate associated with each for an original signal at 144 Kbit/s:

*Table 13.1: Line baud rates for different line codes*

| Code | Line Rate (KBaud) | Remarks |
|------|-------------------|---------|
| AMI | 144 | Alternate Mark Inversion, no change in fundamental frequency. |
| PR4 | 144 | |
| 3B2T | 96 | 3 binary to 2 tenary, fundamental frequency reduced by a third. |
| 2B1Q | 72 | 2 binary 1 quad, fundamental frequency reduced by half. |

British Telecom has opted for a 3B2T line code known as SU32. The transmission system results in a 120 Kbaud ternary signal carrying traffic at a rate of 160 Kbit/s. The 160 Kbits/s signal consists of 144 Kbit/s for the 2B + D traffic from the users' terminals, and 16 Kbit/s of data required for the maintenance and supervisory data required to permit the local exchange to operate the line and the NT, for loop testing and other such procedures.

# 13.7 ISDN Applications

In this final section of the chapter, we take a brief look at some of the ISDN applications which are beginning to emerge. Some ISDN applications, e.g. Video-conferencing, rely heavily on other sophisticated technology to enable them to use ISDN as a bearer medium.

## 13.7.1 Circuit Switched Data

The availability of ISDN2 has resulted in a number of suppliers offering 2 port ISDN terminal adaptors as replacements for modem to modem connections. The highest data rate available over the analogue PSTN using a sophisticated modem was 14.2Kblts/s, although in fact most systems operated at 4.8Kblts/s or 9.6Kblts/s. Because the maximum data rate was low, analogue modems connected by dial up connections through the PSTN were not suitable for applications where large amounts of data were to be transferred between sites. The only viable alternatives were digital leased lines or packet switched networks.

The ISDN terminal adaptors (ISDN TA) now available offer an alternative (ie a circuit switched data) solution. Whilst these TA are capable of operating with existing data terminals which may only be capable of data rates up to 9.6Kbits/s, they are also capable of operating at up to 64Kbits/s and thus will be usable when the data terminals are replaced by high speed devices.

It is now possible to transfer large amounts of data, e.g. for a computer graphics based applications such computer aided design (CAD) between remote work-stations. Not only is the data transfer over ten times quicker than previously possible, it is also far less prone to data transmission errors.

## 13.7.2 Leased Line Back-up

In data communications applications where there is a requirement for a permanent circuit to be available between two sites for a significant period each day, the normal solution is a digital leased line, e.g. a BT Kilostream. Prior to the introduction of ISDN, the only viable way of providing a back-up to the Kilostream circuit was to use dial-up and auto-answer modems, the main disadvantages of which were that modems oeprated at slower data speeds than the kilostream circuit and were more prone to bit errors.

Several manufacturers now offer an ISDN Leased Line Backup Unit, often referred to as a Kilostream Backup Unit or KBU. Figure 13.29 shows how a KBU is connected at each end of the leased line to be protected. Both KBUs monitor the leased line circuit, and should one of them detect a failure, it will, after a short delay to ensure the circuit has not been re-instated automatically, make an ISDN call to the KBU at the remote end of the link. The KBUs establish a path between themselves over ISDN, and then automatically transfer the data traffic to the ISDN route.

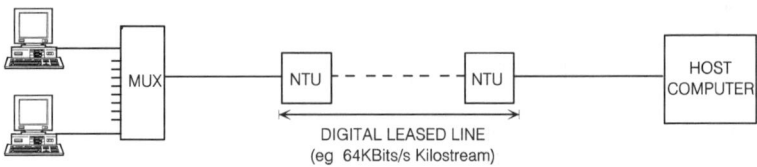

a. Digital leased line without back up

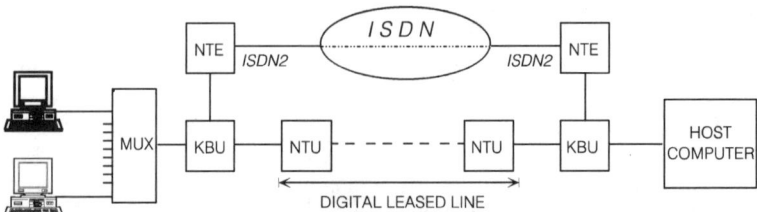

b. Digital leased line with ISDN 2 Back up

*Figure 13.29*

Whilst the traffic is carried over ISDN, call charges accrue, the KBUs therefore continue to monitor the leased line and when it has been restored, the KBUs then arrange to switch the data traffic back on to the leased line route, and then drop the ISDN call.

The advantages of this KBU solution are that the data rate achievable over the ISDN is as high as over the leased line, and the error performance of the ISDN call should be equally as good as that of the leased line.

## 13.7.3 Video Conferencing

Video Conferencing (VC) technology has improved dramatically in recent years. Several years ago, to achieve acceptable quality for a video conference a leased digital link operating at 2MBits/s was required. Today such quality can be achieved with digital circuits operating at rates as low as 128Kbits/s.

Several Video conferencing systems are available which typically use 128, 256 or 384Kbits/s. Without some other technology these systems can not, of course, use ISDN B channels, whose maximum transfer rate is 64Kbits/s, as a bearer medium. In the last year or so inverse multiplexing technology has been introduced for this type of application. The diagram of Figure 13.30 shows a video conferencing unit operating at 128KBits/s. It should be noted that this 128KBits/s signal is not simply a 64Kbits/s Video signal and a 64KBits/s audio signal multiplexed together. Techniques such as video and speech compression are used to produce a highly complex digital signal in which the bandwidth occupied by the video and audio signals are constantly changing.

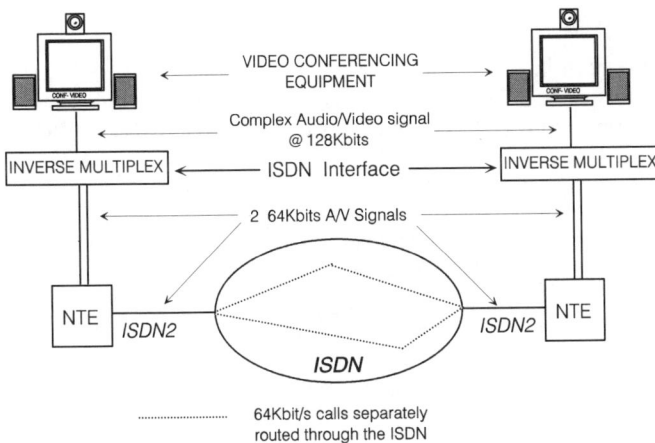

*Figure 13.30*

The ISDN interface unit permits this complex 128KBits/s VC signal to be transmitted as two 64Kbits/s data calls over the ISDN using both B channels of a Basic Rate ISDN access. The ISDN interface splits up the 128Kbits/s VC signal into two

64Kbits/s signals which it then transmits over two separate ISDN calls to the remote ISDN interface unit.

The two ISDN calls are set up independently, and will be treated differently by the ISDN, the two calls will suffer different delays, and may be even connected over different routes. The receiving ISDN interface unit must therefore, contain sufficient buffering and intelligence to allow it to accommodate these delays and reconstitute the original 128Kbits/s VC signal.

The added advantage of such an ISDN based VC system is that there is no longer a requirement for a costly permanent leased circuit between sites, video conferences can simply be dialled up when required and are charges on a pay as you go basis.

Video conferencing systems using this type of technology are becoming increasingly popular especially for trans-atlantic users.

### 13.7.4 Encrypted Speech

The evolution of ISDN, has brought with it the digital telephone. It is now a relatively simple matter to produce a secure speech link between two users, by introducing some form of encryption device between the digital telephone and the B channel over which it is to be connected. D channel signalling messages are not encrypted as they would then be unreadable by the local exchange.

### 13.7.5 High Resolution Colour Facsimile

Group 4 facsimile (G4 fax, see chapter 3), with its ability to produce high resolution images has been available over 64KBits/s leased lines for some time, however, only a small number of specialist users could justify the costs of G4 fax terminals and dedicated point to point lines.

Several G4 fax terminals are now available with a circuit switched option, thus enabling G4 users with ISDN access to transmit to any other G4 user who has ISDN access. As more users move to G4 the costs of these terminals should reduce.

### 13.7.6 Other applications

New ISDN applications being introduced all the time. Typical new applications include local area network (LAN) bridges and ISDN PC cards to permit PC to PC bulk file transfer. Users of LANs operating on geographically dispersed sites are now able to transfer data between each other. The transactions being transparent to the users who do not require to know the location or address of the user to whom they wish to communicate, all these matters are handled by the LAN bridge software which makes use of an ISDN call to facilitate the transfer of data between the two LANs.

# 13.8 Chapter Summary

This chapter has introduced the major concepts of public Integrated Services Digital Networks. The major points discussed were those which impact on users, e.g. the types of user terminal equipment which can be connected to an ISDN, and service distribution within the customers' premises.

Although much of the technical detail regarding network operation is relevant only to the service provider, a basic understanding of the technology involved was provided in this chapter and Chapter 9 on System X. It should be remembered that all System X and AXE 10 exchanges currently being installed into the British Telecom network are capable of providing ISDN service.

Much reference was made to the CCITT I series of recommendations, particularly those referring to the installation, testing and commissioning on ISDN service within a customer's premises. These recommendations describe the various interface points e.g. R, S and T, which are available for the connection of terminal equipment, and describe the functions to be carried out by various network entities such as NT1 and NT2.

The CCITT recommendation does not however specify details for the U interface as this point is generally accepted to belong to the service provider and is therefore subject to his own decisions regarding such matters as line coding, levels, impedances and so on. Within Europe the service provider will provide the Network Terminating Equipment, and ISDN users will connect their equipment to this point via the R, S or T interfaces. Contrary to this however is the decision within the US to specify this interface point, in order to allow competition in the market for the provision of the network terminating equipment.

CCITT recommendations regarding signalling within an ISDN are referred to in the I series but the details are actually contained within the Q series of the CCITT Blue Book.

In line with the European concept of connection to the ISDN at the R, S and T interface points, the requirement for ISDN test equipment has addressed the need to test the user interfaces, the passive distribution system and various ISDN terminal equipment. A section within this chapter used the range of ISDN test equipment produced by Wandel & Golterman to show what functional testing can be undertaken.

# 13.9 Acknowledgment

The diagrams showing ISDN testing arrangements are reproduced by permission of Wandel and Golterman GmbH, whose assistance is gratefully acknowledged.

# 14

# The Digital Transmission Hierarchy

## 14.1 The Digital Transmission Network

### 14.1.1 Review of Primary Rate PCM Systems

Throughout the course of this book we have, until now, considered that the digital links between 30-channel multiplexers, or between digital exchanges in public or private networks, have been established on 2.048 Mbit/s bearers, although in some cases 24-channel, 1.544 Mbit/s systems are used. Both systems are described as Primary Rate Systems and are specified in CCITT recommendations G 731 to G 736.

We saw that the 1.5 Mbit/s primary rate systems, known as T1 systems, in the US, cater for 24 64 Kbit/s PCM Voice or Data channels, with framing and channel associated signalling being transmitted on an extra (the 193rd) bit in the multiplex frame structure. For applications where common channel signalling is required, e.g. ISDN, a 64 Kbit/s second TS is required, thus reducing system capacity to 23 channels. Within Europe the standard is for 2 Mbit/s systems having one framing and alarm channel (TS0), 30 traffic channels (TS1 - 15, and TS17 - 31), and one common signalling channel (TS16).

In medium to large private networks, and in the public networks, the capacity of a single primary rate system is inadequate to meet the heavy traffic demand placed on the network. A simple solution is to connect exchanges with several primary rate links rather than to design multiplexers capable of handling greater numbers of channels. The drawback of this simplistic approach is that the physical transmission media is expensive and thus allocating each primary rate link to its own physical bearer would be prohibitively costly. The CCITT in recognising these problems has recommended a digital transmission hierarchy, not dissimilar in principle to the analogue FDM hierarchy of groups and super groups. The digital hierarchy uses the primary rate multiplex system as a basic building block for high capacity systems capable of handling upwards of 7,680 channels.

During the 60s and 70s the main transmission media was coaxial cable. The available bandwidth of this type of cable precluded high capacity digital systems, and in fact FDM systems were far more efficient in terms of channel capacity per megahertz of bandwidth.

Lately, during the 80s there was a large increase in the use of optical fibre as a transmission medium. Developments in this area have resulted in fibre systems capable of handling a very large bandwidth, typically over a Gigabit/s ($10^3$ Mbit/s) and thus the bandwidth inefficiencies of digital transmission are largely irrelevant.

The remainder of this chapter and the next describe the two closely related subjects of higher rate digital multiplexing techniques and the optical fibre systems often used to provide the physical transmission media for such systems. Optical fibre is by no means the only digital transmission medium. There are still many kilometres of coaxial cable systems, and in areas where it is difficult to lay cable of any type, line of sight microwave radio links are used to provide links with capacities of up to 120 channels.

## 14.1.2 Second Order Multiplexing

High capacity systems can be established in stages, each multiplexing together a number of digital signals from the previous stage to form a single composite digital signal. The concept is illustrated in Figure 14.1 which shows the principle of second order multiplexing. In this diagram it can be seen that a 120-channel link is provided between two sites, and although the diagram shows four primary rate multiplexers (PMUX) at each site, there is no reason why these should not be replaced by four 2Mbit trunk ports of a digital exchange.

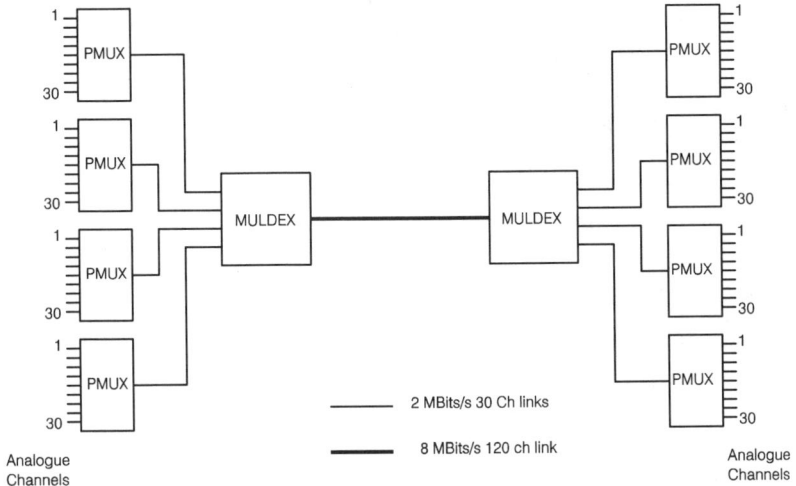

*Figure 14.1 The principle of second order multiplexing*

The 2 Mbit/s outputs of the four 30-channel primary rate multiplexers at each site are multiplexed together to provide an 8 Mbit/s 120-channel link which is in fact the second level of the hierarchy. The equipment used to multiplex digital signals in this form is often referred to as a Muldex (Multiplexer/Demultiplexer). As the equipment actually transmits digital signals in two directions it is necessary to distinguish between the low and high bit rate sides of the equipment. Figure 14.2 shows that the low bit rate inputs and outputs are collectively known as the tributaries, while the high bit rate side is referred to as Muldex inputs and Muldex outputs.

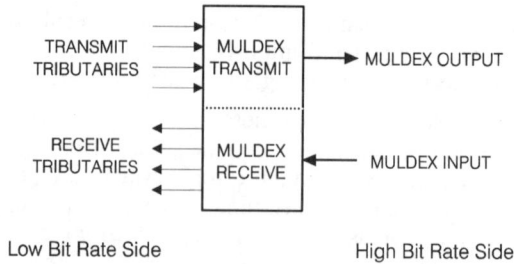

*Figure 14.2  Muldex equipment terminology*

The actual bit rate of this 120-channel link is in fact 8.448 Mbit/s, which is slightly faster than the figure arrived at by multiplying 2.048 Mbit/s by 4. The reason for this is that extra bits must transmitted to provide housekeeping such as framing and other functions which will be described in more detail later.

## 14.1.3 Third and Higher Order Multiplexing

Figure 14.3 shows the third, fourth and fifth levels of the hierarchy. The third level of the digital hierarchy provides capacity for 480 channels. As you might expect from the previous section this is achieved by multiplexing together the outputs of four second order multiplexers. At this level of the hierarchy the digital transmission rate is 34 Mbit/s, not 32 Mbit/s. This is because we are actually multiplying 8.448 Mbit/s by 4 and again adding a few extra bits for housekeeping purposes.

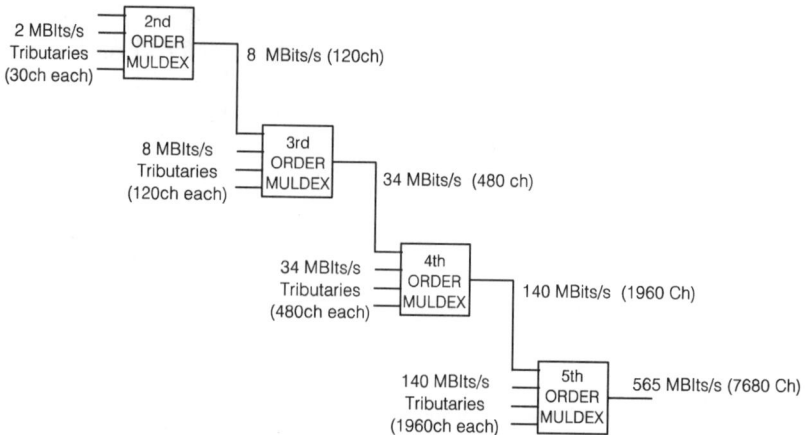

*Figure 14.3 Third and higher Levels of the digital hierarchy*

The fourth level provides capacity for 1,920 channels, and is achieved by multiplexing four third-order systems to produce a composite digital signal which is transmitted at 140 Mbit/s. Although not specified by the CCITT in the 1984 Red Book, fifth order systems are becoming available, especially from the suppliers of optical fibre transmission kit. These systems take four 140 Mbit/s signals and multiplex them to produce a digital signal running at 565 Mbits/s and having the capacity for 7,680 channels. British Telecom has carried out a trial of a 2.4 Gbit/s (30,720 channels) system to exploit the huge bandwidth capacity of its optical fibre network.

In countries such as the USA, which have adopted the 1.544 Mbit/s system as the primary rate, a different digital hierarchy applies. Table 14.1 shows the equivalent output rates and channel capacities for high order systems based on the 1.544 Mbit/s primary multiplex, and summarises the figures already given for systems based on the 2.048 Mbit/s structure.

*Table 14.1  Capacities and bit rates of digital multiplex equipment*

| Hierarchy Level | USA T1 | | CEPT 30 | |
|---|---|---|---|---|
| | Mbit Rate | Channel Capacity | Mbit Rate | Channel Capacity |
| Primary | 1.544 | 24 | 2.048 | 30 |
| 2nd order | 6.312 | 96 | 8.448 | 120 |
| 3rd order | 44.736 | 672 | 34.368 | 480 |
| 4th order | 274.176 | 4032 | 139.264 | 1920 |
| 5th Order | | | 565.148 | 7680 |

*Note: In Japan, where a 24-channel system is also used the third, fourth and fifth levels of the hierarchy are 480, 1440 and 5760 channel systems.*

# 14.2 Digital Multiplexing Techniques

## 14.2.1 CCITT Recommendations Regarding Digital Multiplexing

In order to describe the techniques involved in current digital multiplexers, we will consider the 2nd order Muldex equipment used to multiplex four 2 Mbit/s systems into an 8 Mbit/s output. The same principles will apply to Muldex equipment used at higher levels of the hierarchy.

The principle characteristics of various second order Muldex equipment are specified by the CCITT in recommendations G741 to G 746. The description that follows is based upon recommendations G741 and G742. The recommendations regarding the characteristics of higher order systems are contained in G751 to G754.

## 14.2.2 Multiplexing Methods

Three main methods of multiplexing signals from PMUXs have been identified and are shown in Table 14.2.

*Table 14.2  Multiplexing methods*

| | |
|---|---|
| Synchronous operation | Time Slot interleaving |
| Asynchronous operation | Octet interleaving |
| Plesiochronous operation | Cyclic bit interleaving |

We will discuss the multiplexing schemes for synchronous and plesiochronous systems. I am aware of several plesiochronous systems currently in use, but in the future synchronous operation will probably become more prevalent as this system is suited to operation with digital exchanges to provide high capacity inter-exchange trunk links.

## 14.2.3 Synchronous Digital Multiplexing

In a synchronous system the four PMUXs which are the tributaries to the Muldex are connected to the same external clock source and run completely in step. The fact that everything is in synchronism means it is a relatively simple matter to interleave the signals received in each timeslot from the four multiplexers, especially if the external clock is actually taken from the Muldex equipment.

For example, consider that multiplexing commences from TS0 of the primary multiplexers, and that as the multiplexers are in synchronism, all four are transmitting TS0 at the same time. The output from the Muldex will contain TS0/Trib 1, followed by TS0/Trib 2, followed by TS0/Trib 3 and TS0/Trib 4. The cycle then repeats with TS1s from the four tributaries being transmitted.

Figure 14.4 shows the concept in which the four primary multiplexers and the 2 - 8 Mbit Muldex are all driven from the same reference clock.

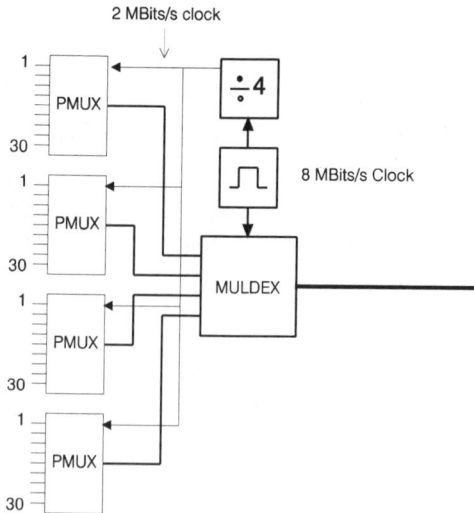

*Figure 14.4 Synchronous digital multiplexing*

The CCITT recommended frame structure for an 8.448 Mbit/s system based on synchronous PMUXs is given in Annex A to G741. This frame structure contains 132 8-bit timeslots, 128 of which are used to transmit every TS in each frame from each of the four PMUXs. The four additional timeslots are required in the Muldex frame structure to take care of extra housekeeping. The assignment of bits within the frame structure is shown in Figure 14.5. As it is not possible to show all 132 TS in the space available, Table 14.3 has been included to show the assignment of those timeslots which are left out of the diagram.

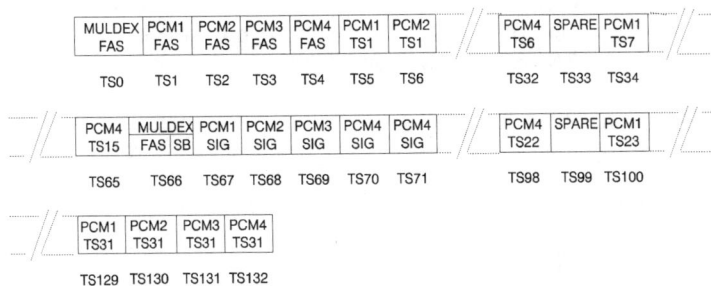

| MULDEX FAS | PCM1 FAS | PCM2 FAS | PCM3 FAS | PCM4 FAS | PCM1 TS1 | PCM2 TS1 | // | PCM4 TS6 | SPARE | PCM1 TS7 | // |
| TS0 | TS1 | TS2 | TS3 | TS4 | TS5 | TS6 | | TS32 | TS33 | TS34 | |

| PCM4 TS15 | MULDEX FAS SB | PCM1 SIG | PCM2 SIG | PCM3 SIG | PCM4 SIG | PCM4 SIG | // | PCM4 TS22 | SPARE | PCM1 TS23 | // |
| TS65 | TS66 | TS67 | TS68 | TS69 | TS70 | TS71 | | TS98 | TS99 | TS100 | |

| PCM1 TS31 | PCM2 TS31 | PCM3 TS31 | PCM4 TS31 |
| TS129 | TS130 | TS131 | TS132 |

*Figure 14.5 8.448 Mbit/s frame structure for synchronous multiplexing*

| TS No. | USE |
|--------|-----|
| 0 | MULDEX FA (1st 8 bits = 11100110) |
| 1 - 4 | PMUX FAS from tribs 1 - 4 |
| 5 - 32 | TS1 - TS7 from tribs 1 - 4 |
| 33 | SPARE TS |
| 34 - 65 | TS8 - TS15 from tribs 1 - 4 |
| 66 (bits 1 - 6) | MULDEX FAS (2nd 6 bits = 100000) |
| 66 (bits 7 & 8) | SERVICE BITS |
| 76 - 70 | SIGNALLING TS from tribs 1 - 4 |
| 71 - 98 | TS17 - TS22 from tribs 1 - 4 |
| 99 | SPARE TS |
| 100 - 131 | TS23 - TS31 from tribs 1 - 4 |

*Table 14.3 Assignment of bits in 8.448 Mbit/s frame structure for synchronous multiplexing*

For the same reasons that the primary rate multiplex frame structure requires a Frame Alignment Signal(FAS), it is necessary to have a Muldex FAS. The FAS consists of 14 bits and can not therefore be transmitted in a single TS. Instead it is distributed through the frame structure with the first eight bits (11100110) being transmitted in TS0, the remaining six bits (100000) are transmitted half way through the frame at TS66. The two spare bits of TS66 are the service bits which are used to transmit alarms etc. Two spare TS are left, TS33 and TS99.

## 14.2.4 Plesiochronous Systems

In some networks it is not possible to synchronise all the multiplexing and transmission equipment. Consider the case shown in Figure 14.6 which depicts a 8 Mbit/s optical fibre link between two sites B and C In this particular case, not all the tributary PMUX are co-located with the Muldex equipments. For example PMUX A is located in a separate building from its associated Muldex and is connected to it by a 2 Mbit system over coaxial cable. At location C, the corresponding PMUX D is some distance from the site and is connected to the Muldex by a microwave radio link. It is possible to install a synchronised network such as this, however in this case the network is not synchronous.

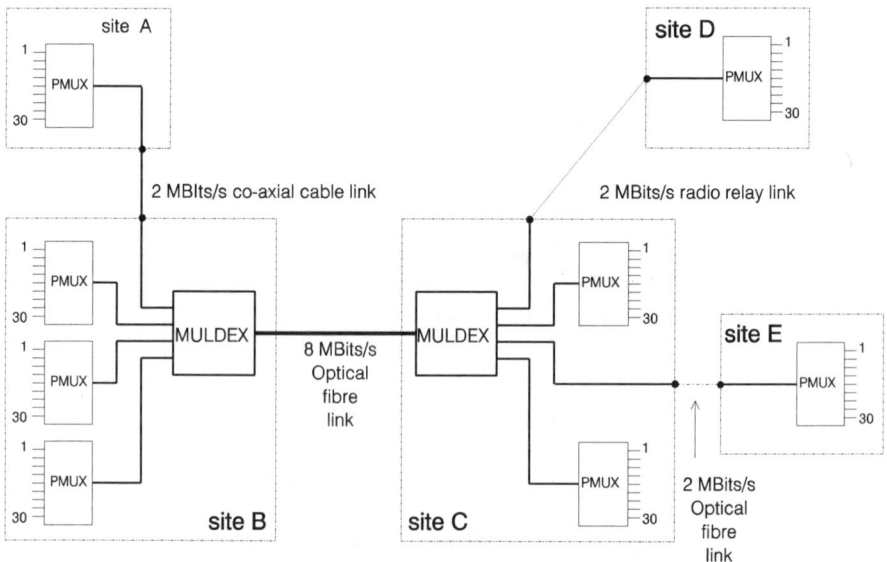

*Figure 14.6  An example plesiochronous network*

The diagram actually depicts eight separate PMUX, each of which has its own internal clock designed to produce a nominal output bit rate of 2.408 Mbit/s. But since the PMUXs are not synchronised to a common clock source, there can be no guarantee that the outputs are all running at exactly 2.048 Mbit/s. In fact the specification for such devices allows for a frequency error of 50 parts per million, the equivalent of 102 bit/s. This is an example of a plesiochronous network.

The definition given in the CCITT vocabulary of G 701 states that two signals are plesiochronous if they are nominally at the same rate, with any variation being constrained within specified limits. This is the case in the example in Figure 14.6.

Each 30-channel system consists of a tributary PMUX, and Muldex equipment at each end of the link. Both PMUX are nominally running at 2.048 Mbit/s, but with a tolerance of 102 bits/s. If the two PMUX were to be connected directly together over a simple 2 Mbit/s transmission path, with no additional multiplexing, the receive side of each PMUX would recover timing from the received digital signal. This recovered timing would be used to control the receive multiplex such that it was always in step with the transmit PMUX at the distant end.

So long as the transmitter output bit rate remains within tolerance, any drift does not affect correct demultiplexing of the signal as the recovered timing in the receiver will contain the same degree of error.

## 14.2.5 Multiplexing of Plesiochronous PMUX Signals

Multiplexing plesiochronous signals presents a problem which does not occur when the signals are synchronous, in that the multiplexing process has to be carried out on signals which will not be exactly at the same bit rate, and even if they were, by chance at the same rate, they may not be in phase.

Figure 14.7 shows a possible case in which the outputs of the four PMUX to be multiplexed are operating nominally at 2.048 Mbit/s but with small errors as shown.

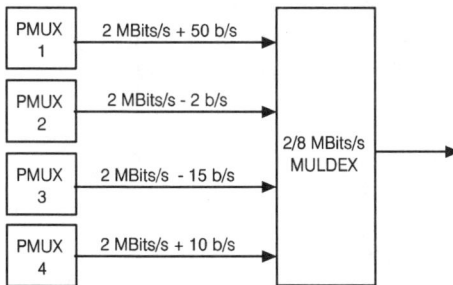

*Figure 14.7 Plesiochronous tributaries with small bit rate errors*

The output bit rate of each tributary is within tolerance, but consider what must happen at each receiving PMUX if the signals are to be correctly demultiplexed. The receiving PMUX must receive a signal from the Muldex which is at exactly the same bit rate as that sent by the corresponding transmit PMUX.

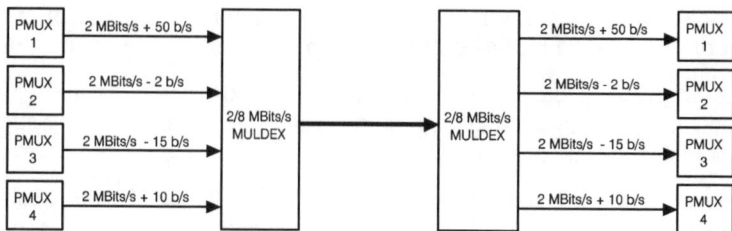

*Figure 14.8  Plesiochonous multiplexing preserving bit rate from PMUX*

Figure 14.8 shows that the transmit Muldex takes account of the incoming error on each tributary when multiplexing the signals together, and then the error is re-inserted in the receive side of the Muldex so the two PMUX can successfully operate together, just as if they were connected by a simple coaxial cable section.

The operation of the Muldex equipment must be completely transparent to each of the four, essentially separate, 2 Mbit/s PMUX links. The CCITT recommendations for plesiochronous multiplexing are contained in G.742. The 8.448 Mbit/s Muldex frame is completely different from that used for synchronous systems, and the two systems are not compatible.

## 14.2.6 Bit Interleaving

Multiplexing is carried out by interleaving single bits from the tributaries, rather than time slots as in the synchronous case. However it is not possible to bit interleave signals which are not at the same rate, as there is no simple way to time the interleaver circuit.

To understand the problem consider a bit interleaving system for synchronous signals as shown in Figure 14.9a. The bit interleaver is basically a 4-input data selector, whose select lines are connected to the output of a 2-bit binary counter which is clocked at exactly four times the incoming tributary rate.

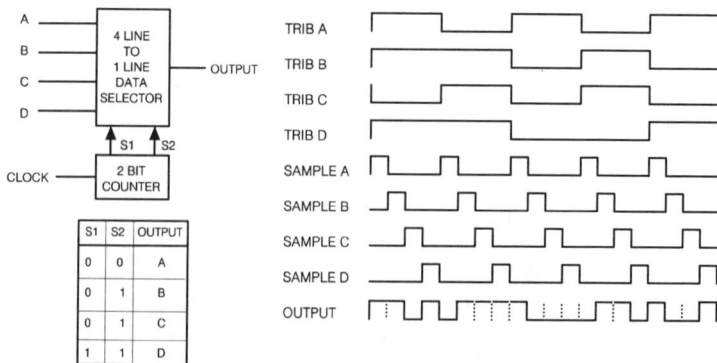

*Figure 14.9 Bit interleaving of synchronous signals*

The data selector cyclically samples the tributary signals on each input line and transmits a corresponding signal for each sample on its output line. Since the signals and sampling are synchronised, there is a constant time relationship between each tributary bit and the sample pulses as shown in Figure 14.9b. Notice that the width of the sampling pulse is T secs, while the pulse width of each bit to be sampled is 4T secs.

When plesiochronous signals are to be interleaved there is no longer a constant relationship between the signals from the tributaries and the sampling pulse.

If the bit interleaver described above is used, and given that the sampling rate is fixed and constant, and has a pulse width of T secs, if the tributary input is fast (i.e. >2.048 Mbit/s), its pulse width will be slightly less than 4T secs (4T - dt secs). On the other hand if the tributary is running slow (i.e. <2.048 Mbit/s) its pulse width will be slightly longer than 4T secs (4T + dt secs). In either case, at some time a form of bit slip will occur, and the digital pulse train will be incorrectly sampled as described in Section 5.5.1.

The PMUX frame structure contains 256 bits. However when the tributary is running fast some bits will not be sampled at all, which will effectively delete a bit from the PMUX frame structure resulting in only 255 bits being transmitted. When the tributary is running slow, some bits will be sampled twice, effectively inserting an extra bit into the PMUX frame structure, resulting in 257 bits instead of 256.

Due to the fact that the PMUX clocks are constrained within certain limits bit slip will not occur every frame. The rate at which slippage occurs is proportional to the degree of frequency error involved. For example a bit rate error of + 1 bit/s will cause far less frequent bit slips than an error of 10 bit/s.

Whenever a bit slippage occurs, the signal received at the distant end PMUX, can not be properly demultiplexed as loss of frame alignment will occur and consequent loss of traffic until frame alignment has been restored.

This situation can not be allowed to occur in a practical network. A process known as justification is employed within the Muldex transmit equipment prior to the actual bit interleaving process to ensure that any error in the output bit rate of a PMUX is taken account of without causing bit slip. A complimentary process, de-justification, is employed within the receive Muldex to restore the original PMUX signal complete with any frequency error that was present in the original signal.

## 14.2.7 Justification

Justification can be defined as the process of changing the bit rate of a digital signal in a controlled manner so that it can be synchronised to a bit rate which is different from its own. Justification usually results in no loss of information from the signal (from CCITT G.701 vocabulary 4022). There are three methods of justification defined in the CCITT vocabulary:

❏ Positive

❏ Negative

❏ Positive/zero/negative.

When positive justification is used the bit rate of the original signal is always increased. The opposite case exists for negative justification, while in the third case the new signal may be at a higher rate, the same rate or a lower rate than the original. This text will concentrate on positive justification.

## 14.2.8 Positive Justification

In order to describe the process of positive justification we will start by describing a hypothetical system in order to get the concept across. After that a real system will be described.

It should be noted that justification is usually used within a multiplex equipment, and it not used as a standalone method of increasing the speed of a signal. However for the purposes of this simple introduction we will ignore the multiplexing functions and concentrate on justification.

Imagine that we have a data signal with a nominal bit rate of 100 bit/s. The tolerance on this signal is such that at any time the actual bit rate may be as low as 99 bit/s or as high as 101 bit/s. It is required to retime this signal so that it is synchronised at 110 bit/s. Positive justification is to be used.

The resulting 110 bit/s signal to be transmitted must contain all the information in the original 100 bit/s data signal. It will also contain its own housekeeping bits. A possible frame structure for the 110 bit/s signal is shown in Figure 14.10.

| FAS | 99 BIT PAYLOAD AREA | SPARE | SPARE | HK |
|-----|---------------------|-------|-------|-----|
| bits 1 - 8 | bits 9 - 107 | bit 108 | bit 109 | bit 110 |

*Figure 14.10  A frame from a hypothetical positive justification system*

The frame (Fx) contains 110 bits and is repeated once a second. An 8-bit FAS has been arbitrarily chosen to enable the equipment at the distant end of the link to correctly retrieve the 100 bit/s data signal which is transported in the 99-bit payload area. The spare bits (SB) and the housekeeping bit (HK) are used to perform positive justification.

Let's assume the data signal is running at exactly 100 bit/s. Starting from time t = 0, the first frame F1 will take 99 bits from the traffic signal into its payload area and transmit them along with the FAS, SB and HK bits at 110bit/s. Thus at this time the 100th bit of the data signal has not been transmitted.

At time t = 1sec, frame F2 will take the 100th bit and the next 98 bits of the data signal into the payload area. The next two bits of the data signal (i.e. the 199th and 200th bits) are moved into the two SB positions. The whole frame now contains 101 bits of the original data signal, and is transmitted at 110 bit/s

Frame F3 will transport 99 data bits just as F1, followed by frame F4 which will contain 101 data bits. Thus over a period of a few seconds the average data rate is exactly 100 bit/s, although the actual bit instantaneous bit rate is 110 bit/s. This is positive justification, because we have increased the actual bit rate from 100 bits/s to 110 bits/s by adding a pair of spare bit timeslots which are used as a form of buffer to match the speed of the original signal to the new bit rate. A point to note in this simplified explanation is that the spare bits can only be used in pairs, i.e. they are both used to carry data traffic, or neither of them is used. In this hypothetical example, a case is not allowed to arise in which one of the spare bits is used and the other not.

Consider what would happen if the speed of the original data signal increased from 100 bits/s to 101 bits/s. The system would cater for this by ensuring that data bits from the original signals would be carried in the two SB bits of every frame transmitted, effectively giving a payload area of 101 data bits per frame. If the data signal speed increased above a rate of 101 bits/s, data bits would be lost as the extra bits can not be transported by this mechanism.

In the case where the data signal speed decreases from 100 bits/s to 99 bits/s, our simple system copes by using only the basic 99-bit payload area, and never sending data in the two SB positions.

Summarising the above, with simple arithmetic:
    Minimum Data Rate = 99 bits/s: Spare bits used in 0% of frames
    Nominal Data Rate  = 100 bits/s: Spare bits used in 50% of frames
    Maximum Data Rate= 101 bits/s: Spare bits used in 100% of frames

From the above information it is possible to appreciate that the system can cope with any bit rate between 99 bits/s and 100 bits/s by using between 0% and 50% of frames for carry traffic in the SBs as appropriate. For example a bit rate of 99.5 bits/s will be accommodated by using the spare bits in only 25% of frames.

For bit rates of between 100 bits/s and 101 bits/s, then between 50% and 100% of frames will be used for carrying data bits in the spare bit positions. e.g a bit rate of 100.5 bits/s will be accommodated by using the spare bit in 75% of all frames.

In the terminology of digital transmission the two spare bits are actually known as Justifiable Digit Timeslots (JDT), which may be defined as a digit timeslot used for justification purposes and may contain an information bit (i.e. a data bit) or a justifying bit which carries no information. From now on we will refer to these bits as the JDT.

Having seen the concept of justification in the transmit equipment, the next step is to appreciate the concept of the dejustification process in the receive equipment. The receive equipment has to synchronise to the incoming 110bit/s signal using the FAS. Having locked on to the frame structure, the receiver has to unload from the payload area the data signal and retransmit this signal at the correct bit rate of 100 bit/s +/- some error.

There are two problems here:

❑ How does the receive equipment know if the JDT contains data bits or not?

❑ How does the receive equipment know the actual bit rate of the original data signal?

The answer to both questions is that the transmitter sends this information in the housekeeping bit of the frame. Using the vocabulary of the CCITT this bit is known as the Justification Service Digit (JSD) which is set by the transmit equipment to indicate whether the JDT bits contain valid data bits, or are simply filling in the frame and contain no valid data.

The JSD is set to 1 if the JDT contains valid data. When the frame is received the condition of the JSD is checked. If this is set to 1, the receiver knows that the data to be unloaded is contained in the payload area and the JDT. If the JSD has been set to 0, the receiver simply unloads the 99 bits from the payload area and completely ignores the two redundant bits in the JDT.

The JSD signal will also reflect the bit rate of the original signal as follows. When the original data signal is running at exactly 100 bits/s the JSD will be set to 1 in 50% of all frames transmitted. If the data signal is at a higher rate the JSD will be set to 1 in a greater proportion of frames. It is thus possible to use the JSD to produce a timing control signal which will ensure that the data signal is reproduced at the correct data rate.

# 14.3 Practical 2-8 Mbit/s Muldex

## 14.3.1 Introduction

This section will briefly describe a practical Muldex equipment which employs positive justification techniques. The first two subsections describe very much simplified diagrams of the transmit and receive equipment to illustrate how the Muldex uses justification prior to multiplexing. The next section describes the Muldex 8 Mbit/s frame structure so that slightly more detail can be provided on the transmit and receive Muldex.

## 14.3.2 Muldex Transmit Equipment

The diagram in Figure 14.11 is a very much simplified schematic showing the transmit side of the Muldex equipment. Each tributary is connected to a justification circuit, the functions of which are:

❑ To retime the incoming 2.048 Mbit/s (nominal) signal to 2.112 Mbit/s

❑ Insertion of Justifiable Digit Timeslot

❑ To decide whether to use the JDT for data or fill

❑ To set the JSD to indicate whether JDT contains valid data.

Although the four inputs are plesiochronous, the outputs from the four justification circuits are synchronous, because all four justification circuits are timed by the same 2.112 Mbit/s clock source. The outputs can be successfully multiplexed by bit interleaving using a simple circuit as that shown in Figure 14.9.

The output of the bit interleaver is fed to a data selector along with the output from a Frame Alignment Signal Generator and Service Bits. The FAS is transmitted at the beginning of every Muldex frame, and is followed by two service bits which are used for the transmission of alarms and so on.

Figure 14.11 showing Muldex transmit equipment block diagram with:
- Tributary Inputs 2.048 MBits/s Plesiochronous feeding four JUSTIFICATION CIRCUIT blocks
- BIT INTERLEAVER producing Synchronised signals at 2.112 MBits/s
- DATA SELECTOR receiving SERVICE BITS
- MULDEX OUTPUT 8.448 MBits/s
- FAS GENERATOR
- TIMING CIRCUIT
- 8.448 MHz CLOCK
- ÷4 CIRCUIT producing 2.112MHz Clock

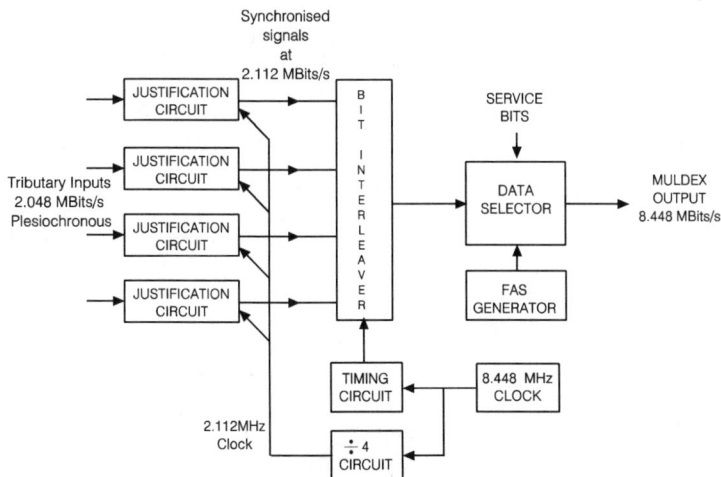

*Figure 14.11 Simplified diagram of Muldex transmit equipment*

## 14.3.3 Muldex Receive Equipment

A simplified diagram of the Muldex receive equipment is shown in Figure 14.12. A FAS detector ensures that frame alignment is established and maintained. The service bits are decoded to determine the alarm status of the distant transmit Muldex.

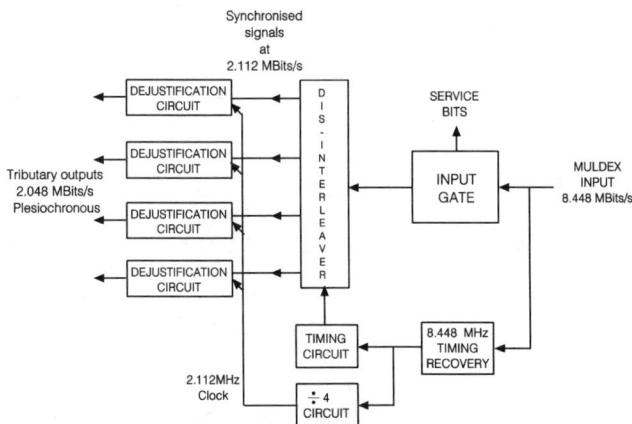

Figure 14.12 showing Muldex receive equipment block diagram with:
- MULDEX INPUT 8.448 MBits/s feeding INPUT GATE with SERVICE BITS
- DIS-INTERLEAVER producing Synchronised signals at 2.112 MBits/s
- Four DEJUSTIFICATION CIRCUIT blocks
- Tributary outputs 2.048 MBits/s Plesiochronous
- TIMING CIRCUIT
- 8.448 MHz TIMING RECOVERY
- ÷4 CIRCUIT producing 2.112MHz Clock

*Figure 14.12 Simplified diagram of the Muldex receive equipment*

The remaining bits are then fed to a demultiplexer which separates the incoming signal into four tributary signals, which are then fed to dejustification circuits. Each dejustification controller retimes the tributary signal to a nominal 2,048 Mbit/s and performs dejustification by examining the JSD to determine whether the JDT has been used or not.

## 14.3.4 8.448 Mbit/s Muldex Frame Structure

The CCITT recommendations for this type of Muldex frame structure are given in G.742. The 8.448 Mbit/s Muldex frame structure is divided into four sub-frames, or sets. Each set consists of 212 bits, thus the frame itself contains 848 bits.

Table 14.4 shows the assignment of each bit in the frame.

| Set No | Bit Position | Assignment |
|---|---|---|
| 1 | 1 - 10 | FRAME ALIGNMENT SIGNAL |
|  | 11 & 12 | SERVICE BITS |
|  | 13 | Bit from TRIB 1 |
|  | 14 | Bit from TRIB 2 |
|  | 15 | Bit from TRIB 3 |
|  | 16 | Bit from TRIB 4 |
|  | 17 - 212 | Interleaved bits from tribs as above (50 bits per Trib) |
| 2 | 1 | Bit 1 of JCW for TRIB 1 |
|  | 2 | Bit 1 of JCW for TRIB 2 |
|  | 3 | Bit 1 of JCW for TRIB 3 |
|  | 4 | Bit 1 of JCW for TRIB 4 |
|  | 5 - 212 | Interleaved bits from tribs (52 bits per Trib) |
| 3 | 1 | Bit 2 of JCW for Trib 1 |
|  | 2 | Bit 2 of JCW for Trib 2 |
|  | 3 | Bit 2 of JCW for Trib 3 |
|  | 4 | Bit 2 of JCW for Trib 4 |
|  | 5 - 212 | Interleaved bits from tribs (52 bits per Trib) |
| 4 | 1 | Bit 3 of JCW for TRIB 1 |
|  | 2 | Bit 3 of JCW for TRIB 2 |
|  | 3 | Bit 3 of JCW for TRIB 3 |
|  | 4 | Bit 3 of JCW for TRIB 4 |
|  | 5 | JDT for TRIB 1 |
|  | 6 | JDT for TRIB 2 |
|  | 7 | JDT for TRIB 3 |
|  | 8 | JDT for TRIB 4 |
|  | 5 - 212 | Interleaved bits fromtribs (51 bits per tri |

*Table 14.4 Assignment of bits in 8.448 Mbit/s frame structure*

The Frame Alignment Signal is the 10-bit sequence 1 1 1 1 0 1 0 1 0 0 0 0, and is always transmitted in the first 10 bits of Set 1. Bit 11 is used to indicate the presence of alarms to the distant Muldex equipment, while bit 14 has been reserved for national use, and it may used for transmission of other alarms or network management functions as decided by individual national PTTs.

The Justification Control Word has the same function as the Justification Service Digit. For each tributary a 3-bit JCW indicates the status of the JDT in set 4. The 3-bit JCW consists of one bit in each of sets 2, 3 and 4. For example, for tributary 4, the JCW consists of bit 4 in sets 2, 3 and 4. These three bits are set to indicate whether or not the JDT for tributary 4, assigned to bit 8 of set 4 is be used to transmit a data bit or not. If the JDT is used for valid data the three bits of the JCW are all set to zeroes, otherwise they will be set to ones.

A 3-bit JCW is used in preference to a single JSD to provide more reliable reception of the justification status indication. The corruption of a single JSD would result in incorrect treatment of the received JDT, and subsequent loss of frame alignment in the receive tributary PMUX equipment. A 3-bit JCW in which all three bits are transmitted with the same value is used, as the risk of corruption to two or more JCW bits is substantially less than that to a single JSD. Majority vote logic is used on all three bits of the JCW to determine the value of the JCW transmitted.

## 14.3.5 Muldex Transmit Block Diagram

Figure 14.13 shows a simplified block diagram of the transmit cards of the Muldex equipment. The diagram shows one of four identical tributary cards and a transmit common equipment card; it does not show the power supply and alarm cards.

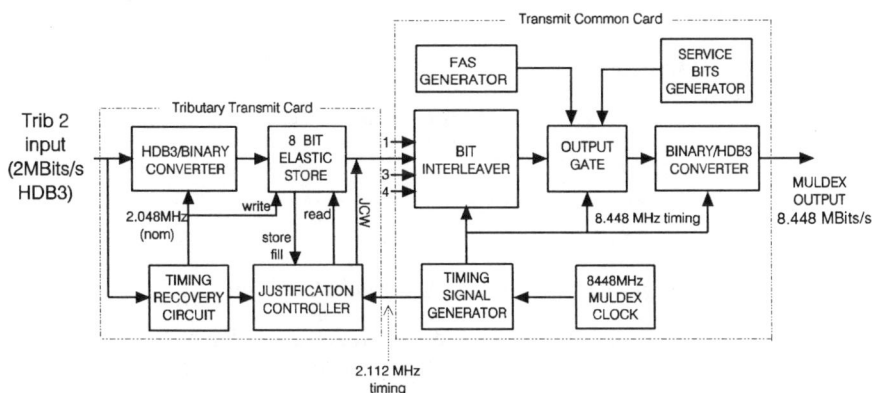

*Figure 14.13 Block diagram of Muldex transmit tributary and common cards*

The output bit rate and all transmit functions within the Muldex are controlled by a timing generator which produces the necessary control signals from a single 8.448 MHz master oscillator. The maximum permitted deviation from the nominal frequency of this master oscillator is 30 parts per million, equivalent to +/-225Hz.

The 2.048 Mbit/s (nominal) input to each card is in HDB3 form in line with CCITT

G.703. A timing recovery circuit produces a tributary clock signal with the same frequency as the original PMUX, while an HDB3 to Binary converter changes the format of the incoming bipolar signal to unipolar binary.

The incoming digital signal, after conversion to binary form, is written into a shift register which forms a special type of serial in, serial out device known as an elastic store. The recovered clock signal is used as a write clock to ensure that the signal is correctly written into the store without the loss of any bits.

All tributary cards are fed from the Muldex clock master oscillator and timing circuit which produces a pulse sequence (read clock) to clock data out of the shift register. Thus the individual tributary stores are filled at their respective incoming signal rates (nominally 2.048 Mbit/s), but emptied at a common rate of 2.112 Mbit/s. The actual read out rate will in fact depend upon the frequency of the master oscillator, since the read clock is derived by dividing the master oscillator frequency by 4.

It is a requirement that no bits are fed out of the shift register during the transmission of the FAS, service bits and JCW, This is achieved by inhibiting the read clock during these periods as shown in Figure 14.14.

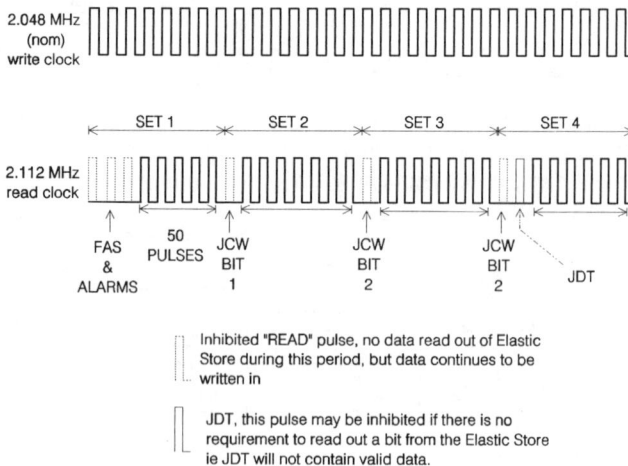

*Figure 14.14 2.112 Mbit/s read clock signal*

The elastic store is typically an eight stage shift register in which the output can be taken from any stage dependent upon the current state of the fill of the register. For example, consider that prior to the transmission of the FAS the output was being taken from stage 3. While the FAS and alarm bits are transmitted by the Muldex no bits are being read out of the shift register. But three new bits will have arrived from the tributary PMUX and must be written into the shift register. All the preceding bits still in the store will move along three stages and thus the next bit to be read out must come from stage 6.

During the remainder of the frame bits are being read out faster than they are being written in, and eventually the output will need to be taken from stage 5, then stage 4 etc. The ability of the shift register to store a variable number of data bits acting as a buffer between input and output gives rise to the term *elastic store*.

A detector circuit monitors the fill status of the store and transmits a suitable signal to the justification controller. If the store fill is below a given threshold it is not necessary to read out a stored bit during the JDT, and so the justification controller will also inhibit the read clock during the JDT bit period. The justification controller will also insert the necessary bits of the JCW and a fill bit in the JDT of the 2.112 Mbit/s sequence.

The store fill is determined by comparing the phase of the read and write signals. When the store fill has passed the required threshold, the controller determines that it is necessary to use the JDT to send a valid data bit and enables the read clock during the JDT period. The controller will also insert the required JCW, but not the JDT as the data to be transmitted in this timeslot will come from the shift register. Note that since all three bits of the JCW are transmitted before the JDT, the decision regarding use of the JDT must be taken at the beginning of the frame.

The output of all tributary cards are synchronous signals at a rate of 2.112 Mbit/s. These signals are fed to the input of the bit interleaver to produce a composite multiplexed signal which is then passed to an output gate. A timing signal causes the output gate to select the appropriate input from the multiplexed signal, the FAS generator or the service bit generator. The total output signal is then passed still in binary format, to a binary to HDB3 converter prior to transmission.

## 14.3.6 Muldex Receive Block Diagram

A simplified block diagram of the Muldex receive equipment is shown in Figure 14.15. This shows the receive common card and one of four receive tributary cards. In typical equipment the transmit and receive functions for each tributary would be on single card.

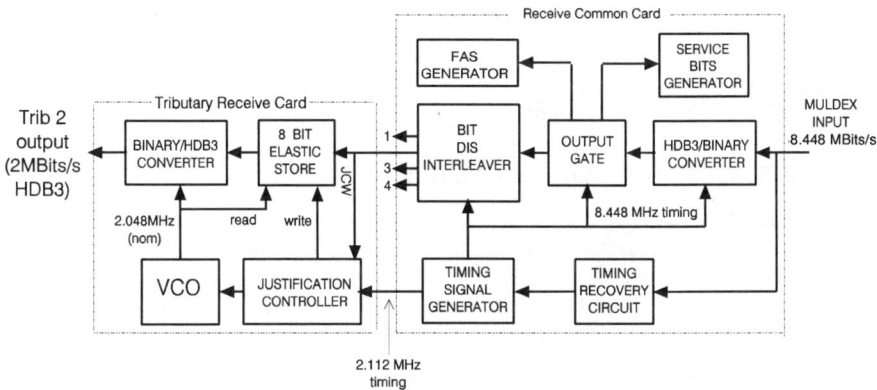

*Figure 14.15 Block diagram of Muldex receive tributary and common cards*

Since the Muldex equipment is also operating plesiochronously, the incoming 8.448 Mbit/s signal from the distant transmit Muldex must be passed to an 8.448 MHz timing recovery circuit which produces a clock signal at the same frequency as the master oscillator in transmit section of the distant Muldex. All timing for the receive equipment is derived from this recovered clock signal by a timing signal generator.

The incoming 8 Mbit/s signal is in HDB3 format and is passed to a converter to change the format to binary. The binary signal is fed to an input gate which directs the signal to one of three outputs according to the current bit position within the frame. Initially, when frame alignment has not been established, the input signal is gated to the FAS detector, so that all incoming bits can be examined for the FAS sequence. When frame alignment has been gained the service bits can be extracted and the remaining bits of input signal can then be successfully disinterleaved

The signal sent to the tributary receive cards is nominally at 2.112 Mbit/s, the actual bit rate being dependent upon the frequency of the recovered clock signal, which is itself dependent upon the frequency of the distant master oscillator. Within the tributary cards the 2.112 Mbit/s write signal from the dejustification controller is used to clock the incoming traffic signal into the receive store. The write signal is inhibited during the JCW bit periods, and instead the JCW is received by the dejustification controller. As the JCW indicates the status of the JDT, the write signal can be inhibited if the JDT contains only a fill bit. If the JDT contains valid traffic, the write signal is not inhibited during the JDT period and the data bit carried is thus written into the receive store.

Although the write signal has an instantaneous frequency of 2.112 Mbit/s, due to the fact that it is inhibited during the reception of the FAS, service bits and JCW, its average frequency will be the same as the master oscillator in the transmit equipment of the PMUX connected to the tributary card at the distant Muldex.

The dejustification controller produces a reference voltage which is used to determine the actual frequency of a voltage controlled oscillator (VCO) whose nominal frequency is 2.048 MHz. The output signal from the VCO is used as a read clock to time the transmission of all bits from the receive store to the tributary PMUX. Prior to transmission to the PMUX receive equipment the signal format is changed to HDB3 from binary.

## 14.4 The Transmultiplexer

### 14.4.1 Interfacing Analogue and Digital Networks

Although there are many new digital networks emerging, many organisations still have a large capital investment in analogue FDM plant which may still have several years of economic life left. While this FDM plant still exists there will be a requirement to produce circuits which have one end terminating in a digital network while the other end of the circuit terminates in an analogue system.

The simplest method of achieving this connectivity, and certainly the cheapest when only a few circuits are required, is to physically connect the audio output of a PCM channel to the audio input of an FDM channel at the point where the two networks interface. This approach is illustrated in Figure 14.16, and while it is satisfactory for a few interfaces, it is very clumsy and unsatisfactory when a large number of circuits are required.

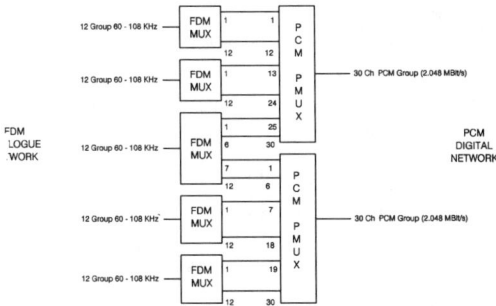

*Figure 14.16 Interconnecting audio circuits between analogue and digital networks*

## 14.4.2 The Role of the Transmultiplexer

Consider the case of an international trunk telephone system between two countries, of which one has a well established digital network while the other is just starting to convert its own system from analogue to digital. The number of telephone circuits crossing the border between these two countries could conceivably be as high as 120 channels and it is obvious that the method shown in Figure 14.16 would be expensive in terms of FDM and PCM multiplex equipment. A solution to the problem may be found in the transmultiplexer, or TMUX, a complex piece of equipment which provides an interface between analogue and digital systems at group level rather individual channel level. A diagram showing how a TMUX may be used as a 60-channel network bridge is shown in Figure 14.17.

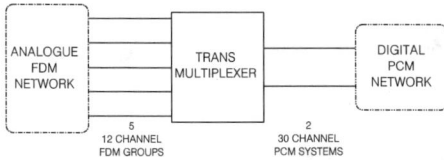

*Figure 14.17 The role of the transmultiplexer*

On the analogue side of the TMUX, up to five 12-channel FDM groups each in the frequency band 60 KHz to 108 KHz are connected, typically on coaxial connectors.

Within the TMUX complex signal processing is used to produce two 30-channel PCM systems, each operating at the standard rate of 2.048 MBits. These two PCM system are then connected into the digital network. The arrangement is easily adapted to cater for 120 channels by the introduction of a second TMUX. The four PCM systems involved may, if required, be connected to a 2/8 Mbits/s Muldex. The TMUX must also cater for various types of signalling convention.

## 14.5 Chapter Summary

This chapter commenced with a brief review of 24 and 30-channel PCM systems as these systems are the building blocks of high capacity digital links capable of carrying up to 7,680 64 Kbit/s channels.

Within the European environment the digital hierarchy is built up by multiplexing four 2 Mbit 30-channel primary rate systems to produce a 120-channel system operating at 8 Mbit/s. The next stage in the hierarchy uses a similar concept to produce a 480-channel system operating at 34 Mbit/s.

By using successive multiplexing stages it is now possible to produce a PCM system capable of handling 7,680 channels and operating at 565 Mbit/s, and trials with even larger systems have already taken place, with a view to exploiting the extremely large bandwidth available on optical fibre systems.

Outside Europe, and particularly in the US, the digital hierarchy is similar in concept, but is based on a 24-channel primary rate multiplexer. Comparisons regarding channel capacities and bit rates were made between the two systems in Table 14.1.

Section 14.2 described synchronous and plesiochronous multiplexing systems and showed that synchronous multiplexing involved interleaving timeslots from primary multiplexers that were synchronised to the Muldex clock.

In plesiochronous systems, where the primary multiplexers have their own individual master oscillators such a multiplexing scheme could not be possible. In these systems, multiplexing is carried out by speed buffering the signals from the primary multiplex in a process known as justification. The four justification circuits within the Muldex are synchronised, and so multiplexing then involves a bit interleaving process. This section contained a description of the speed changing process using a positive justification example.

In either system, the output bit rate of the Muldex is not simply four times the individual input bit rates as there is a bit overhead due to the requirement to transmit Muldex frame alignment signals and alarms.

Section 14.3 contained a brief description of the block diagrams of a 2-8 Mbit/s plesiochronous Muldex. The main emphasis of this section concerned the justification circuits to provide a greater understanding of how these systems operate.

The last section dealt with the transmultiplexer, a device for providing a 60 channel group level interface between analogue and digital networks.

## 14.6 Further reading and references

*PCM and Digital Transmission;* GH Bennet published by Marconi Instruments

*Digital Multiplexing;* V W Wilson, supplement to British Telecommunications Engineering Vol 5 Part 1 April 1986

*Multiplexing for a Digital Main Network;* ER Brigham, MJ Snaith, DM Wilcox

*The Post Office Engineering Journal*

*15*

# Optical Fibre Transmission Systems

## 15.1 Introduction to Optical Fibre Transmission

The aim of this chapter to is introduce the main concepts of optical fibre transmission. This means of digital transmission is becoming more important as the demand for communications systems increases and outpaces the ability of radio systems such as microwave radio relay and satellite communication to meet the demand. British Telecom has invested heavily in fibre optic systems for its all-digital trunk network, while Mercury Communication's own digital system is almost entirely based on optical fibre. This chapter is concerned with the technology of long haul large capacity systems and will not deal with the short low capacity systems which are replacing traditional intra-site data communications systems.

### 15.1.1 The Use of Optical Fibre Systems

Many current communications systems are based on optical fibre cables rather than conventional copper cables. Typical optical fibre applications include connections between terminals and computers, local area networks, connections between subsystems of a digital exchange and connections between digital exchanges.

Optical fibres are manufactured from very pure silica, and are in fact extremely thin fibres of glass. Information can be transmitted along these fibres in the form of light pulses in much the same way as digital information is transmitted along conventional cable as electrical pulses.

The purpose of this chapter is to give a basic introduction to the concepts of optical fibre transmission, particularly in its application to higher order multiplexed systems. There are already many good text books covering this subject in greater depth should you wish to pursue the subject further.

The introduction of digital transmission into telephony and data networks has been largely possible due to the availability of optical fibre as a transmission medium. It is true that the traditional transmission media, e.g. coaxial cable, microwave radio and increasingly satellite communications are capable of carrying digital signals, but unfortunately these traditional systems have limited bandwidths which tend to preclude their use for the transmission of digital signals above 140 Mbit/s.

Other factors which work against the use of traditional media include the increasingly high cost of copper for metallic cables, and the lack of available space in the frequency spectrum for microwave links.

The main benefit of using optical fibre is that it offers tremendous bandwidths, and thus may be used for the transmission of extremely high bit rate digital signals. As an example of the bandwidths that are achievable, systems operating at up 2 Gbit/s have recently been trialled in British Telecom's digital trunk network.

At the present time, there are increasing demands for telecommunications resources, particularly for transmission capacity on trunk routes. By exploiting the bandwidth opportunities of optical fibre, network operators can satisfy this demand with a system which has the ability to meet the needs of the present time, but which also has inbuilt, the ability to meet the increased demands of the next decade or so.

The next few sections describe the major benefits of using optical fibre as a transmission medium, although it should be noted that some of the factors discussed will not be relevant in all applications.

## 15.1.2 Bandwidth of Optical Fibre Systems

The large available bandwidth may be exploited by using a digital multiplexing system such as that described in the previous chapter. It is possible to transmit well over 7,000 64 Kbit/s channels over a single fibre, although in many applications it will not be necessary to use the full available capacity of the optical fibre cable that is installed.

Should demand for capacity over the route increase in the future, the increase in demand can be met simply by installing upgraded multiplexing equipment at each end of the link, without the need to replace the optical fibre cable.

Currently optical communications systems are based on the transmission of a single high speed digital signal, which would normally be produced by some form of multiplexing equipment. Research is being carried out with a view to increasing the available bandwidth of optical fibre by making it possible to transmit several high speed digital signals along the same fibre completely independently of each other, without the problem of one signal interfering with another.

## 15.1.3 Low Optical Loss

Repeaters are required approximately every 2 Km in a coaxial line system to overcome the effects of signal attenuation. These repeaters are often buried underground, and must of course be fed with electrical power normally from the nearest exchange.

Optical fibres exhibit such very low loss characteristics that many routes can be accomplished without the need for repeaters, and their incumbent problems. Repeaters which may be required on longer routes can be located within the building of an exchange which is on the route. This obviates the need for repeater burial and the requirement for external power feeding.

## 15.1.4 Size and Weight Considerations

An optical fibre is extremely small and light. To use optical fibre in a practical environment the fibre must be made into a cable, often with several other fibres. The whole cable is protected from the environment and from the stresses of installation. Despite the protection optical fibre cables are considerably smaller and lighter than copper cables capable of the same transmission capacity.

In any application where space and weight are at a premium, optical fibre can make a contribution. For example within aeroplanes and ships considerable space and weight savings can be made by using optical fibre in preference to copper cable for communications purposes.

Cables are usually routed around buildings in ducts. There are many instances in which the ducts have been filled to capacity by conventional cables but the expense of providing new ducts has been avoided either by installing small fibres between the existing cables or by replacing several cables by one fibre cable.

## 15.1.5 Lack of Interference in Optical Fibre Systems

Information is transmitted along optical fibre cable in the form of pulses of optical energy. The optical energy is contained completely within the fibre cable, with only minute amounts of leakage. This contrasts with the transmission of information along copper cables using electrical energy, these systems do not completely contain the electrical energy and much is lost through electromagnetic (EM) radiation.

The EM radiation from a pair in an electrical cable will cause unwanted crosstalk on adjacent pairs in the cable unless transmission levels are strictly controlled. In the commercial and military worlds, sensitive applications are prone to attack as the information that is transmitted along a cable may easily be intercepted by picking up the radiation from the cable. As this activity is passive, it can be carried out with little risk of being detected.

The use of optical fibres for this type of application does not completely eliminate the problem of interception, however it is extremely difficult to intercept information transmitted along optical fibres without being detected.

From the opposite standpoint, electrical systems are prone to interference by picking up the electromagnetic radiation from other sources which radiate electrical energy, e.g. electric motors, fluorescent lighting and so on. These sources of electrical noise do not interfere with optical systems and so optical fibres find applications for voice and data transmission in electrically noisy environments such as factories, and close to high power radio and radar installations.

EM radiation can also cause Electromagnetic Compatibility (EMC) problems when siting different electrical systems close to another, as radiation from one system can cause interference problems in others. Such EMC problems do not occur on the optical portions of communications systems.

Taking these factors together, optical systems are immune from the following problems which tend to affect all electrical communications systems to some degree:

❏ No crosstalk between one fibre and another in the same cable

❏ No EMC problems

❏ No interference in electrically noisy environments

❏ Communications are relatively secure because it is not easily possible to pick up optical radiation from the cable and recover the information that is being transmitted.

### 15.1.6 Electrical Insulation

Optical fibre cables may be constructed without the use of any metallic conductors or strength members and thus can be used to convey information between two points of differing electrical potential. Equally no earth loop current problems are encountered and the isolation ensures that electrical faults can not propagate from one part of the system to another.

The lack of electrical conductors permits optical fibres to be used in hazardous environments such as mines, where the risk of explosion from electrical sparks must be avoided.

### 15.1.7 The Problems

In the early days of optical fibre transmission there were several problems which had to be overcome. One area of particular concern was the jointing of one length of fibre to another. Skilled staff were required to prepare the fibres and to perform the delicate fusing operation that is necessary. Today these problems have largely disappeared thanks mainly to better manufacturing methods and sophisticated fusing equipment.

## 15.2 Basic Optical Fibre Transmission Principles

### 15.2.1 Optical Transmission Systems

Just as with any transmission media, an optical link must consist of three major items:

❏ The transmitter, which includes an electrical interface and an optical source which is normally an LED or Laser Diode.

❏ The transmission path, i.e. the fibre optic cable

❏ The receiver, which includes some form of optical detector, often a PIN or Avalanche photodiode, with a suitable electrical interface.

Only a one-way, or simplex, communications path is possible with these items. For a normal two-way or duplex system, two of each item are required as shown in Figure 15.1.

Several manufacturers are experimenting with a form of optical hybrid which will permit both a transmitter and a receiver to be simultaneously connected to one end of a single optical fibre. By using such a combination at both ends of the fibre duplex communications over a single fibre can be achieved.

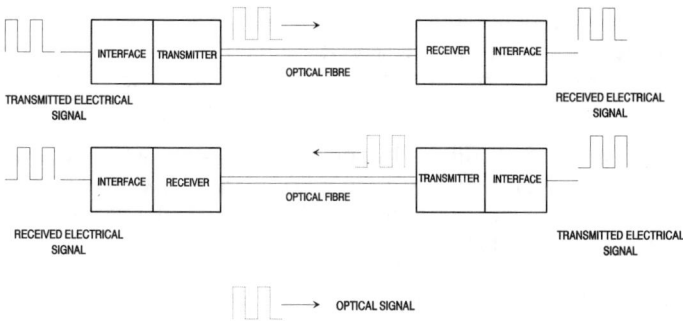

*Figure 15.1 Basic Duplex communications using optical fibre*

## 15.2.2 How is Light Transmitted Along an Optical Fibre

Light travels at different speeds in different mediums e.g. air, glass and water. It travels faster in less dense mediums such as air than in optically dense mediums such as glass. The speed of light in a particular medium relative to its speed in a vacuum is given by the refractive index of the medium. The refractive index of a vacuum is 1, while that of glass is approximately 1.5. The formula below shows the relationship which links the optical velocity and the refractive index of a medium.

Refractive index (n) = <u>Speed of light in a vacuum</u>
of the medium      Speed of light in the medium

## 15.2.3 Refraction and Reflection

Light passing across the boundary between two media having unequal refractive indices will be subject to a refraction, or bending, as shown in Figure 15.2. This is the phenomenon which causes a stick to appear to be bent when partially immersed in water. The degree of refraction will depend upon the refractive indices of the two materials, and is given by a formula known as Snell's Law.

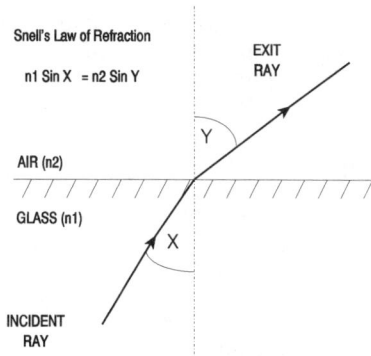

*Figure 15.2 Refraction of light at the boundary between two materials*

Snell's law is only true for a limited range of incident angles. If the angle of incidence is increased, a point is reached where the light is no longer refracted as it passes across the boundary, but is totally reflected at an angle equal to the incident angle, as if from a mirror.

The incident angle at which this reflection starts to occur is known as the critical angle, and it, also, is dependent upon the ratio of the refractive indices of the two materials concerned. It is this Total Internal Reflection (TIR) characteristic which is exploited in optical fibre systems.

## 15.2.4 Basic Fibre Construction

Figure 15.3 shows the construction of a basic optical fibre. In the central area there is a cylindrical core of very pure silica having a diameter of around 50 microns (1 micron = 1 millionth of a metre). The silica of the core has a refractive index of approximately 1.45 and is surrounded by a cladding of silica which has been doped to give it a slightly lower refractive index than that of the core.

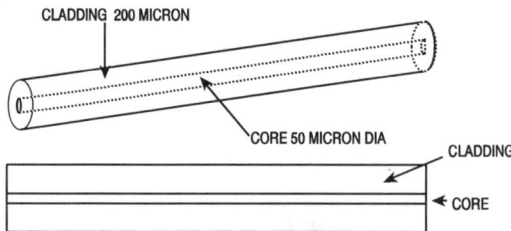

*Figure 15.3 Basic Optical Fibre Construction*

This is an example of one of the earliest types of practical optical fibre and is known as stepped index fibre, due to the step in refractive index that occurs at the core/cladding boundary. Typical dimensions for this type of fibre are:

core diameter: 50 micron; cladding diameter: 125 micron to 200 micron

Many other sizes of stepped index fibre are available, although the 50/125 stepped index fibre is reasonably common. In a later section we will see that nowadays little use is made of this type of fibre in trunk communications systems due to its limited bandwidth.

## 15.2.5 Light Propagation in an Optical Fibre

Light travels along the fibre by being repeatedly reflected from the core cladding boundary as shown in Figure 15.4. This process is extremely efficient and only a minute amount of light escapes across the core/cladding boundary to be lost.

Light will be launched into the fibre at many angles but the reflection at the core cladding boundary, and thus propagation along the fibre, will only occur for that proportion of the light that is launched into the fibre at an angle which is greater than the critical angle.

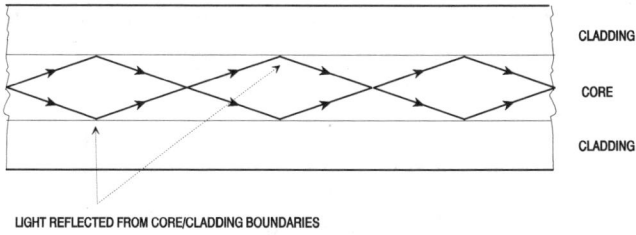

LIGHT REFLECTED FROM CORE/CLADDING BOUNDARIES

*Figure 15.4 Transmission of light along an optical fibre*

As light enters the end of the fibre it passes across the boundary between the optical source and the optical fibre, if these two have different refractive indices, the light will also be refracted at this boundary.

## 15.2.6 Numerical Aperture

The requirement for light to be launched into the fibre such that it is incident upon the core/cladding boundary at an angle greater than the critical angle, and the fact that light is refracted as it enters the fibre give rise to the concept of a light acceptance cone or numerical aperture illustrated in Figure 15.5.

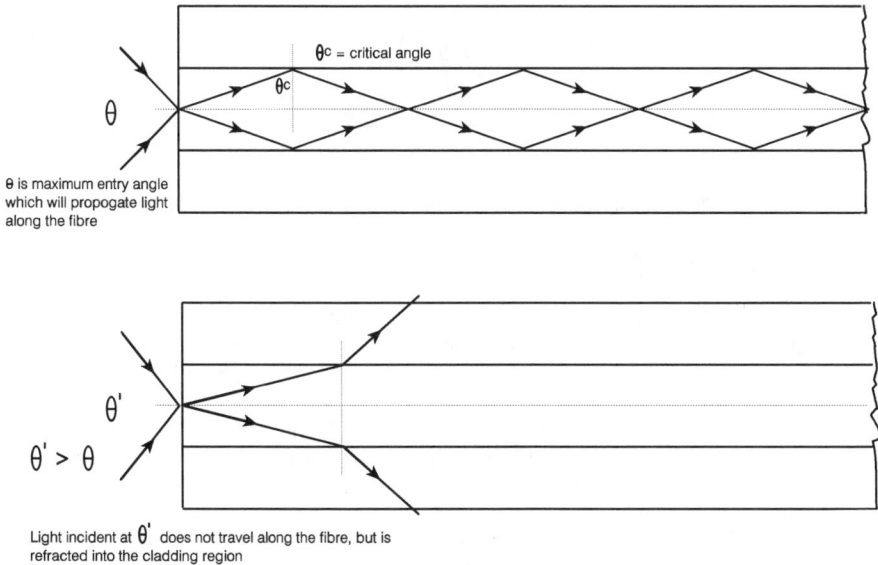

*Figure 15.5 Light acceptance cone and numerical aperture of an optical fibre*

The Numerical Aperture (NA) of a fibre is measure of ability of the fibre to accept light from a source and typical values lie in the range 0.15 to 0.5.

The value of NA is dependent upon the refractive indices of the cladding and core materials and is given by the formula:

$$NA = n_{core}^2 - n_{cladding}^2$$

## 15.2.7 Loss in Optical fibre

Not all the light energy entering the fibre emerges at the distant end and some light is lost en route because of scattering and absorption in the core. These losses are proportional to the fibre length and are analagous to the power losses on audio and RF transmission lines. It has become standard practise to express these losses in terms of decibels per kilometre length of fibre. Recent improvements in fibre manufacturing technology has resulted in the fibres of today having extremely low loss and attenuation figures as low as 0.1db/km are now the accepted norm for fibres used in long haul telecommunications applications.

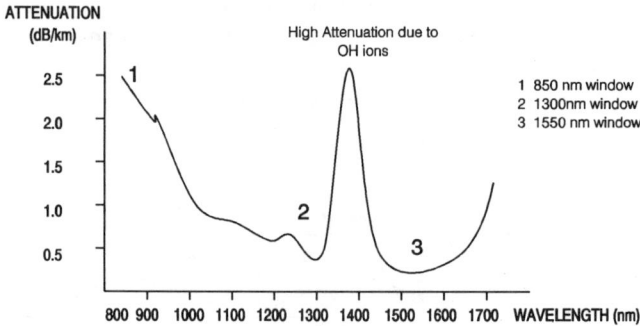

*Figure 15.6 Spectral response of a typical optical fibre*

## 15.2.8 Spectral Response

The devices which are used as optical sources transmit optical energy of different wavelengths, typically 850nm, 1300nm and 1550nm. It is therefore important to know the attenuation characteristics of a fibre at various optical wavelengths. Manufacturers often show this information in the form of a spectral response diagram accompanied by the actual attenuation figures at specific wavelengths of interest. Figure 15.6 shows the spectral response of a typical optical fibre between the wavelengths 800nm and 1600nm.

Below 800nm, attenuation is increased due to ultraviolet absorption and a phenomena known as Rayleigh Scattering, while above 1600nm attenuation is increased due to infrared absorption. There is also an attenuation peak around 1400nm due to absorption by certain ions trapped in the silica.

The diagram shows three transmission windows which are the wavelengths of commonly available optical sources. In the early years the most common operating wavelength was around 850nm, but with the minimum possible attenuation in this region being around 2db/km, interest soon moved to the 1300nm region. The minimum attenuation in this region is under 1db/km and today many long haul systems operating at 1300nm are in use.

Today, there is growing interest in the 1550nm region and systems which exploit the extremely low attenuation possible in this region are becoming available. In some cases fibre systems are being designed to operate at 1300nm and 1550nm. Initially 1300nm transmitters and receivers are being installed, but these systems are being engineered so that in the future 1550nm devices can be used.

## 15.2.9 The effect of pulse dispersion on the bandwidth of optical systems

Although it has been stated that optical fibres have extremely large bandwidths, the bandwidth is not infinite, and is in fact limited by pulse dispersion.

Pulse dispersion causes a reduction in available bandwidth and as it is related to the length of the optical fibre, the effect is that the bandwidth of a fibre reduces as the fibre length is increased. In other words, if a 20 Km system has a bandwidth of 100 Mhz, a 40 Km system using exactly the same components will have a bandwidth of approximately 50 Mhz.

Pulse dispersion is caused by a number of factors; in this section these factors will be briefly discussed and we will show how different optical fibres have evolved to counter the problems.

## 15.2.10 Pulse dispersion in stepped index multimode fibres

The optical fibre introduced in an earlier section (15.2.4) was referred to as 50/125 stepped index fibre when in fact the correct name is actually 50/125 Stepped Index Multimode Fibre (SIMF). Simply put, light travels along this fibre over a multiplicity of paths, each path having a different angle of incidence at the core/cladding boundary as shown in Figure 15.7.

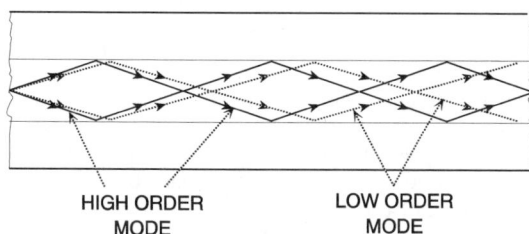

HIGH ORDER
MODE

LOW ORDER
MODE

*Figure 15.7 Multipath propagation in a stepped index multimode fibre*

Consider what happens to a very narrow pulse of light launched in to one end of the fibre as it propagates along the length of the fibre. The available light energy divides into a number of modes some of which tend to travel along the fibre with relatively few reflections as the angle of incidence is large. Those modes whose incident angle is close to the critical angle have considerably more reflections and thus have a greater actual distance to travel. The effect of these path length differences is that the light pulse which arrives at the far end of the fibre is no longer narrow but has been spread in time, or has been dispersed.

If the length of the original launched pulse was t microseconds, it has been increased to t + dt microseconds. It is the factor dt which is dependent upon fibre length. Intuitively we can visualise that over a short length of fibre the difference in path lengths will be small, but as the length of the fibre is increased, so the difference in path lengths will become more significant.

To realise how this pulse dispersion affects bandwidth, consider a constant stream of light pulses with a 50% duty cycle, i.e. t microseconds on, t microseconds off. At the distant end of the fibre an optical receiver must be able to correctly detect all these pulses by detecting the presence or absence of a pulse of light.

If the fibre is short, the situation will be similar to that shown in Figure 15.8(a), although there is a degree of pulse dispersion, it is not significant relative to the pulse width (i.e. dt < t) and the receiver will have no problem detecting the presence and absence of pulses.

In Figure 15.8(b) the length of fibre has been increased and pulse dispersion has also increased to a point where the received pulses are just beginning to merge (i.e. dt = t). It now becomes more difficult for the receiver to detect each pulse and some bit errors will occur. If the fibre length is increased above this limit, (Figure 15.8(c)), the pulses will become more merged and it will be impossible for the receiver to detect the pulses without a significant number of bit errors occurring.

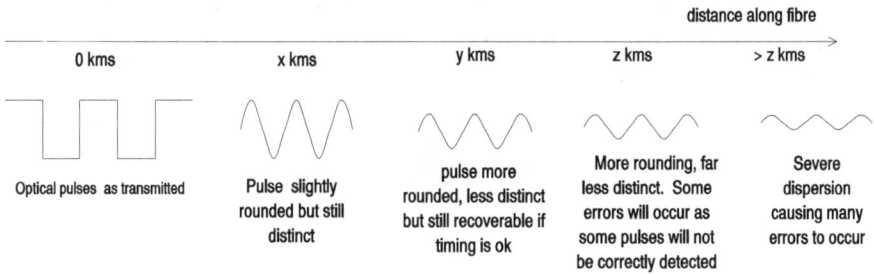

| | | | distance along fibre | |
|---|---|---|---|---|
| 0 kms | x kms | y kms | z kms | > z kms |
| Optical pulses as transmitted | Pulse slightly rounded but still distinct | pulse more rounded, less distinct but still recoverable if timing is ok | More rounding, far less distinct. Some errors will occur as some pulses will not be correctly detected | Severe dispersion causing many errors to occur |

*Figure 15.8 Effect of Pulse Dispersion On Fibre Bandwidth*

Fibre manufacturers may quote the information required to determine bandwidth in a number of ways. One method is to quote the amount of pulse dispersion that occurs per kilometre of cable. As pulse dispersion itself is in no way dependent upon pulse width, the system designer can calculate the amount of dispersion for the required length of fibre and compare this with the pulse width at the bit rate to be used.

Another method is to quote the bandwidth-distance product for the fibre. This figure is a constant, for example, a particular fibre may be quoted as having a bandwidth of 40Mhz.Km. (Note this is not quoted as Mhz/Km.) This means that the product of distance in Km and bandwidth in MHz for any length of the fibre is always 40. The bandwidth-distance products of typical stepped index multimode fibres lie in the range 20 – 50 MHz.Km

The system designer can use the bandwidth distance product to determine the bandwidth of the length of fibre required. For example, if the fibre length is to be 5 Km, the bandwidth will be 8 MHz as 8 MHz x 5 Km = 40 MHz.Km. If the fibre length is 10 Km, the bandwidth reduces further to only 4 MHz as 4 MHz x 10 Km = 40 MHz.Km. Conversely, given the required bandwidth for a system, it is possible to calculate the maximum length of fibre that can be used without repeaters.

For example, if the 40 MHz.Km fibre is to be used for a 2 MHz 30-channel link, the maximum fibre length that can be used without repeaters is 20 Km.

If the calculated bandwidth for the length of fibre required is not sufficient for the application, repeaters can be used to improve the situation. As these devices are effectively able to re-shape and retime the received pulses prior to retransmission over a second, or subsequent sections of fibre. The bandwidth can now be calculated for each section of fibre to be used.

The bandwidth of the whole system will depend upon the bandwidth of all components in the system, transmitter, repeaters, fibre sections and receiver. Although a full treatment is not possible here, suffice to say that if a particular system bandwidth is required, the bandwidth of each component in the system must be greater than the required system bandwidth.

## 15.2.11 Graded Index Multimode Fibre

In the last section it was stated that the bandwidth-distance product for a stepped index multimode fibre was typically between 20 – 50 MHz.Km. As the digital hierarchy is based on bit rates of approximately 2 Mbit, 8 Mbit, 34Mbit, 140 Mbit etc, it can be seen that only very short distances are achievable at the higher bit rates if this type of fibre is to be used. For example with a 50 MHz.Km fibre the maximum workable length at 34 Mbit/s is just under 1.5 Km.

Graded Index Multimode fibre was developed to allow the higher bit rates to be used on longer lengths of fibre without the need for repeaters. Typical bandwidth distance products for this type of fibre range from 200 – 1500 MHz.km.

Using a 1500 MHz.km graded index fibre at 34 Mbit/s gives a reach of approximately 44 Km. This is a far more practicable length than the 1.5 Km using stepped index multimode fibre as the economics of such a low repeater spacing make it an unattractive proposition.

The principle of graded index fibre is far easier to explain than it is to achieve in practice. In concept the fibre does not contain separate core and cladding regions. Instead the refractive index of the fibre is at a maximum value in the centre, reducing gradually away from the centre. Figure 15.9(a) is an attempt to show the concept. The diagram is effectively a graph showing the value of the refractive index of the material varies along a cross section of the fibre.

The bandwidth-distance product of this type of fibre is considerably better than that of the stepped index multimode fibre because pulse dispersion is significantly reduced.

(a) Basic construction of a graded index fibre

(b) light propagation in a GI fibre due to refraction
caused by changing refractive index across the fibre

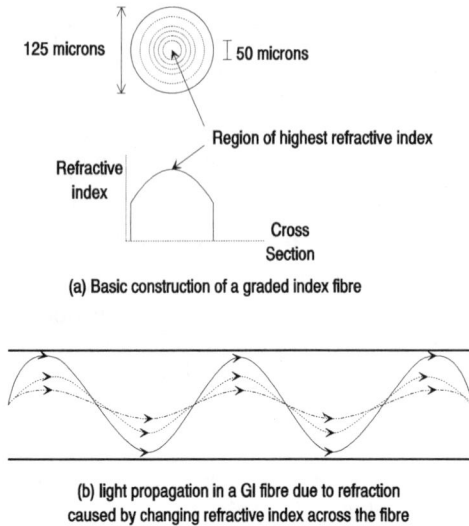

*Figure 15.9 Diagram to show the concept of Graded Index Multimode fibre*

As there is no distinct core/cladding boundary propagation is not achieved by reflection. Propagation in this type of fibre occurs as the light signal is continually refracted as it passes across the material moving in regions of changing refractive index. This concept is simply shown in Figure 15.9(b) by using just two light rays. In fact this is also a multimode fibre and several paths are possible.

Although this is a multimode fibre, there is less pulse dispersion than in a stepped index fibre. This is due to the fact that when light travels in the centre of the graded index fibre, it is travelling in the region of highest refractive index, and will thus travel more slowly in this region than when it is travelling in the outer regions where the refractive index is lower. Although the differences in path length will remain, the overall effect is that the differences in transit time over the different paths are not as great as those which occur in the stepped index multimode fibre and thus pulse dispersion is reduced.

The value of refractive index of the fibre material depends to some extent on the wavelength of the light being transmitted along the fibre. It is therefore necessary to manufacture graded index fibres optimised for a specific wavelength. A typical graded index fibre optimised for use at 1300nm is quoted as producing pulse dispersion of 0.44ns/km, or a bandwidth.distance product of 1000 MHz.km. Pulse dispersion increases dramatically if the fibre is not used at the specified wavelength as illustrated in Figure 15.10.

## 15.2.12 Stepped Index Single Mode Fibre

The main cause of pulse dispersion, and thus reduced bandwidth in both types of fibre mentioned so far, is multimode propagation within the fibre. This type of pulse dispersion, known as mode dispersion, does not occur in stepped index single mode fibre.

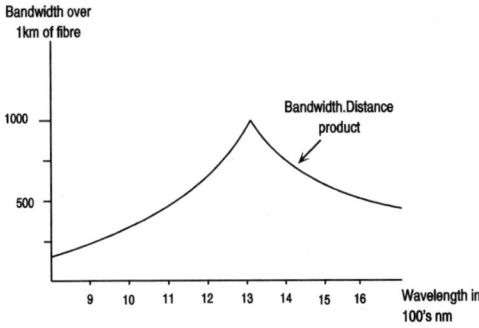

*Figure 15.10 Bandwidth of a graded index fibre optimised for use at 1300nm*

In a stepped index multimode fibre the diameter of the core is typically 50micron and is considerably greater than the optical wavelengths to be used (0.85micron to 1.55micron). The concept of single mode fibre, as shown in Figure 15.11 is to reduce the diameter of the core region to around 5 to 8 microns, so that it is now comparable to the optical wavelength, and only one mode, the axial ray is possible. As there is only one path there can be no difference in path lengths, therefore no difference in transit times and thus mode dispersion does not occur.

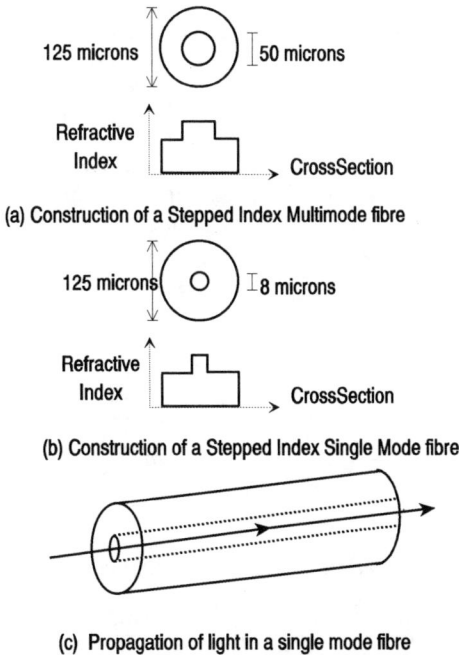

(a) Construction of a Stepped Index Multimode fibre

(b) Construction of a Stepped Index Single Mode fibre

(c) Propagation of light in a single mode fibre

*Figure 15.11 The concept of Stepped Index single mode fibre*

In fact a very small amount of pulse dispersion does occur due to two factors known collectively as chromatic dispersion as they are strictly related to the operating wavelength of the light source.

Chromatic dispersion consists of:

❏ Material Dispersion

❏ Waveguide Dispersion

It should be noted that chromatic dispersion does occur in multimode fibres but it is of such little significance when compared to mode dispersion, that its effects can be ignored.

## 15.2.13 Material Dispersion

It has already been stated that the refractive index of the fibre material is related to the operating wavelength, and therefore the velocity of light within the fibre is also dependent upon the wavelength of the optical source.

The LEDs and laser diodes used in the optical transmitter do not produce light of a single wavelength. Typically a 1300nm LED will produce light energy between 1200nm and 1400nm, although most of the power is concentrated between 1250nm and 1350nm as shown in Figure 15.12. The diagram also shows the output spectrum of a typical laser diode to show that these devices produce a much narrower spectrum than LEDs.

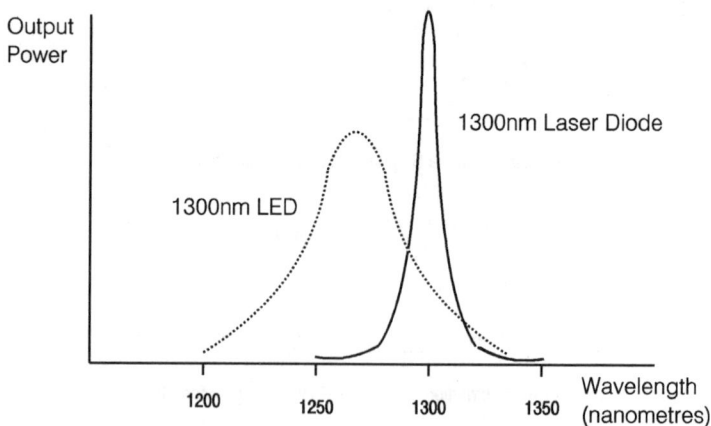

*Figure 15.12 Output power spectra of typical Optical Sources*

This wide power spectrum is the cause of material dispersion. The different wavelength components travel along the fibre at different velocities and thus cause pulse dispersion due the difference in transit times for each wavelength component. To give you some idea of the scale, manufacturers of multimode fibre quote mode dispersion in terms of nanoseconds/km (e.g. .44ns/km), whereas material dispersion is quoted in terms of picoseconds. In fact because material dispersion is caused by the spectral width of the source, the material dispersion parameter is often quoted in terms of the amount of dispersion (in psecs) per nanometre of source spectral width per kilometre of fibre. The figure quoted for a typical fibre is:

d(material)= 8ps/nm.km

This means that the actual material dispersion in any application will depend upon the spectral width of the optical source and the fibre length.

To appreciate the effect of material dispersion, consider that a single mode fibre designed for use at 1300nm is quoted as having a dispersion of 6ps/nm.km. This fibre may be used with either a LED having a spectral width of 100nm, or a laser diode with a spectral width of 10nm. The length of fibre to be used is 20km.

In the case of the LED, the amount of material dispersion over this length of fibre is calculated as:

d(material)= 6ps/nm.km x 100nm x 20km
          = 1200ps or 1.2ns

The same calculation for the laser diode source gives:

d(material)= 6ps/nm.km x 10nm x 20km
          = 120ps or 0.12ns

These dispersion figure must be compared with the pulse width of the signal to be transmitted, however they do show that the use of laser diodes rather than LED causes less material dispersion and thus permits greater system bandwidth.

## 15.2.14 Waveguide Dispersion

The other component of chromatic dispersion, waveguide dispersion, is due to the fact that some of the optical power will travel in the cladding region of the fibre, in so called cladding modes. As the refractive index of the cladding is less than that of the core, the power in these modes will travel at a greater velocity than that in the core, and thus give rise to pulse dispersion.

The effect of waveguide dispersion is similar in degree to material dispersion. However at wavelengths around 1300nm, material dispersion and waveguide dispersion are of approximately equal amplitude but opposite sign and thus tend to cancel producing an overall minimum figure for chromatic dispersion at this wavelength. Optical fibres can be also be designed to have a chromatic dispersion minimum at 1550nm.

# 15.3 Optical Transmitters

## 15.3.1 Requirements of Optical Sources

Light emitting diodes (LEDs) and semiconductor lasers come close to meeting the requirements of an optical source that is to be used in the transmitter of a fibre optic communications system. These requirements are summarised below:

❑ High Radiance, i.e. high optical output

❑ Small emission area, comparable to diameter of fibre core

❑ Small emission cone angle, so that good optical coupling to the fibre can be achieved

❑ Close access to emitting surface, also for good optical coupling

❑ Optical wavelength compatible with fibre transmission windows and operating wavelength of available detectors

❑ Small spectral width, i.e. power produced over a limited range of wavelengths produced

❑ Fast response to electrical modulation so that high bit rate signals can be transmitted

❑ Long life, to keep maintenance costs to a minimum

❑ Reasonably low cost, to keep initial system costs to a minimum.

## 15.3.2 Light Emitting Diodes (LEDs)

These are semiconductor PN diodes manufactured from materials such as gallium arsenide (GaAs). When these diodes are forward biased, they emit light having wavelengths in the region 500nm to 900nm (i.e. 0.5 micron to 0.9 micron). These wavelengths are compatible with the first fibre transmission window at around 850nm and with several types of optical detector.

One of the disadvantages of LEDs when compared with laser diodes is that although basic LEDs produce relatively large amounts of optical power, the radiance of these devices is not high. The reason for this is that the optical power tends to radiate within the LED in all directions, whereas the requirement is for a narrow beam of optical energy. Although a significant amount of power does radiate in the required direction from the surface of the LED, power also radiates in all other directions. The power level at any angle is proportional to the cosine of the angle of radiation. This means that not all the optical power produced can be coupled into a fibre.

Another disadvantage of basic LEDs is the large output spectral width, i.e. they transmit light of a wide range of wavelengths, typically in the order of 100nm.

## 15.3.3 High Radiance LEDs

In recent years, high radiance LEDs have been developed for both 850nm and 1300nm operation. These are more complex structures which produce greater optical power in the required direction and also have spectral widths of around 50nm, as opposed to the 100nm or so associated with the basic LED.

High radiance LEDs may be classified as one of two types: surface emitters or edge emitters. The Surface Emitter LED (SLED) most commonly found in communications applications is known as a Double Hetrostructure Burrus LED and is shown in Figure 15.13. These devices incorporate a well which permits a multimode fibre to be mounted immediately above the emitting surface thus achieving good optical coupling. In the device shown in Figure 15.13 optical coupling is further increased by mounting a lens in the well between the emitting surface and the fibre.

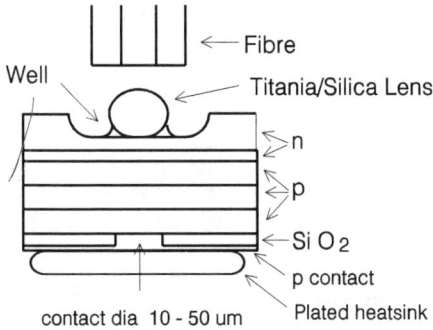

*Figure 15.13 Surface emitting LED: The double hetrostructure Burrus LED (Reproduced by permission of GEC Marconi Materials Technology Ltd)*

Edge emitting LEDs (ELED) produce a more directional beam from a narrow stripe on the edge of the device. This permits a far greater coupling efficiency to be achieved, and the optical power levels claimed by manufacturers are generally higher for this type of device. The structure of a typical ELED is shown in Figure 15.14.

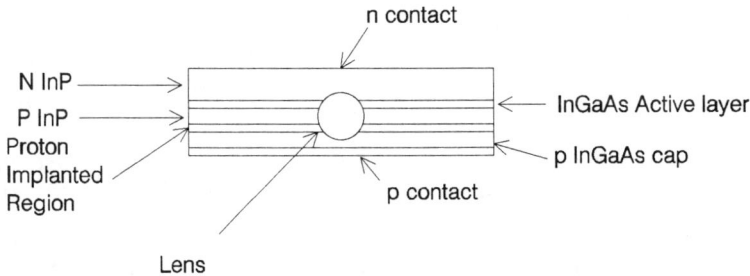

*Figure 15.14 Edge Emitting LED (Reproduced by permission of GEC Marconi Materials Technology Ltd)*

## 15.3.4 Semiconductor Lasers

The term LASER is a mnemonic for Light Amplification by Stimulated Emission of Radiation. Semiconductor laser diodes are found in many applications, particularly those involving long links or very high data rates. Many such applications are based on a combination of 1300nm laser diode and single mode fibre.

The laser consists basically of an isotropic optical source, i.e. one which produces spontaneous emission in all directions, located between two parallel reflecting surfaces. Most of the spontaneous light produced by this source escapes after reflection from the edge of the reflectors, but some will be continuously reflected back and forth, confined between the reflectors.

If the initial energy is high enough, lasing occurs due to the confined radiation stimulating the production of more light energy, most of which is emitted with the same frequency, phase and direction. This process is cumulative, in that the stimulated light energy released causes even more light to be produced. This leads rapidly to a highly intense collimated and monochromatic (single wavelength) beam between the reflectors. A proportion of the beam is allowed to escape from the edge of the reflectors, and is thus available for transmission into an optical fibre.

Although semiconductor lasers do not produce beams which are as well defined as gas lasers (e.g. Helium Neon Lasers), their small emission areas mean that very high optical power levels can be launched into even the smallest fibre.

The characteristic of a semiconductor laser has two distinct regions. The device behaves rather like an LED at low values of forward bias, light output increasingly linearly with increasing forward current. At a certain threshold, losses within the device are overcome, and lasing takes place. This is marked by a dramatic increase in light output and efficiency. Figure 15.15 shows the structure of a semiconductor laser and its output characteristic at various operating temperatures.

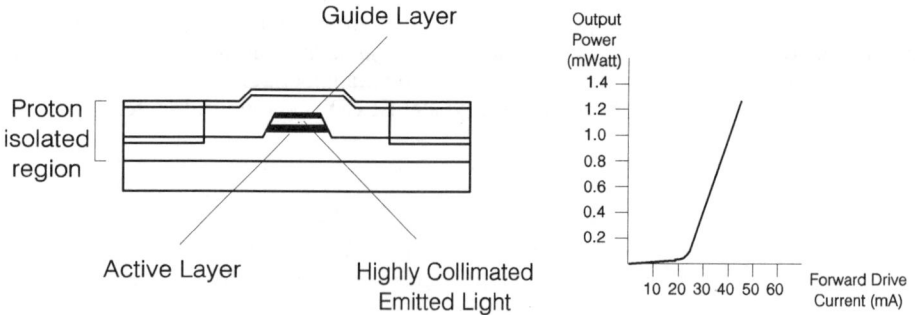

*Figure 15.15 Structure of a semiconductor laser and output characteristic*
*(Reproduced by permission of GEC Marconi Materials Technology Ltd)*

Note that more drive current is required to produce the lasing effect at higher temperatures. Lasers have several advantages over LEDs, two of the more significant advantages being:

❑ Greater output power; typically up to 10mW

❑ Narrow Spectral Width; typically 10nm.

Lasers are normally operated over the lasing region of their characteristic. The digital modulating signal causes the laser to switch between a low output state and a high output state. This is in contrast to the operation of LEDs in which the modulating signal causes the LED to switch between off and on states.

One of the disadvantages of using laser diodes is that they require more sophisticated drive circuitry. This is because the significant effects of temperature change make it necessary to carefully control the operating point. To achieve this, most laser diodes are manufactured with an integral optical detector, the purpose of which is to monitor the high output from the laser and produce a signal which can be used in a feedback circuit to automatically control the biasing of the laser diode.

## 15.3.5 Safety Aspects of Laser Diodes

Another major disadvantage of using laser diodes is that strict safety procedures have to be designed into systems. Because semiconductor laser diodes produce a narrow spectral width and output powers, typically in the region of 10mW or above, there is a danger that damage to the eyes can result from exposure to the optical radiation from these devices. It should be pointed out that as optical fibre systems operate in the infrared band, i.e. over 750nm, the light produced by both lasers and LEDs is invisible to the human eye. However exposure to high optical power levels at these wavelengths will cause thermal damage to the cornea.

British Standard 4804 contains safety regulations governing the use of optical sources to reduce the risk of injury to operations and maintenance personnel. Within the regulations optical systems are classified according the perceived danger. Systems which involve low power and are considered safe under all conditions are classified as class 1 systems. Class 2 systems require caution, while class 3 and 4 systems, involve high power devices and present a danger, making safety measures mandatory.

# 15.4 Optical Receivers

## 15.4.1 Introduction

The function of the optical receiver is to receive optical energy from the fibre system and produce an electrical signal which can then be processed to recover the traffic signal that was originally transmitted. Optical receivers thus consist of two elements, the first of which is a photodetector which produces a very low level electrical signal. The second element is a low noise preamplifier stage to increase the level of the electrical signal prior to further demodulation.

Two factors impact on optical receiver designs. The first of these is the demand for very high transmission speeds with low error rates, typically now in the region of 565 Mbits/s and above, at error rates at $1 \times 10^{-10}$ or better. The second factor is the demand for longer repeaterless optical links. Both of these factors give rise to the requirement to use receiver components with high sensitivities.

There are two main types of photodetector in use for telecommunications applications: the PIN diode and the Avalanche Photodiode (APD).

The PIN diode tends to be used for short distance applications, such the local exchange digital network, operating at the lower transmission speeds of 2 – 34 Mbits. The more sensitive APD is used for higher speed applications, such as the digital trunk network, where the required range is far greater.

## 15.4.2 PIN Diodes

A semiconductor photodiode in its simplest form is a PN junction operated under reverse bias conditions upon which light is incident normal to the plane of the junction.

The absorption of light at the junction causes the formation of electron-hole pairs which are subject to the electric field in the depletion region of the junction. The electrons and holes drift across the junction under the influence of the electric field and thus give rise to an electric current which can flow in an external load circuit. This concept is illustrated in Figure 15.16.

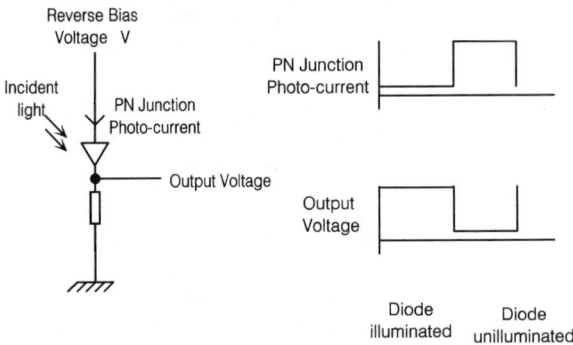

*Figure 15.16 The operation of a simple PN junction photodiode*

The performance of a simple PN junction diode is improved by incorporating a high resistance layer of undoped silicon between the p and n doped region of the junction. The undoped silicon is referred to as intrinsic silicon and hence the term PIN (P material – Intrinsic – N material) Photodiode. Figure 15.17 is an illustration of a typical silicon PIN photodiode, in cross section, to show the highly doped p and n material regions sandwiching a layer of intrinsic silicon.

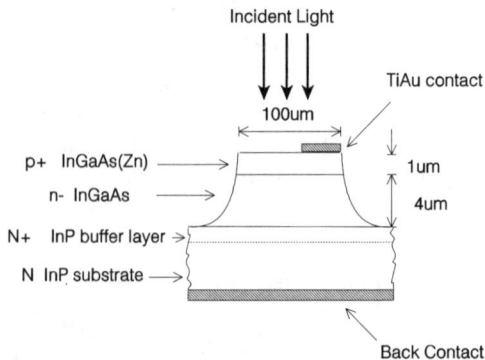

*Figure 15.17 Cross section of a silicon PIN photodiode (Reproduced by permission of GEC Marconi Materials Technology Ltd)*

Because some form of amplification stage is always required to boost the output of the PIN diode to usable levels, several manufacturers supply integrated PIN photodiode and FET high input impedance amplifier devices known as PIN-FETS.

These devices exhibit performances which are almost as good as APDs, and are particularly suited to applications at 1300 and 1550nms which make optimum use of the fact that fibre attenuation at these wavelengths is very low.

## 15.4.3 Avalanche Photodiodes (APDs)

When a reverse bias voltage is applied to a pn junction no current should flow, though in reality a small leakage current will in fact be present. The principle of the APD is based on the fact that when the reverse bias voltage is increased past a certain critical point, the avalanche effect occurs. This avalanche effect is caused by the initially small number of free electrons in the depletion region travelling with so much energy that they collide with other electrons which are then also released and move through the depletion region colliding with and releasing more electrons. The effect is cumulative and results in a large current flowing in the direction which is opposite that which would flow naturally if the diode was forward biased.

The operation of the APD in simple terms involves reverse biasing the pn junction almost to critical point of junction breakdown. The effect of light incident upon the junction is to cause the formation of a few free electrons, and associated holes which are subject to the intense electric field caused by the reverse bias voltage. Because of the applied electric field these electrons move very quickly across the junction, and collide with other electrons giving rise to an avalanche effect. A large photocurrent results which can then flow in the external load circuit.

The breakdown of the junction is not however destructive, and when the incident light is removed, electrons and holes recombine and the photocurrent no longer flows across the junction.

The amount of photocurrent which flows in both the PIN diode and the APD is related, if not proportional to the intensity of the light illuminating the PN junction. However in the case of optical systems used for digital transmission the photodiode simply has to detect the presence or absence of a pulse of light rather than to respond to the instantaneous level of optical energy. This rather simplifies the point, because in systems where long lengths of fibre are in use, the level of optical energy available at the receiver is low.

The sensitivity of a photodiode is its ability to respond to low level pulses of optical energy, and correctly detect the incoming pulse stream without error. The sensitivity of a photodiode is usually quoted as the minimum incident optical power which can be detected at a particular bit rate with a specified BER. Typically a manufacturer will quote sensitivities for a range of bit rates and possibly a range of BER. Photodiode sensitivity is reduced as bit rate is increased and it is also reduced for better BER performances.

## 15.4.4 Comparison of PIN and Avalanche Photodiodes

Avalanche photodiode are generally more sensitive than their PIN counterparts. However, there are two operating disadvantages that have to be taken into account during the design of APD receiver systems. In some ways these points are similar to those of semiconductor laser diodes in that APDs require very high voltages, typically up to 300volts to reverse bias the diode correctly, and achieve the correct level of sensitivity. For example, an APD that is insufficiently reversed bias will not respond to even high levels of optical power. The operating point is also very temperature conscious and thus relatively complex temperature stabilisation circuits are required.

These problems contrast sharply with the relatively simple PIN diode which is not sensitive to large changes of incident optical signal level and operating temperature and which requires a reverse bias voltage of only around 12volts.

# 15.5 Designing Optical Systems

## 15.5.1 Factors Affecting System Design

The design of an optical system will depend to a great extent on the following four major factors:

❏ Length of the optical link

❏ The proposed data rate

❏ The maximum acceptable bit error rate

❏ Economics.

These factors will greatly influence the choices of the types of optical components for the system.

Short links, up to say 1km, for low data rate applications e.g. connecting a 2.4Kb terminal to a host computer or local area network, may be constructed from low cost optical transmitters and receivers, and use cheap Plastic Clad Silica (PCS) Fibres which tend to have relatively high losses and low bandwidths. These systems are simple to design and implement and it is unlikely that this type of application will involve the use of repeaters.

On the other hand, long telecommunications links running at up to 565 Mbit/s may involve expensive laser diodes, high quality single mode fibre and very sensitive receivers. Often this type of long link also requires optical repeaters.

Another factor which some organisations are beginning to take into account is a requirement for a degree of standardisation. Due to the wide variety of components available there is concern that systems will be incompatible and that initial and ongoing costs will be unnecessarily high due to the large number of different components, installation and test equipment that are required.

Standardisation on one type of fibre for example permits standardisation on one type of connector, and one type of installation and test equipment. Safety factors regarding the use of laser diodes may also need to be taken into account.

## 15.5.2 The Design Choices

The system designer must choose the most appropriate combination of optical components for each individual system. The choices to be made will include:

**Optical Fibre?**
      Single mode or Multimode
      Bandwidth and dispersion characteristics
      Operating wavelength
      Optical loss

**Transmitter?**
      LED or semiconductor laser
      Spectral width
      Operating wavelength
      Output power

**Receiver?**
      PIN diode or Avalanche Photodiode
      Operating wavelength
      Sensitivity/error rate

The system design will involve the choice of suitable components, based on an optical power budget analysis and a system bandwidth analysis. There is not sufficient space to cover these analyses in detail, however a brief outline of the concepts is given below.

## 15.5.3 Optical Path Loss Budget Analysis

Path loss budgets are not unique to optical systems; they are used in the design of coaxial systems and microwave links (both terrestrial and satellite systems).

The purpose of the path loss budget analysis is to ensure that a given combination of transmitter, receiver and fibre will operate successfully at the data rate required. The analysis must also take into account changes that may occur to the system during its lifetime – component ageing, and repairs to the fibre are the most often quoted examples.

The analysis presupposes that the components chosen will all operate at the same wavelength.

The following parameters are required:

❑ Output power of the optical transmitter (Po)

❑ Maximum permissible receiver input power (Pr max)

❑ Minimum permissible receiver input power i.e. sensitivity (Pr min)

❑ Attenuation of the Optical Fibre (in dBs/Km)

❑ Length of the optical route (in Kms)

❑ Losses due to optical connectors and cable splices

The output power of the optical transmitter at the desired operating wavelength must be known, along with the operating conditions, such as drive voltage, that are required to produce this output. Due to the difficulty of aligning a fibre to a transmitter for optimum coupling some manufacturers build their devices with a short length of fibre (often called a pigtail) connected, and aligned for optimum coupling. The manufacturer will then quote the output power that can be delivered into a given type of fibre from the pigtail.

Sensitivity figures quoted for receivers are usually quoted at specified data rates and bit error rates. It is important to ensure that the actual input power incident at the receiver is greater than the minimum specified for the data rate and bit error rate of the application e.g. typically telecomms applications require data rates of 140 Mbits/s at BERs of $10^{-9}$ or better.

While the power incident at the receiver must be greater than the required minimum, there is in many cases also an upper limit to the amount of power which the receiver can handle without being overloaded.

The manufacturer's quoted figures for attenuation at the wavelength concerned can be used to calculate the total optical loss over the route. To this figure must be added all losses due to connectorised and spliced connections between cable length. The designer must also allow a margin to take account of other factors such as component ageing and cable repair. This is usually done by regarding these as an additional loss in the system.

## 15.5.4 Example Optical Path Loss Budget

In this hypothetical example, we show that there is more than one possible solution and that in a real situation a final choice would have to be made by taking account of cost factors which are neglected for the purpose of this example.

An optical link is required between two sites 55 Km apart. The link is to run at 8.448 Mbits/s and will be used with appropriate PCM multiplex equipment to provide a 120 telephone channel system. The target BER for this link is better than $1 \times 10^{-10}$.

Two systems have been identified as possible solutions, the parameters of which are given in Table 15.1. In both cases, the electrical interface at 8 Mbits is in HDB3 format conforming to CCITT recommendation G.703 which is compatible with the PCM 2-8 Mbits/s Muldex equipment.

*Table 15.1 Main parameters of example optical systems*

| Component | Parameter | Option A: 850 nm System | Option B: 1300nm System |
|---|---|---|---|
| FIBRE | Type | 50/125 Graded Index Multimode | 8/125 Stepped Index Single mode |
| | Loss/Km | 3.0 dB | 0.6 dB |
| | Splice Loss | 0.5 dB | 0.1 dB |
| | BW.Distance product | 50 MHz.Km | 5000MHz.Km |
| OPTICAL TX | Type | High Radiance LED | Semi-Conductor LASER |
| | Output Power (Po) | -19 dBm | -3 dBm |
| OPTICAL RX | Type | PIN Photodiode | PIN-FET Combination |
| | Sensitivity | -53 dBm | -50 dBm |
| | Max Input Power (Pr Max) | - 33 dBm | -30 dBm |
| MANUFACTURERS RECOMMENDED SYSTEM MARGIN | | 6 db for BER < 1 x 10E-10 | 6 db for BER < 1 x 10E-10 |

## 15.5.5 Solution A: Optical Path Loss Budget for the 850nm System

The first step involves calculating the Path Loss Budget (PLB). This is the maximum optical power loss that can be tolerated between the transmitter and receiver without a degradation of BER performance below that specified for the link. In this case a target BER of $1 \times 10^{-10}$ has been specified, and so the figures quoted by the receiver manufacturer for this BER have been used in Table 15.1.

The maximum permissible loss is calculated using:

Po: The transmitter output power
Pr min: The receiver sensitivity
Recommended System Margin

Initial Path Loss Budget (IPLB)= Po – Pr min
$$IPLB = -19dBm - (-53dBm)$$
$$IPLB = 34dB \qquad (1)$$

Path Loss Budget $= IPLB$ – Recommended System Margin
$$PLB = 34dB - 6dB$$
$$PLB = 28dB \qquad (2)$$

The PLB of 28 dB represents the maximum loss that can be tolerated on each optical fibre section due to fibre attenuation, splice losses and connectorised joints.

The next step of the optical power budget involves calculating the maximum length of each optical section to ensure that the PLB figure of 28dB is not exceeded. This calculation involves the following:

PLB    28dB

Lc     Loss due to the connectorised joint at each end of the optical fibre cable. Note connectors are used to facilitate end to end testing of the optical fibre. 1dB is a typical figure for connectorised joints at 850nm.

Ls     Loss due to the splice which is required to join two sections of optical fibre. Fibre splicing gives a permanent low loss reliable joint. Losses of 0.5db per splice being typical at 850nm.

Maximum permissible loss (Lf max) = PLB − Lc
$$= 28db − ( 2 \times 1dB)$$
$$Lf\,max = 26dB \qquad (3)$$

Since fibre is supplied in lengths of 1Km and attenuation figures quoted are 3dB/Km, the additional loss (Ls) due to splicing each kilometre length gives a figure of 3.5dB/km. This neglects the fact that there will always be one less splice than the number of fibre sections, but since the figure quoted for splice loss is an average typical figure the effect of this error on the calculation is negligible, and in any case the error is on the safe side.

The maximum distance (D max) of an optical section is simply obtained by dividing the allowable loss (3) by the fibre loss per kilometre (L).

Distance  (D max)  =  $\dfrac{Lf\,max}{L}$

D max  =  $\dfrac{26db}{3.5db/km}$

D max  =  7.4 Kms (approx.)

The required link is actually 55 Kms in length, and so this system must use eight optical sections with seven intermediate repeaters. Figure 15.18 shows a simple block diagram of a repeater to illustrate that it consists of two receivers, each connected to a transmitter. Each receiver extracts timing information from the received optical signal to ensure that the optical signal to be retransmitted by the repeater is at exactly the same clock rate as the original signal.

The optical signal incident at the repeater receiver will be a low level distorted version of that originally transmitted. The function of the transmitter in the repeater is to reconstitute the signal so that it becomes a replica of the original signal in terms of amplitude and pulse shape. This reconstituted signal can then be transmitted along the next section of the fibre.

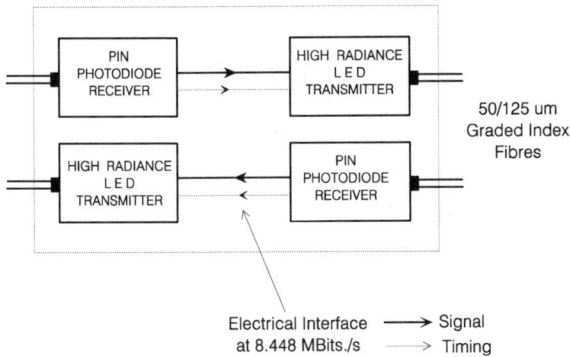

*Figure 15.18 Simplified block diagram of an optical repeater*

Figure 15.19 is a diagram of the total system, showing PCM multiplex equipment, Muldex equipment and the optical components necessary for the link.

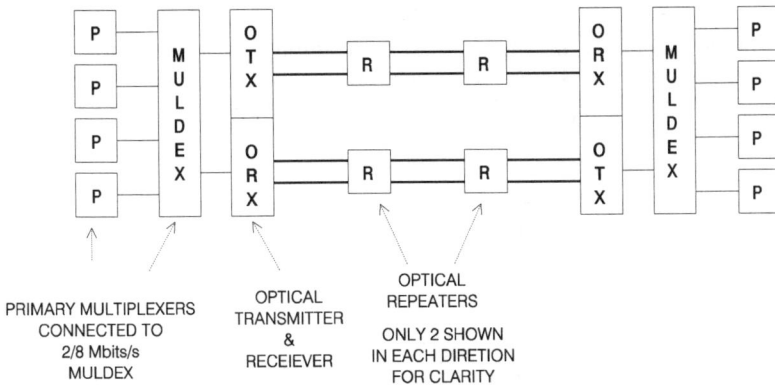

*Figure 15.19 Example 850nm optical system*

## 15.5.6 Solution B: Optical Path Loss Budget for the 1300nm System

The same calculations are carried out as for the 850nm system, using the parameters shown for the 1300nm components.

Initial Path Loss Budget (IPLB)= Po – Pr min
IPLB = -3dBm – (-50dBm)
IPLB = 47dB (1)

Path Loss Budget (PLB)= IPLB – System Margin
PLB = 47dB – 6dB
PLB = 41db (2)

Connector Loss (Lc) = 2dB

Maximum permissible loss
on optical fibre (Lf max)   $= PLB - Lc$
$= 41dB - 2dB$
$= 39dB$

Quoted Fibre Attenuation (L/km)   $= 0.6db/km$
Splice loss at 1300nm (Ls)   $= 0.1dB$
Overall loss per kilometre (L)   $= 0.7dB$

Maximum distance (D max)$= \dfrac{Lf\,max}{L}$

$= \dfrac{39dB}{0.7db/Km}$

D max $= 55.4$ Kms

Since the figure obtained for D max exceeds the length of the required link this system can be implemented without the use of repeaters. One method of presenting the information obtained in the path loss analysis is to produce a graph representing optical power level against distance. The figures used in the 1300nm system have been used to produce an example in Figure 15.20.

These graphs are rarely drawn to scale and simply serve as an aid. The area of importance on the graph is the area between Pr max and Pr min as this represents the possible range of incident power at the receiver for satisfactory operation.

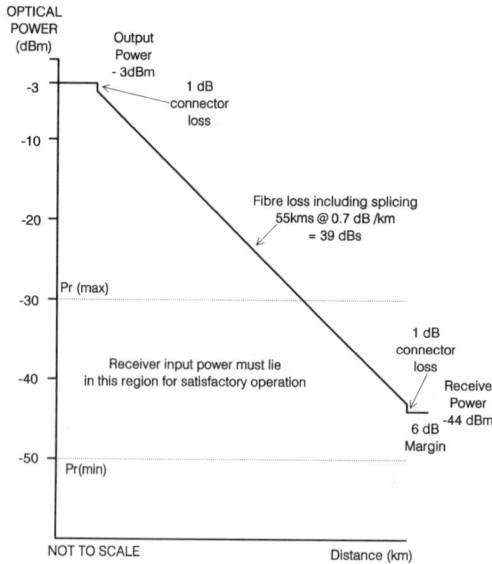

*Figure 15.20 Graphical presentation of optical path loss budget for the 1300nm system*

Commencing at the transmitter, the graph shows the output power from the laser (-3dbm), and then power level decreases through connectorised joints and normal fibre attenuation to become the power incident upon the receiver (-44dBm).

This incident power level is inside the required operating band (-30dBm to – 50dbm) and is 6dBs above Pr min, and thus there is a system margin of 6dBs, i.e. the situation can deteriorate by up to 6dBs before the system will fail.

Although generally the cost of 1300nm components is significantly greater than that for comparable components for use at 850nm, in this case the extra cost of the repeaters required for 850nm operation, coupled with the reduced reliability due to the increased complexity of the 850nm system make it the less preferred option. Other factors against the use of an 850nm system include:

❏ Cost of power to repeaters

❏ Increased difficulty of isolating faults in the system (see later)

❏ Very limited future expansion possibilities.

## 15.5.7 The System Bandwidth Analysis

Having carried out the path loss budget to determine that sufficient power will be available at the receiver, the next analysis is to ensure the system has sufficient bandwidth to handle the required data rate. In this example a simple method only is illustrated. More detailed methods involving the calculation of optical pulse dispersion and comparison with overall pulse durations can also be carried out.

### Option B Bandwidth analysis

The fibre is quoted as having a bandwidth.distance product of 5000 MHz.Km. The bandwidth that can be expected from a 55 Km section of this fibre is estimated by dividing the bandwidth.distance product by the length of the optical section. In this case:

$$\text{BW estimated} = \frac{5000 \text{ MHz.Km}}{55 \text{ Kms}}$$
$$\text{BW} = 90\text{MHz approx.}$$

As this figure compares very favourably with the required data rate of 8 Mbit/s there is no need for further analysis. Note that this optical system would also be suitable for a 34 Mbits/s (480ch) link, but would not have sufficient bandwidth for a 140 Mbits/s (1920ch) link unless a repeater was used.

In cases where the estimated fibre bandwidth approaches the required system bandwidth a more detailed analysis of pulse dispersion will be required. Such analysis will also take account of the pulse rise times attributable to the optical transmitter and receiver components.

### Option A Bandwidth Analysis

Carrying out the same calculation for option A; The fibre bandwidth.distance product is quoted as 50 MHz.Km. The estimated bandwidth of 7.4 Kms of this fibre is only 6.2 MHz, this is insufficient to support the 8 Mbits/s application and in this case the repeater spacing would have to be reduced to about 5 Kms to permit the required bandwidth to be achieved.

These examples also illustrate how individual optical systems can be described as being loss limited or bandwidth (i.e. dispersion) limited. Using Option B we found that the maximum length of an optical section, i.e. the repeater spacing, was determined by the path loss budget, there being sufficient bandwidth for the system concerned with the spacing of 55.4 Kms (in fact no repeaters were required on this link). Systems such as this are described as loss limited.

With option A we found that the maximum repeater spacing was determined by the bandwidth analysis, as the distance calculated by the path loss budget provided insufficient bandwidth for the application concerned. Option A is therefore an example of a bandwidth limited system.

## 15.5.8 System Design Summary

Optical system design involves the selection of suitable components to achieve the required transmission performance in terms of data rate and bit error ratio over the specified link.

Two analyses are required: A path loss budget is used to ensure there will be sufficient optical power at the receiver, while a bandwidth analysis ensures that pulse dispersion does not reduce the system bandwidth below the required operating rate. One of these two analyses will determine the maximum possible repeater spacing and so systems may be described as being loss limited or bandwidth limited.

Many systems are now being designed to take account of future communications requirements. For example, a 140 Mbit/s system being designed today will probably be designed so that it has sufficient bandwidth to be used at 565 Mbits/s in the future, simply by adding the required extra multiplex equipment and changing the optical transmitters and receivers in the system.

# 15.6 Testing Optical Systems

## 15.6.1 Introduction to Optical System Testing

As with any transmission system, it is necessary to test optical links after installation to ensure that the required performance has been achieved. Testing is also required at intervals during the system lifetime to ensure the system performance continues to meet its required specification, to detect fall in performance and to isolate faults which may have occurred.

There are three basic types of test which can be used to check optical systems:

❑ Path Loss Test

❑ Bit Error Ratio (BER) test

❑ Fault Location.

## 15.6.2 The Path Loss Test

This test is carried out only on the fibre link and does not involve the system's active optical components such as transmitters, receivers and repeaters. In systems where repeaters are used it is necessary to test each section of the link separately.

As its name suggests, the path loss test determines the actual loss on the fibre route. The test involves connecting a stable optical source of known output power to one end of the fibre, and a calibrated optical receiver to the other as shown in Figure 15.21.

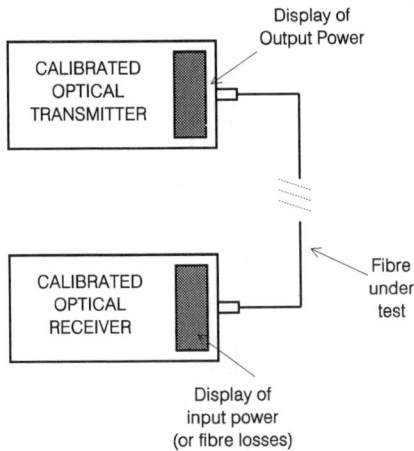

*Figure 15.21 Measuring the optical path loss*

Normally the two pieces of test equipment would be calibrated together prior to the test. The actual fibre loss can then be read directly from the panel of the receiver and compared with the design figures.

Results of these tests should be recorded so that they can be used as a comparison for future tests to detect a fall off in performance.

## 15.6.3 Bit Error Ratio Test (BERT)

The BER test is carried end to end over the whole optical system including transmitter, receiver, fibre and all repeaters as a single entity.

The test determines whether the whole system is capable of operating within specified performance limits at the required data rate. Test equipment used for this test can also be used for end to end testing of other digital transmission media.

The diagram in Figure 15.22 shows how the BERT is carried out using a pseudorandom (pr) pattern generator operating at the required system bit rate which is connected to the electrical interface of the optical transmitter at one end of the link.

At the distant end of the link a pseudorandom pattern receiver is connected to the electrical interface of the receiver.

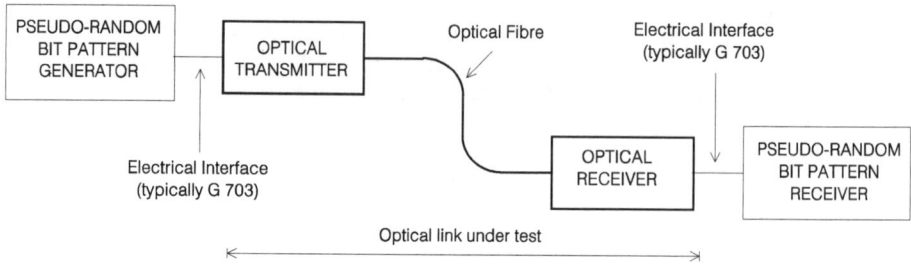

*Figure 15.22 Bit error ratio testing of the optical system*

In the type of systems discussed in this book the pr generator and receiver interface will conform to CCITT recommendation G.703 and the test equipment is in effect simulating the multiplex equipment.

The pr pattern receiver monitors the incoming digital signal and locks its own internal pattern generator to this received signal. The pr pattern receiver is then able to compare the received signal with the expected signal derived from its internal pattern generator, on a bit-by-bit basis, and thus detect any errors which have occurred during transmission over the optical system. The resulting BER is then be displayed on the front panel of the receiver

The test equipment normally consists of a pr generator and receiver, thus enabling end to end testing in both directions simultaneously.

## 15.6.4 Fault Location using Optical Reflectometry

The purpose of fault location is to determine the position of a break or fracture in the optical fibre. Testing is carried out using a technique known as optical reflectometry to determine the position of a fault to a high degree of accuracy even on cables exceeding 50 Kms in length.

## 15.6.5 The Principles of Optical Reflectometry

The principle behind the reflectometry technique is based on the fact that when light travels along the fibre, small quantities are reflected in all directions from all parts of the fibre. This phenomenon, known as back scatter, results in a very small quantity of light being reflected back to the source and this light can be detected and measured, and the result used to determine the path loss over every section of the fibre.

Several general purpose test equipment manufacturers produce microprocessor driven devices know as Optical Time Domain Reflectometers (OTDR). One advantage of these machines over the test equipment required for the path loss test is that they need only be connected to one end of the fibre, and are thus suitable for operation by a single technician.

A typical OTDR will include an aperture to which the fibre under test is connected, a display, a printer and a keyboard for control of the equipment, and to facilitate the input of data regarding the type of fibre under test. Having being initially set up by the user, with data such as the refractive index of the fibre, these devices launch a very high power short duration pulse of optical energy into the fibre.

The OTDR then switches to receive mode and records the level of optical energy received over a short period of time following the pulse. Since the time taken for back scatter to arrive back at the OTDR is dependent upon the speed of propagation in the fibre, and thus the refractive index, the microprocessor can then use this data to calculate the loss over the length of the fibre. This information is normally shown on a display screen, as a trace which represents the attenuation characteristics of the fibre under test. Figure 15.23 shows a trace that would be obtained from an ideal fibre. Note that the horizontal scale is calibrated in Kms and the vertical scale is calibrated in dBm.

Most practical fibre systems will not be ideal, and the trace will not normally be quite like that shown in this diagram.

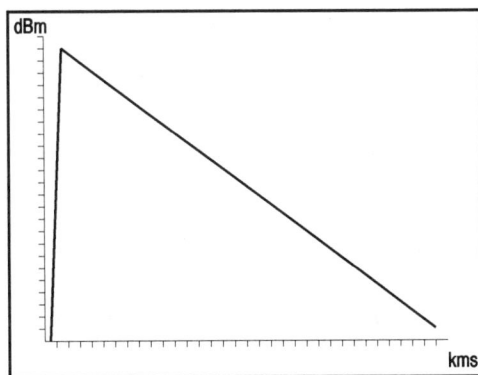

*Figure 15.23 Display of an optical time domain reflectometer with an ideal fibre under test*

## 15.5.7 Use of the Optical Time Domain Reflectometer to Locate Faults

Discontinuities in the fibre such as those caused by connectorised joints, splices and fractures will cause levels of reflection which are different from those caused by back scatter and so these discontinuities can easily be detected on the OTDR display.

When fibres are prepared for jointing it is necessary to ensure that they are cut (or cleaved) square to the fibre axis and then polished. A perfect cleave or polished termination causes a point of high local reflectivity (approx. 4% of the available light being reflected). The reflection from these points is significantly greater than that caused by the back scatter in the adjacent regions of the fibre, and will be displayed on the OTDR as a sharp peak in the trace. (It should be noted that although it appears that there is a gain in optical power at this point, this is most definitely not the case).

Figure 15.24 shows a simplified typical display to highlight a few points.

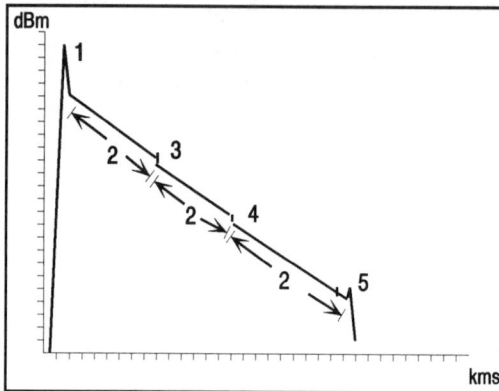

*Figure 15.24 Optical time domain reflector example display*

At point 1 there is a large peak in the trace caused by the reflection from the connector joint between the OTDR and the fibre under test.

Over the region marked as 2 there is a steady decrease in optical power caused by the normal attenuation of the fibre. The trace will be steeper for fibres having higher attenuation figures.

At point 3, the small peak is typical of that caused by reflection from a connector joint. However, the trace does not continue at the previous level, there being a slight drop due to the attenuation of approximately 1dB across the connector.

At point 4, there is a slight drop (exaggerated in this diagram) due to an imperfect spliced joint. In a typical system it is normally very difficult to detect good quality spliced joints unless they have become broken in some way.

Point 5 represents the end of the fibre. The small peak is caused by the reflection from the fibre end, while the very steep fall off indicates the end of the fibre as there is no further backscatter. The distance to the end of the fibre can be read directly from the display using a cursor control. However the OTDR will also have a alphanumeric display normally at the top of the screen to display data such the refractive index of the fibre, total loss over the whole fibre, length of the fibre, fibre attenuation in dBs/Km etc.

A technician is able to pin point faults by observing unexpected discontinuities in the trace and then moving a cursor to the point of interest on the display. The OTDR will then calculate the distance to the fault and the loss across the fault, if it is not a complete fibre break. It is then a relatively simple matter to have the fibre repaired and retested.

The built-in printer can be used to produce a copy of the screen display which can then be filed with the cable records, and used as the basis for comparisons to detect further impairment when subsequent tests are carried out.

## 15.5.8 Summary of Fibre Systems Testing

The three basic tests give a good indication of the current state of the fibre cable, and the ability of optical system including transmitter, receiver and repeaters to operate under the required traffic conditions. It should be pointed out that the optical tests, i.e. path loss and OTDR tests should always use test equipment that operates at the same optical wavelength as the intended optical components.

It is no use testing an 850nm system with an OTDR which operates only at 1300nm or 1550nm, as the results obtained will be completely invalid and bear no relation to the state of the system at 850nm. However as was pointed out in the summary of the last section, some systems are designed to be used at different wavelengths in the future. It is therefore sensible to test such systems at all intended operating wavelengths.

# 15.7 Chapter Summary

The aim of this chapter was to introduce the major concepts involved with optical fibre transmission systems used for long haul large capacity links such as those used in the digital trunk network. Although optical systems offer many advantages over traditional communications media such as coaxial cable and radio systems, it is true to say that there are actually few alternatives to fibre as the cost of copper increases and the availability of radio frequencies decreases.

Optical systems depend upon the fact that optical energy is guided along an enclosed fibre. Light emitting diodes and semiconductor lasers operating at typical wavelengths of 850nm, 1300nm and 1550nm are used as the transmitter in these systems, while optical receivers effective at these operating wavelengths are based on PIN-FET or Avalanche photodiode detectors.

Several types of fibre are available, although almost without exception, long haul systems are based on single mode fibre operating at 1330nm or 1550nm.

Optical system design involves ensuring that sufficient optical power reaches the optical detector so that the original electrical digital signal can be reproduced without errors, while also ensuring that the bandwidth requirements of the system are met. Any particular design will produce a system which is either loss limited or bandwidth limited.

Optical systems are subject to testing upon commission and regularly during their operation. Methods of testing such optical loss measurement and bit error ratio testing are employed to monitor operation. When faults are discovered an optical time domain reflectometer can be used to locate the fault on the fibre.

# INDEX